THE ELEMENTS

10	11	12	13	14	15	16	17	18

								VIIIA
								2 **He** 4.00260
			IIIA	IVA	VA	VIA	VIIA	
			5 **B** 10.81	6 **C** 12.011	7 **N** 14.0067	8 **O** 15.9994	9 **F** 18.9984	10 **Ne** 20.179
	IB	**IIB**	13 **Al** 26.9815	14 **Si** 28.0855	15 **P** 30.9738	16 **S** 32.066(6)	17 **Cl** 35.453	18 **Ar** 39.948
28 **Ni** 58.69	29 **Cu** 63.546	30 **Zn** 65.39	31 **Ga** 69.72	32 **Ge** 72.59	33 **As** 74.9216	34 **Se** 78.96	35 **Br** 79.904	36 **Kr** 83.80
46 **Pd** 106.42	47 **Ag** 107.868	48 **Cd** 112.41	49 **In** 114.82	50 **Sn** 118.71	51 **Sb** 121.75	52 **Te** 127.60	53 **I** 126.905	54 **Xe** 131.29
78 **Pt** 195.08	79 **Au** 196.967	80 **Hg** 200.59	81 **Tl** 204.383	82 **Pb** 207.2	83 **Bi** 208.980	84 **Po** (209)	85 **At** (210)	86 **Rn** (222)
110 **Uun** (267)	111 **Uuu** (272)	112 **Uub**	113 **Uut**	114 **Uuq**	115 **Uup**	116 **Uuh**	117 **Uus**	118 **Uuo**

64 **Gd** 157.25	65 **Tb** 158.925	66 **Dy** 162.50	67 **Ho** 164.930	68 **Er** 167.26	69 **Tm** 168.934	70 **Yb** 173.04	71 **Lu** 174.967
96 **Cm** (247)	97 **Bk** (247)	98 **Cf** (251)	99 **Es** (252)	100 **Fm** (257)	101 **Md** (258)	102 **No** (259)	103 **Lr** (260)

The History and Use of Our Earth's Chemical Elements

A Reference Guide

Second Edition

Robert E. Krebs
Illustrations by Rae Déjur

GREENWOOD PRESS
Westport, Connecticut • London

Library of Congress Cataloging-in-Publication Data

Krebs, Robert E., 1922
The history and use of our earth's chemical elements : a reference guide / Robert E. Krebs ;
illustrations by Rae Déjur. — 2nd ed.

   p.   cm.
  Includes bibliographical references and index.
  ISBN 0–313–33438–2 (alk. paper)
  1. Chemical elements. I. Title.
QD466.K69 2006
546—dc22   2006012032

British Library Cataloguing in Publication Data is available.

Library of Congress Catalog Card Number: 2006012032

ISBN: 0–313–33438–2

First published in 2006

Greenwood Press, 88 Post Road West, Westport, CT 06881
An imprint of Greenwood Publishing Group, Inc.
www.greenwood.com

Printed in the United States of America

The paper used in this book complies with the
Permanent Paper Standard issued by the National
Information Standards Organization (Z39.48–1984).

10 9 8 7 6 5 4 3 2 1

To Carolyn, my wife, proofreader, pre-editor,
constructive critic, and friend.

Contents

---

[*] The following elements belong to the subseries of transuranic elements as well as to the actinide series: Neptunium, Plutonium, Americium, Curium, Berkelium, Californium, Einsteinium, Fermium, Mendelevium, Nobelium, Lawrencium.

Alphabetical List of the Elements

Name of Element	Chemical Symbol	Atomic Number	Atomic Weight	Page
Actinium	Ac	89	227.028	307
Aluminum	Al	13	26.981538	178
Americium	Am	95	243	321
Antimony	Sb	51	121.760	217
Argon	Ar	18	39.948	267
Arsenic	As	33	74.92158	215
Astatine	At	85	210	257
Barium	Ba	56	137.327	78
Berkelium	Bk	97	247	324
Beryllium	Be	4	9.012182	65
Bismuth	Bi	83	208.98038	220
Boron	B	5	10.811	175
Bromine	Br	35	79.904	225
Cadmium	Cd	48	112.41	143
Calcium	Ca	20	40.078	73
Californium	Cf	98	252	326
Carbon	C	6	12.01115	189
Cerium	Ce	58	140.116	279
Cesium	Cs	55	132.90546	59
Chlorine	Cl	17	35.453	248
Chromium	Cr	24	51.996	95
Cobalt	Co	27	58.9332	105
Copper	Cu	29	63.546	111

Name of Element	Chemical Symbol	Atomic Number	Atomic Weight	Page
Curium	Cm	96	247	323
Dysprosium	Dy	66	162.50	294
Einsteinium	Es	99	252	328
Erbium	Er	68	167.259	297
Europium	Eu	63	151.964	289
Fermium	Fm	100	257	330
Fluorine	F	9	18.9984	245
Francium	Fr	87	223	62
Gadolinium	Gd	64	157.25	290
Gallium	Ga	31	69.723	181
Germanium	Ge	32	72.64	198
Gold	Au	79	196.967	165
Hafnium	Hf	72	178.49	147
Helium	He	2	4.002602	261
Holmium	Ho	67	164.903	295
Hydrogen	H	1	1.0079	40
Indium	In	49	114.818	159
Iodine	I	53	126.9044	254
Iridium	Ir	77	192.217	159
Iron	Fe	26	55.847	100
Krypton	Kr	36	83.798	268
Lanthanum	La	57	138.9055	277
Lawrencium	Lr	103	262	335
Lead	Pb	82	207.19	203
Lithium	Li	3	6.941	47
Lutetium	Lu	71	174.96	302
Magnesium	Mg	12	243.1	69
Manganese	Mn	25	54.9380	97
Mendelevium	Md	101	257	332
Mercury	Hg	80	200.59	168
Molybdenum	Mo	42	95.94	127
Neodymium	Nd	60	144.24	283
Neon	Ne	10	20.179	265
Neptunium	Np	93	237.0482	316
Nickel	Ni	28	58.6934	108
Niobium	Nb	41	92.906	124
Nitrogen	N	7	14.0067	207
Nobelium	No	102	259	333

Name of Element	Chemical Symbol	Atomic Number	Atomic Weight	Page
Osmium	Os	76	190.2	157
Oxygen	O	8	15.9994	223
Palladium	Pd	46	106.42	137
Phosphorus	P	15	30.97376	212
Platinum	Pt	78	195.078	162
Plutonium	Pu	94	239.11	318
Polonium	Po	84	210	241
Potassium	K	19	39.0983	53
Praseodymium	Pr	59	140.9075	281
Promethium	Pm	61	145	285
Protactinium	Pa	91	231.0358	311
Radium	Ra	88	226.03	81
Radon	Rn	86	222	272
Rhenium	Re	75	186.207	155
Rhodium	Rh	45	102.906	135
Rubidium	Rb	37	85.4678	57
Ruthenium	Ru	44	101.07	133
Samarium	Sm	62	150.36	287
Scandium	Sc	21	44.9559	87
Selenium	Se	34	78.96	237
Silicon	Si	14	28.0855	194
Silver	Ag	47	107.868	140
Sodium	Na	11	22.9898	50
Strontium	Sr	38	87.62	76
Sulfur	S	16	32.065	234
Tantalum	Ta	73	180.948	150
Technetium	Tc	43	98.9062	130
Tellurium	Te	52	127.60	239
Terbium	Tb	65	158.925	292
Thallium	Tl	81	204.383	186
Thorium	Th	90	231.0381	309
Thulium	Tm	69	168.9342	299
Tin	Sn	50	118.710	200
Titanium	Ti	22	47.88	90
Tungsten	W	74	183.85	153
Uranium	U	92	238.0291	312
Vanadium	V	23	50.9415	93
Xenon	Xe	54	131.293	270

Name of Element	Chemical Symbol	Atomic Number	Atomic Weight	Page
Yittrium	Yb	70	173.04	300
Ytterium	Y	39	88.9059	119
Zinc	Zn	30	65.39	114
Zirconium	Zr	40	91.224	122
Rutherfordium				342
(Unnilquadium)	Rf (Unq)	104	⁻253 to 263	342
Dubnium				343
(Unnilpentium)	Db (Unp)	105	⁻262	343
Seaborgium				345
(Unnilhexium)	Sg (Unh)	106	⁻263	345
Bohrium				346
(Unnilseptium)	Bh (Uns)	107	⁻272	346
Hassium				347
(Unniloctium)	Hs (Uno)	108	⁻277	347
Meitnerium				349
(Unnilennium)	Mt (Une)	109	⁻276	349
Darmstadtium				350
(Ununnilum)	Ds (Uun)	110	⁻281	350
Röentgenium				352
(Unununium)	Rg (Uuu)	111	⁻280	352
Ununbium	Uub	112	⁻285	353
Ununtrium	Uut	113	⁻284	354
Ununquadium	Uuq	114	⁻289	358
Ununpentium	Uup	115	⁻288	359
Ununhexium	Uuh	116	⁻292 (in dispute)	361
Ununseptium	Uus	117	Unconfirmed (undiscovered)	362
Ununoctium	Uuo	118	Unconfirmed (in dispute)	363

Note: Superactinides and super heavy elements (SHE) are elements beyond lawrencium 103. All are artificially produced, unstable, and radioactive and have very short half-lives. Most are made in small amounts, even one atom at a time.

How to Use This Book

The History and Use of Our Earth's Chemical Elements is organized to reflect the chemical and physical properties of the elements that are depicted in the periodic table of chemical elements. This book uses the same general format as the periodic table. Familiarity with how the periodic table is organized and its terminology is not required to understand this book. However, the beauty of the table's organization will become apparent as this book assists you in understanding how the elements are classified according to their atomic numbers, atomic weights, and other chemical and physical properties.

There are several ways to use this book. First, the book is not divided into chapters, but rather into sections. The first five sections include background information that will provide an understanding of basic chemistry and how the rest of the book is organized. If you already are familiar with basic chemistry, you may wish to skip some of the beginning sections. There are three ways to proceed if you are just interested in knowing more about a particular element.

1. One way is to look up the name of the element in which you are interested in the "Alphabetical List of the Elements" located immediately following the table of contents.
2. Another way to start is to look up the element you wish to know more about in its **period** or **group** as listed in the table of contents. The Contents is really the guide for presenting the major classes or categories of elements, that is., periods and groups of elements.
3. Still another way is to look up the page for the element in which you are interested in the index at the end of the book.

The first sections of this reference book set the stage for the presentation of the elements. First is the section "How to Use This Book" followed by a short introduction. Next is "A Short History of Chemistry," the narrative of which progresses from prehistoric times to the Age of Alchemy and then to the Age of Modern Chemistry. Next is the section titled "Atomic Structure," which traces the history of our knowledge of the structure of the atom; some theoretical models, including quantum mechanics; the discovery of subatomic (nuclear) particles and radiation; and the **chemical bond**. Following "Atomic Structure" is the section "The Periodic Table of the Chemical Elements," which describes the table's organization and how to use the table. The rules for placing the elements within the matrix structure of the periodic table are given in this section.

"A Guide to the Elements," which is the major section of the book, presents the elements as they are arranged in the periodic table. Each element is presented as a separate entry that includes information on

- the element's symbol, period, and group and where it is found in the table, its atomic number, its atomic mass (weight), its valence and oxidation state, its natural state, the origin of its name, and the element's isotopes;
- a figure representing the electron configuration for each element up to and including element 103 (lawrencium);
- the important chemical and physical properties of each element;
- the important characteristics of each element;
- each element's abundance and its source on Earth;
- the history of each element, including the discoverer(s) of the element and how and when it was discovered;
- a section for each element's important uses;
- examples of common compounds; and
- an explanation of each element's potential hazards to humans and the environment.

As previously stated, the reader can locate specific elements by checking the Contents for the elements' placement in the periodic table or by consulting the "Alphabetical List of the Elements" that follows the Contents.

The following notations are used in this book:

BCE = Before the Common Era (instead of B.C.)

CE = Common Era (instead of A.D.)

BP= Before the Present (time)

c. = Approximate date (e.g., approximate birth or death dates)

ˉ = Approximate amount, quantity, or figure

amu = atomic mass units, or for heavy elements average mass units.

aw = atomic weight

mw = molecular weight

$_0n$-1 = a single neutron (no charge with atomic weight of 1)

α = alpha particle (helium nucleus; $_2$He-2 or He^{++})

β = beta particle (high-energy electron)

λ = gamma radiation (similar to high-energy X-rays)

SF = spontaneous fission

SHE = abbreviation for super heavy elements

Z = atomic number ($_{92}$uranium, or $_{92}$U, can be shown as Z-92)

IUPAC = International Union of Pure and Applied Chemistry (England)

ACS = American Chemical Society (U.S.A.)

SHIP = Separator for Heavy Ion Products (Germany)

JINR = Joint Institute for Nuclear Research (Russia)

ppm = parts per million

ppb = parts per billion

% = percent

Δ = heat

$\uparrow$ = gas produced

$\rightarrow$ = yields

Terms set in bold type throughout the text can be found in the Glossary of Technical Terms.

Introduction

This volume is a reference for students and other readers interested in chemistry as well as for school and public library use.

It provides the background of how we came to know and understand the chemical nature of our planet and everything in the universe, including ourselves. Early humans had limited knowledge and use of their chemical environment. Through the ages humans progressed from a practical to a spiritual and finally to a rational approach to the nature of the chemical **elements** found on Earth, as well as those in the entire universe.

As science developed, our accumulation of knowledge about the structure of **atoms** and **molecules** was an achievement of early philosophers and scientists. These men and women did not use scientific procedures, but they did build the foundation of our current understanding of the structure of **matter** and how different species of matter interact. This history has led to our current understanding of the theoretical and practical nature of the chemical elements.

This reference book describes the chemical elements according to such characteristics as structure, size, weight, activity (energy), abundance, usefulness, and hazards. Each element's structure relates to its "fit" within the periodic table of the chemical elements. The book is about chemical elements found on Earth as well as in the entire universe. Interestingly, the proportions (by weight or number) of elements found on Earth are not the ratios found in the rest of the universe. For instance, although hydrogen is the most common element in the universe (90+%), the most common element on Earth is oxygen (49%), which was formed by phototropic bacteria and later green plants long after the Earth was formed about 4.5 billion years ago.

Chemistry is a physical science that studies the structure and properties of elementary matter. Matter interacts with other substances. Matter can be defined as something that occupies space, has mass, and cannot be created or destroyed, but can be changed from one form to another. Matter is "stuff" that can be perceived by one or more of our senses, as opposed to something intangible such as an idea, the mind, or spirits (ghosts and angels). We usually think of matter as the chemical elements composed of atoms and the **compounds** composed of molecules, which are combinations of atoms. Chemistry is the science of how and what happens when two or more atoms combine to form myriad compounds that make up all physical things we can see, touch, and taste. Although chemistry is universal, it is not spiritual or mysterious. We study the chemical and physical nature of what makes up atoms and the interactions of atoms and molecules in the stars, the Earth's matter, living **cells,** and everything that exists. Chemistry is also a science of energy. The formation of molecules by the combination of atoms involves energy. **Chemical reactions** involve energy during the combination or separation of atoms and molecules. When energy is released, the reaction is known as an exothermic chemical reaction. Other chemical reactions that require an input of energy to complete the reaction are known as endothermic reactions. In addition to the nuclei of atoms that are mainly composed of protons and neutrons, there are electrons orbiting the nuclei of atoms. These electrons exist in a specific state, or level of energy. Atomic structure is discussed in the section on atomic structure.

Almost everything we live with on Earth is made up of about 100 different chemical elements that, in various combinations, form compounds. But only a few of these elements are essential to explain life on Earth. The air we breathe, the food we eat, the clothing we wear, the cell phones we use, and our bodies all consist of chemicals. The trillions upon trillions of atoms and molecules in our bodies consist of only six major elements: carbon, oxygen, hydrogen, nitrogen, phosphorus, and sulfur. In addition, there are over 40 trace elements in our bodies that are important for our well-being. Some of these are copper, iron, chromium, magnesium, calcium, and zinc. The current market price for all the elements in your body is about $10.95.

What makes chemistry so interesting is that each specific chemical element is related to its own kind of atom. Elements with specific characteristics have unique atoms. Each type of atom is unique to that element. If you change the basic structure of an atom, you change the structure and properties of the element related to that atom. Also of interest is what happens when two or more different atoms combine to form a molecule of a new substance. Once they form a molecule of a new compound, the original atoms no longer exhibit their original properties.

All this is dependent on the electron arrangement in the **shells**[*] around the nucleus of the atom. How atoms interact—that is, how they combine to form molecules—is dependent on the arrangement of their electrons. The electron is one of the three major **subatomic particles.** It has a negative charge and a negligible mass (only 1/1837 the weight of a proton). Although **orbital electrons** are continually circling the central positively charged **nucleus** of the atom, it is not possible to determine exactly where they are within their orbit (shell) at any point in time. (See quantum mechanics in the section on atomic structure) The outer ring, shell, or orbit of electrons consists of a specific number of electrons for each element. For most elements, the electrons in the outermost shell may be thought of as **valence** electrons because they partially determine the chemical properties of atoms and how they combine with each other. Atoms of some elements may have more than one valence number. Valence is a whole number that represents the combining power of one atom to another and provides the relative amounts of each of the interacting elements in the new molecule of a new compound. Valence is discussed in detail in the section on atomic structure.

---

[*] The illustrations that depict the electron configurations of the atoms of each element are based on the Bohr model of quantum energy shells.

This reference work uses the periodic table of the elements as the basis for organizing the presentations of the elements. Once you learn how this remarkable chart is organized, you will be able to relate the characteristics of many elements to each other based on the structure, which can be determined by their placement on the periodic table.

The section "How to Use This Book" provides details and specifics concerning the organization of this user-friendly reference work.

Robert E. Krebs

A Short History of Chemistry

The Beginnings of Science

There are two major theories of how science developed over the ages. One states that early humans, being curious and having some intelligence, began to explore nature by using trial and error. To continue existing, humans learned what to eat, how to protect themselves, and, when time permitted, how to cope with their environment to make life easier and more understandable. This is the "continuum" or "accumulative" approach to science and discovery—which is still ongoing.

The other theory, as presented by Alan Cromer in *Uncommon Sense—The Heretical Nature of Science* (1993), postulates that science was not a natural sequence or continuum of inventions and discoveries from ancient to modern times. He states that science and technological developments occurred in "spurts" of periods of discovery interspersed with periods of ignorance and status quo. He also states that what we think of as science developed in early Greece, possibly because of the democratic nature of Greece's culture, which included rational inquiry and debate. From Greece, science spread to Egypt, China, India, and the Mesopotamian region, where it flourished for a time. As Muslims conquered many countries, the Arab world introduced science to southern Europe as far west as Spain. In the Middle Ages, these Arabic texts of Greek science were retranslated first into Latin and later into the vernacular languages of western Europe.

The libraries of Alexandria (in Egypt) became depositories of knowledge and were major contributors to the advancement of science and intellectual studies in many other countries.

Modern science is very different from the descriptions of early systems of thought. Early philosophers and theorists lacked the objective methodologies and rational investigative processes required for the controlled experiments that led to modern science. They were more concerned with seeking universal cures for sickness, **transmutation** of base metals into gold, and mysticism in general. Most, but not all, ancient philosophers depended more on the written words of "experts" than on their own observations and insights.

Origins of the Earth's Chemical Elements

Humans have espoused all manner of theories as to the origins of the universe—including philosophical, scientific, theological, and mystical myths, as well as just plain speculation—

from the time they first became cognizant of their environment. Even though it is impossible to have complete certainty, during the twentieth century and continuing into the present, science has accumulated a tremendous amount of data and knowledge to support the belief that a cataclysmic event now known as the "Big Bang" resulted in the conditions necessary for the creation of the universe and our solar system.

The most accepted theory of several proposed ideas for the origin of the universe, the Big Bang theory states that it all started 13 to 15 billion years ago with the "explosion" of an incredibly small, dense, and compact point source of matter and energy. This resulted in the creation and rapid expansion of space and the beginning of time, as well as the formation of the universe and all that is contained within in it. Astronomers' observations indicate that the universe continues to expand at an accelerated rate, and this original unknown primordial mass is the source of all the chemical elements and energy existing in the universe. The formation of the chemical elements began in just seconds. Hydrogen was the first element to be formed and remains the most abundant element in the universe. In the heat of the explosion, hydrogen nuclei, having one **proton,** fused to form helium, which has two protons in its nucleus. Together, hydrogen and helium make up 98 to 99% of all the atoms in the universe. As these gaseous elements condensed, they formed billions of galaxies consisting of trillions of stars as well as dark matter that we can't see but that constitutes a large portion of the universe. In addition, many of the stars in the galaxies have accompanying planets, comets, meteorites, and cosmic dust as well as all the chemical elements.

Another popular theory speculates that there was no beginning and there will be no end to the universe—it is infinite. Another theory is that there is continuous death and rebirth of stars and matter in the universe. Others postulate that it all started by some spontaneous and unknown force—possibly supernatural.

We know a great deal about the nature of the universe. For instance, the element hydrogen makes up about 75% of all the **mass** in the universe. In terms of number, about 90% of all atoms in the universe are hydrogen atoms, and most of the rest of the atoms in the universe are helium. All the other heavier elements make up just one to two percent of the total. Interestingly, the most abundant element on Earth (in number of atoms) is oxygen (O_2). Oxygen accounts for about 50% of all the elements found in the Earth's **crust,** and silicon, the second most abundant element, makes up about 25%. Silicon dioxide (SiO_2) accounts for about 87% of the total Earth's mass. Silicon dioxide is the main chemical compound found in sand and rocks.

Early Uses of Chemistry

Early beliefs about and uses for the chemical elements are not well recorded. Obviously, early humans learned, probably through trial and error, how to use several of the Earth's chemical elements for survival. Fire is one of the earliest examples of the use of a chemical reaction. Humans were obviously aware that fire gave off light and heat and caused death, but in time they learned how to use fire to their advantage, most probably for cooking and warmth. Oil lamps that used liquid fat were invented in about 70,000 BCE. Solid fat to form candles for light originated about 3000 BCE. The burning of a candle involved a great deal of chemistry, but many centuries passed before humans understood the science involved.

An interesting application of early chemistry was the use of fire to make pottery. Early pots were formed from soft clay that, of itself, cannot hold much weight or water. Early humans had baskets, wood containers, and leather bladders—yet there was a need for something in which to hold water and to cook food. Around 13,000 BCE, people learned, either by accident or by trial and error, to bake the soft clay over fire, causing it to harden. Once fired, the clay pot became a waterproof ceramic, resulting in a more useful container. This may have been the first time humans used fire for a purpose other than for cooking, heat, and protection.

Another example of an early use of chemistry is the discovery of **fermentation:** overripe fruit or honey, if left uncovered, turned into wine after coming into contact with an airborne organism. During fermentation, large glucose molecules (sugar) break down into the smaller molecules of carbon dioxide and alcohol. The fermentation process that produces alcohol dates back nearly 10,000 years. Wine-making dates back about 7,500 years, evidenced by the residue at the bottom of a clay pot, dated to 5400 BCE, recently found in the mountains of Iran. Wine-making is an ancient chemical process.

A related, and possibly accidental, discovery was bread-making. Presumably some ancient human noticed that when flour pounded out of wild grains was moistened and then exposed to air, it "rose," forming leavened bread. Airborne yeasts are responsible for the fermentation that produces the gas carbon dioxide (CO_2) in the dough. As the dough rises and then is baked, the carbon dioxide gas produces an open texture in the final product that is far more edible. A standard practice was to save a small portion of the unbaked dough, which contained some of the yeast, as a "starter" for future baking. It was kept separate and cool until mixed with fresh dough for the next batch of bread. Then a new starter was put aside for future baking, and so on. If you heat or bake a starter sample, you run the risk of killing the yeast, which is exactly what happens during the baking process after the bread loaf has been raised by the action of the carbon dioxide and heat.

There are numerous examples of the early uses of chemistry in metalworking. Copper and some other **metals** were found as nuggets or exposed native deposits. Sometime around 8000 BCE, it was discovered, probably by accident, that when heated, copper ore combines with air to form a gas (CO_2) and metallic copper (Cu). Thus, humans did not need to rely on stone and wood for tools and weapons, given that metal was superior for these purposes and was now available. A later discovery made circa 3600 BCE involved the mixing of copper and tin ores, resulting in the **alloy** called bronze. This was the beginning of the Bronze Age. Bronze is stronger and holds a sharper edge on tools and weapons than pure copper.

High-grade iron exists in meteorites and in some iron ores. Early humans found these sources of iron but were unable to make use of them except as ornaments. Most iron on Earth exists in ores where it is combined with other substances (often oxygen). It cannot be melted down by wood fires because wood fires do not produce temperatures hot enough to separate the iron from its impurities. In about 1500 BCE, humans learned how to convert wood to charcoal (another chemical reaction) and found that charcoal produces a higher temperature than wood when burned, thus making it possible to smelt the iron from its ore. Iron made sharper, stronger, and more durable tools and weapons than did bronze, thus the beginning of the Iron Age.

These examples, as well as other early uses of chemistry, all involved chemical changes that are well known today. It was many years before humans began studying how chemical changes occur and how to explain and control these reactions.

In 340 BCE, Aristotle (384–322 BCE) published *Meteorologica,* in which he postulated that the Earth's matter is composed of four elements—earth, water, air, and fire. His speculations led to the idea that the Earth is composed of "shells," which is a rather modern concept in earth science. This was the extent of humankind's understanding of the composition of the Earth's chemical elements for several centuries—until the Age of Alchemy.

The Age of Alchemy

Various forms of alchemy were practiced from about 500 BCE into the seventeenth century. It was not until about 320 BCE, at the time of Alexander the Great, that alchemists made serious studies of chemical changes. Alchemy encompassed Greek and Egyptian as well as Arabian and Chinese concepts of matter and energy. The early alchemists were not scientists, and they also did not use scientific procedures, as we think of these terms today. Mostly they were philosophical and theological about their trial-and-error procedures. They were unable to decipher the nature of chemical reactions, but they did make some discoveries that advanced knowledge. In the year 300, Zosimus of Egypt made the first attempt to summarize the accumulated knowledge of alchemists.

Unfortunately, most early alchemists are unknown, considering that they were very secretive about their methods and left little in the way of written history. Their goals were mystical, economic, secret, unpublished, and unshared. Alchemic practices were also related to medicine as well as religion during some periods of time and in some countries. The alchemists' main search was for the "philosopher's stone" that could unlock the secrets of transmutation—that is, the secrets of how to transform base metals and chemicals into different, more useful and valuable products, such as gold and silver. This also led to the futile search over many centuries for the *elixir vitae* that would be both the universal "cure" for all illnesses and the way to achieve immortality.

For over 2,000 years, alchemy was the only "chemistry" studied. Alchemy was the predecessor of modern chemistry and contributed to the slow growth of what we know about the Earth's chemical elements. For example, the alchemists' interest in a common treatment for all diseases led to the scientific basis for the art of modern medicine. In particular, the alchemist/physician Paracelsus (1493–1541) introduced a new era of medicine known as iatrochemistry, which is chemistry applied to medicine. In addition, alchemists' elementary understanding of how different substances react with each other led to the concepts of atoms and their interactions to form compounds.

The Age of Modern Chemistry

It is both difficult to determine an exact date for the beginning of modern chemistry and impracticable to bestow the designation of "father of chemistry" on any one individual. Some historians date the end of alchemy and the beginning of modern chemistry to the early seventeenth century. Over the years many men and women of different races and from many countries have contributed to our current knowledge and understanding of chemistry. A few examples follow.

In 1661 Robert Boyle (1627–1691), an early chemist from Great Britain, published a book titled *The Skeptical Chymist,* which was the beginning of the end of alchemy. His book ruled the perceptions and behavior of early scientists for almost 100 years. Two of his contributions were the use of experimental procedures to determine properties of the chemical elements and the concept that an element is a substance that cannot be changed into something simpler. Robert Boyle is best known for Boyle's Law, which states that the volume of a gas varies inversely with the pressure applied to the gas when the temperature remains constant. In other words, as you squeeze a container of gas, as occurs when the piston compresses the air and gas mixture in an internal combustion engine, the volume of the gas decreases. In the reverse, if you increase the amount of gas in a closed container, the pressure on the inside of the container becomes greater (similar to pumping up a bicycle tire). On the other hand, if you increase the volume of the container but not the amount of gas, the pressure inside the container will be reduced. In the early 1700s, Georg Ernst Stahl (1660–1743), a German chemist, developed a theory that when something burned, *phlogiston* (from the Greek "to set on fire") was involved. His idea was that burnable things had a limited amount of phlogiston and that when burned they lost their phlogiston, leaving residues that would not burn because they no longer contained this so-called substance. Although this theory was considered viable for many years, it could not be sustained as the physical science of chemistry advanced with new concepts for **combustion.** Through experimentation, it was shown that different products resulted from combustion, depending on the particular substances that burned in the atmosphere.

Most historians credit the French chemist Antoine-Laurent Lavoisier (1743–1794) with the death of alchemy and the birth of modern chemistry. His many contributions to the profession included the important concept that one must make observational measurements and keep accurate written records. Lavoisier mixed substances, burned common materials, and weighed and measured the results. His work led to the discovery of more than 30 elements. He described **acids, bases,** and **salts** as well as many organic compounds. Through a unique experiment with water (H_2O), he determined that it is made up of the gases hydrogen and oxygen, with oxygen having a weight eight times that of hydrogen. This led to a later theory of the Law of Definite Proportions, which states that a definite weight of one element always combines with a definite weight of the other(s) in a compound. (It should be noted that Lavoisier was unaware that two atoms of hydrogen combine with one atom of oxygen to form a water molecule. Therefore, the actual ratio of weight [atomic mass] is 1:16 instead of 1:8.) Up until this time, no standard nomenclature (names and symbols) was used for the elements, compounds, and reactions. In 1769, Lavoisier and others published a book titled *The Methods of Chemical Nomenclature* that proposed a logical systematic language of chemistry. Even with modifications by the Geneva System of 1892 and additional reforms by the International Union of Pure and Applied Chemistry (IUPAC) in 1930, Lavoisier's nomenclature of chemical names and symbols is still in use today.

Jöns Jakob Berzelius (1779–1848), a Swedish chemist, is also considered one of the founders of modern chemistry. He prepared, purified, and identified more than 2,000 chemical elements and compounds. He also determined the atomic weight (mass) of several elements and replaced pictures of elements with symbols and numbers, which is the basis of our chemical notations today.

Elements and Compounds

Essentially all of us have used the terms "weight" and "mass," and most people use these terms interchangeably. However, there is an important scientific distinction. The mass of an object is the amount of matter (stuff) contained in a particular body or volume of a substance, *regardless of the object's location in the universe.* The mass of an object is constant and always the same, no matter the planet or galaxy in which it is located.

The weight of an object relates to its size (mass) and distance from the gravitational pull of another body, such as Earth or any other large mass in the universe. In other words, weight is gravity's effect on the mass of an object. Thus object *A*'s weight depends on two factors: first, the size (mass) of the two bodies (for instance object *A* and the Earth); and second, the square of the distance separating the two bodies. Mass is constant in the universe, whereas weight may be thought of as the strength with which gravity "pulls" objects *(A)* to Earth, or any other object in the universe whose gravity might affect "A."

The concept of an "atom" is very old. Leucippus of Miletus in Greece (c. 490–430 BCE) proposed that all matter is composed of very minute **particles** called atoms, from the Greek word *atomos,* meaning indivisible. Atoms are so small that nothing can be smaller, and they cannot be further divided and still be characteristic of the elements they form. One of Leucippus' students, the philosopher Democritus (c. 469–370 BCE), is credited with further developing his teacher's concept of the atom. Another philosopher, Zeno of Elea (c. 495–430 BCE), used paradoxes to present his hypotheses that distance and motion, as well as matter, could be divided into smaller and smaller units in perpetuity. His most famous paradox is the fable of the race between the hare and tortoise wherein the tortoise is given a head start to the halfway point of the total distance of the race before the hare has started. Next, the tortoise progresses to half of the remaining distance—1/4—and then to 1/8, 1/16, and so on, so that no matter how the race progressed, the tortoise would always lead the hare. Democritus challenged Zeno's concept with the analogy that if a handful of dirt is divided by half, and then that half is divided into half ad infinitum, there would be a final limit to the point of divisibility, and this would be the indivisible atom of the dirt (matter). Other Greek philosophers and scientists, including Aristotle and Epicurus (c. 341–270 BCE), accepted Leucippus' theory of the atom and described it as logical and rational and claimed that it could be used to explain reality and eliminate superstition. Epicurus also stated that atoms were always in constant motion, as well as perceivable, and thus deterministic. Although these ancient scientists lacked experimental evidence, they did speculate philosophically about the concept of atoms, which was not well accepted until much later. It would be many centuries before Epicurus' idea was developed into the concept of kinetic energy, heat, and thermodynamics.

As mentioned previously, many people contributed to the development of modern chemistry. An important advancement in our understanding of the Greek philosophers' concept of the indivisible atom occurred in the early nineteenth century. After conducting many experiments, the English chemist John Dalton (1766–1844) published his book *New Systems of Chemical Philosophy* in 1808. In essence, he said that the atoms of a specific element are exactly alike and have the same weight; that the atoms of that specific element are different from the atoms of every other element; and that combinations of the elements are merely combinations of the atoms in simple or multiple units. At this time, Dalton did not know the exact number of atoms that could combine to form molecules. However, through experimentation, he knew the relative weights of the elements that formed compounds. From this information, he developed the first table of **atomic weights.** Despite the fact that Dalton's table was inaccurate, his concept was another step that advanced our knowledge of chemistry.

It became increasingly clear that atoms were the smallest, indivisible units of elementary substances, but it was important to determine the atomic weights (mass) and sizes of different atoms. Since atoms were too minute to see with nineteenth-century microscopes, it was impossible to count and weigh them as we do with larger objects. Therefore, a different standard was needed, and this is where the concept of "relative weights" became useful. Using this relative-weight concept makes it possible to determine the mass of an individual atom of one element relative to the mass of an atom of a different element. Using this system to ascertain atomic weight (mass), chemistry advanced rapidly. In a sense, all standards used for measurements of weight (pounds and kilograms), distance (miles and kilometers), volume (gallons and liters), temperature (Fahrenheit and Celsius), and so forth are arbitrarily related to a specific physical characteristic. Elements that are gases and consist of a single atom are called **monatomic** elements (Ar, Ne, Xe). Gas molecules that are composed of two atoms of an element are called **diatomic** (O_2, H_2, Cl_2), and three atoms of the same gaseous element can combine to form what are known as **triatomic** molecules (O_3 Ozone). The combinations of three or more atoms of different elements form **polyatomic** molecules (PO_4, H_2SO_4). Thus, there are several distinct types of molecules. When atoms of elements combine to form molecules, the chemical and physical properties of the new molecules are different from the characteristics of the original elements. For instance, the compound NaCl is unlike the two elements that make up this molecule. Na is sodium, a very reactive metal, and Cl_2 is chlorine, a reactive gas, whereas the molecule NaCl is well known as table salt. (There is further discussion of atomic structure later.) Based on what we have learned, let's define elements and compounds.

Atom: The smallest unit of each of the more than 100 known elements and different basic types of matter that either exist in nature or are artificially made. All atoms that compose a specific element have the same nuclear charge and the same number of electrons and protons. Atoms of some elements may differ in mass when the number of neutrons in that atom's nucleus is different (such atoms are called isotopes). This is discussed in the next chapter.

Compound: A molecular substance composed of atoms and formed by a chemical reaction of two or more elements. The molecules of the new compound have properties very different from the properties of the elements that formed the compound. An example is when two diatomic molecules of hydrogen gas ($2H_2$) combine with one diatomic molecule of oxygen gas (O_2) to form two water molecules ($2H_2O$), which have none of the properties of the original two substances.

Allotrope: An **allotrope** is formed when an element or compound exists in more than one form. Carbon is an example of an element found in different forms (e.g., carbon black, graphite, and diamonds). Oxygen has three allotropes: monoatomic or nascent oxygen (O); diatomic oxygen (O_2), the gas we breathe; and triatomic oxygen (O_3), which is known as ozone.

Atomic Structure

Early Ideas of Atomic Structure

As mentioned in the previous section titled "A Short History of Chemistry," many scientists identified elements, determined their characteristics, similarities, and differences, and designed symbols for them. Using unique experiments, scientists devised ways to define the structure of atoms and determine atomic weights, sizes, and electrical charges as well as energy levels for atoms.

Many of these men and women recognized the existence of some order in the manner in which chemicals relate and react to each other. Although these scientists could not see the atoms themselves, they were aware that the structure of each element's atoms has something to do with these characteristics. There were several attempts to organize the elements into a chart that reflected the particular nature of the atoms for these elements. Before the periodic table of the chemical elements was developed as we know it today, several relationships had to be established. (See the next section for more on the periodic table of the chemical elements.)

The concept of electrons had been known for many years, but determining how these negatively charged particles react required experimentation and analysis of data. In about 1897, Joseph John Thomson (1856–1940) sent streams of electrons through **magnetic fields,** which resulted in the dispersion or spreading of the electrons. Thomson's experiments, and those of others, led him to speculate that the atom was a positively charged "core" and that negatively charged particles of energy surrounded and matched the positive charge of this core or nucleus. Further, when these electrons were excited or "stirred up" with strong light, electricity, or magnetism, some of them were driven from the outer regions of the atom. This was one of the first experimental evidences for the structure of the atom. Many refinements of this concept were made by the work of several scientists, including the French chemist Marie Sklodwska Curie (1867–1943), the British physicist Baron Ernest Rutherford (1871–1937), the American physicist Robert Andrew Millikan (1868–1953), and others of many nationalities.

Two of Rutherford's students conducted a classic experiment to determine the structure of an atom. They beamed **alpha particles** through a sheet of gold foil that was 1/50,000 of an inch thick. Since the thickness of this thin foil was only about two thousand atoms of gold, it should have allowed all the alpha particles to pass through and be detectable on the other side of the foil. The experimenters recorded the pattern of alpha particles with a detecting instrument located behind the foil. Rutherford noticed that most particles went straight through the foil as if nothing interfered with them. However, unexpectedly, a few seemed to be diverted from the target. As it turned out, one out of every 10,000 alpha particles, as it passed through the foil, was deflected sideways, away from the center of the target located behind the foil, similar to a billiard ball glancing off another ball when struck. During the experiment, one out of every 20,000 particles bounced back from the foil toward the source of the alpha particles and thus did not pass through the gold foil at all. This indicated that some alpha particles were being deflected by something in the foil. After making some calculations, Rutherford concluded that this backward and side scattering of the alpha particles was evidence of a few collisions with something that had almost all its mass concentrated in a central, very small "nucleus." He also determined that this tiny nucleus had a positive charge. Since the vast majority of the alpha particles passed straight through the foil to hit the center of the target used to detect the particles, the atoms of gold must be composed of mostly empty space. He stated, "It was almost as incredible as if you fired a 15-inch shell at a piece of tissue paper, and it came back to hit you."[*] One way to understand the great distance between an atom's nucleus and its surrounding electrons is to imagine a baseball on the floor of your living room as the nucleus of an atom. The electrons that surround the nucleus are about the size of peas that are orbiting in three dimensions at about a 10-mile radius from the baseball. This means that an atom with a 20-mile diameter would have a nucleus about the size of a baseball. The conclusion may be that there is more empty space than there is solid matter in matter and the universe, including in you.

Robert Andrew Millikan (1868–1953) continued the work of J. J. Thomson by conducting a classic experiment to determine the exact charge and mass of the electron. He constructed two flat round brass plates with a one-centimeter space between them. He drilled a tiny pinhole through the top plate and arranged them so that the top plate carried a positive charge and the bottom plate a negative charge. Placing a microscope at the edge of the opening between the plates, he then atomized thin oil droplets over the top plate. As the oil droplets drifted through the tiny hole and moved between the plates, Millikan measured the time it took the droplets to descend from the top to the bottom plate. By varying the electric charge on the droplets, he could control their rate of descent across the space between the charged plates. This enabled him to determine the negative electrical charge of electrons. His figure was close to the accepted figure of $-1.602177 \times 10^{-19}$ coulombs. From this data, he then calculated the mass of the electron to be 1/1,835 that of the hydrogen atom. The hydrogen atom is essentially composed of a single proton with a single negative charge, or electron.

Some questions still required answers before a logical organizational chart of the elements could be completed, among them how to determine the energy, position, and number of electrons in atoms of different elements. When electrons are "excited" by an input of energy, they jump to a higher energy level, orbit, or shell—that is, from a shell (energy level) that is closer to the nucleus to a higher energy level further from the positive nucleus—and then they return to their original energy level closer to the nucleus, or they are driven off the outer region of

---

[*] Rutherford, Ernest. Qtd. in Aaron J. Ihde. *The Development of Modern Chemistry.* New York: Dover Publications, 1984.

the atom. When the energized electrons return from their higher energy position to their original lower energy position, they emit the acquired energy in unique electromagnetic frequencies, across the **electromagnetic spectrum.** This energy release occurs in uniform "packets" of energy known as photons (**quantum** units of electromagnetic radiation). This jump of the electron(s) changes the atom and is known as a quantum transition or a quantum leap of radiant energy, units of which cannot be divided into smaller bits. Contrary to the media's use of "quantum leap" when referring to some large change or action, a quantum leap is a very, very small bit of matter or energy.

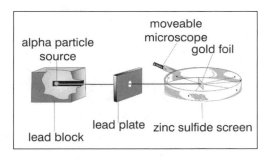

Figure 2.1: Rutherford's experimental apparatus to determine the structure of the atom.

Max Karl Ernst Ludwig Planck (1858–1947), following some theories of electromagnetic radiation, devised a mathematical equation to explain the relationship between energy and wavelengths of light. Now referred to as Planck's constant, it is an invariant that, when multiplied by the frequency of electromagnetic radiation, describes the amount of energy contained in one quantum. It is an extremely small measurement, but the concept is so fundamental that it heralded the beginning of modern physics in the early 1900s.

Werner Karl Heisenberg (1901–1976), the founder of quantum mechanics, described the behavior of very small particles and their interactions based on some higher mathematical concepts. The theory is somewhat easier to understand in a descriptive sense. His most unusual theory changed classical physics and chemistry and is now referred to as indeterminacy or the "uncertainty principle." In essence, it states that it is impossible to know both the exact position and the momentum (mass) of a subatomic particle at the same moment. Thus, the ultimate nature of matter is not susceptible to objective measurement (indeterminacy). However, an accurate prediction (statistical average) of the nature of matter is possible when based on statistical probabilities. An interesting aspect of the Heisenberg uncertainty principle is that merely viewing and attempting to measure subatomic particles may alter their positions and other characteristics. It might be mentioned that this physics concept of quantum uncertainty has nothing to do with our macro-world's philosophical concepts of being "uncertain" or of sociological relativity.

Wolfgang Pauli (1900–1958), an American physicist, was awarded a Nobel Prize in 1945 for developing the "exclusion principle." In essence, it states that a particular electron in an atom has only one of four energy states and that all other electrons are excluded from this electron's energy level or orbital. In other words, no two electrons may occupy the same state of energy (or position in an orbit around the nucleus). This led to the concept that only a certain number of electrons can occupy the same shell or orbit. In addition, the wave properties of electrons are measured in quantum amounts and are related to the physical and, thus, the chemical properties of atoms. These concepts enable scientists to precisely define important physical properties of the atoms of different elements and to more accurately place elements in the periodic table.

The terms "**shell,**" "**orbital,**" and "**energy** level" are sometimes used interchangeably. In this text we use the term "shells" most often because it is descriptive and conveys the image of three-dimensional layers or structures of electrons surrounding the nucleus. The term "orbit" is more closely related to the image of two-dimensional concentric rings, similar to an archery target. The term "energy level" describes the energy that electrons possess, depending on their distance from the atom's nucleus. The term "orbital" is the more descriptive term and describes the original four distinct energy levels detailed in Pauli's exclusion principle based on the angular momentum quantum numbers. Because the original four energy levels for electrons surrounding the nucleus were identified by Wolfgang Pauli and others, three more orbitals (energy levels) have been added to accommodate new artificial elements as they have been discovered. Shells, orbitals, and energy levels make up the levels of electron energy surrounding the nuclei. Similar to "shell," "orbital," and "energy level," the notions of subshells or orbitals are used interchangeably. This electron structure is depicted by the figure for each element listed in the section, "Guide to the Elements." The patterns for electron configuration for each of the following seven energy levels (shells or orbits) follow:

1. The first shell or energy level out from the nucleus is called the "K" shell or energy level and contains a maximum of two electrons in the "s" orbital—that is, K = s2, where the "K" represents the shell number (or principle quantum number), the "s" describes the orbital shape of the angular momentum quantum number, and the "2" is the maximum number of electrons that the "s" orbital can contain. This particular sequence is "K = s2," which means K shell contains 2 electrons in the "s" orbital. This is the sequence for the element helium. Look up helium in the text for more information.

2. The second shell or energy level is "L" and may contain a maximum of eight electrons; its orbital is called "p" and can contain a maximum of six electrons. Therefore, its sequence would be K = 2 (s2) and L = 8 (s2 + p6). This sequence of 10 electrons in the first two shells (K and L) represents the element neon, which has an atomic number of 10.

3. The third shell or energy level is "M" and may contain a maximum of 18 electrons; its orbital is called the "d" subshell, and it may have a maximum of 10 electrons: for example, K = 2 (s2), L = 8 (s2 + p6), and M = 18 (s2 + p6 + d10). See the elements in the text for more details.

4. The fourth shell or energy level is "N," which may contain a maximum of 32 electrons; its orbital is called "f" subshell: for example, K = 2 (s2), L = 8 (s2, p6), M = 18 (s2, p6, d10), and N = 32 (s2 + p6 + d10 + f14). Refer to the elements in the text for more information.

5. The fifth shell or energy level is "O," also with a possible maximum of 32 electrons; its highest orbital is also the "f" subshell: K = 2 (s2), L = 8 (s2, p6), M = 18 (s2, p6, d10, f14), and O = 32 (s2, p6, d10, f14).

6. The sixth shell or energy level is "P," with a maximum of 10 electrons. Remember that the numbers of electrons in each shell (K, L, M, N, O, and P) for each individual element are added together to find the total number for a particular atom, which is also the element's atomic number.

7. The seventh shell or energy level is "Q," with a maximum of two electrons, and is represented as Q = 2 (s2) with just two electrons in it first orbital. (This sequence holds until the element ununtrium-113, where Q = 3 (s2, p1), and those heavy elements beyond 113, where the "Q" shell may contain more than three electrons.)

Note: All elements with a depiction of higher energy levels or shells are both synthetically produced and radioactive, and electrons are added to inner shells rather than the usual outer shell of the elements. As these heavy elements' atomic numbers increase, their half-lives decrease to a fraction of a second, and they are produced one or few atoms at a time. This concept is depicted with the data presented for each element.

Some Theoretical Atomic Models

The earliest concept of atomic structure dates back to Greece in the fifth century BCE, when Leucippus and Democritus postulated that tiny particles of matter, which they called *atomos,* were indivisible.

Over the centuries, many other concepts were proposed to explain the nature of matter—many of them extensions of the Greek concept of an ultimately indivisible and indestructible elementary bit of matter. But it was not until J. J. Thomson proposed his model of the atom, consisting of a sphere with an agglomeration of particles with negative electric charges somehow positioned randomly inside a very small ball of matter, that the modern structure of the atom began to take shape.

The Rutherford model of the atom is a significant improvement over the Thomson model. Baron Ernest Rutherford incorporated the background and understanding of many scientists as he developed experiments designed to show that the atom has a central and small, but relatively heavy, nucleus. His experiments verified that this positively charged dense nucleus has negatively charged electrons surrounding it at a very great distance compared to the size of the nucleus. This concept resembles the planets revolving around the sun, including with regard to the laws of motion and energy.

The Bohr model of the atom took shape in 1913. Niels Bohr (1885–1962), a Danish physicist, started with the classic Rutherford model and applied a new theory of **quantum mechanics** to develop a new model that is still in use, but with many enhancements. His assumptions are based on several aspects of quantum theory. One assumption is that light is emitted in tiny bunches (packets) of energy call **photons** (quanta of light energy).

The Bohr model continues:

First, the orbital or quantum theory of matter assumes that the electron is not a particle, as we normally think of particles. **Orbital theory** considers the electron as a three-dimensional wave that can exist at several energy levels (orbitals), but not at the same time.

Second, the electrons are in constant motion around the nucleus, even though it is not easy to determine the position of a particular electron in its shell at any particular moment.

Third, the electrons are revolving at different distances from their nuclei; these pathways are called shells, orbitals, or energy levels.

Fourth, there are more than a few shells, orbitals, or energy levels in most atoms.

Fifth, the electrons can continue to move in a specific shell without emitting or absorbing energy.

Sixth, an electron at a specific energy level will remain in its shell until it either loses energy and "jumps" to an outer shell or gains energy and proceeds to a higher energy level (shell). Seventh, if the electron is excited by external energy, it can "jump" to a different shell or higher energy level.

Eighth, when the electron returns to its former shell or lower energy level, it will emit the energy (photon), which represents energy the electron acquired to raise it to the higher energy level.

Bohr's idea led to the comparison that likened the structure of the atom to the structure of an onion. The outer layers of skin on an onion are the "shells" where the electrons exist.

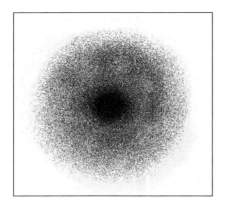

Figure 2.2: Artist's depiction of the "fuzzy ball" atom whose electrons had no sharp boundaries but exhibited limits to the number of energy levels.

At the center of the onion is a dense, tiny BB-like shot for a nucleus. Unlike the onion, however, the area between the nucleus and the electron's shells is merely space. There is no distinct boundary for the shells. The electrons assigned to a particular shell are in constant motion, so the shell does not seem to have a sharp definition. It is something like a "fuzzy ball," with no distinct edges.

A similar model proposed at that time was related to "raisins in a plum pudding," with the electrons resembling raisins spread throughout the pudding. The fuzzy-ball or raisin-in-plum-pudding model of the atom has no sharp boundaries, but does have definite limits as to the number and energy levels for the electrons residing within each shell. The fuzzy ball depicts the energy concept of the "electron cloud," which considers the electrons as energy levels around the nucleus. The concept states that the atom is spherical, and the electrons are all over the atom at any one time. The electron cloud concept statistically depicts the probable distribution of electrons as they exist at any particular distance from the nucleus at any one particular time. The fuzziness is the image one might see while viewing the atom over an extended period, such as a time exposure with a camera.

The number of electrons in each shell is partially dependent on how far the shell is from the nucleus. In addition, electrons assigned to a specific shell stay in position until forced out by an input or loss of energy. Even more important, a particular shell or energy level "dislikes" having any electrons missing. If a shell does not have its complete quota of electrons, the atom will "bond" with other atoms by "taking in electrons" or "giving up electrons" or "sharing electrons" in order to maintain a complete outer shell, thus forming molecules. This is the essence of how atoms of elements form molecules of compounds, and it is the essence of chemical reactions.

There are seven possible shells or energy levels for electrons surrounding the nucleus at a relatively great distance. The lightest atoms have only one shell, which is the innermost shell closest to the nucleus. Other atoms have multiple shells, and the largest and heaviest atoms have all seven shells of electrons. All the electrons in a particular shell have the same energy. The electrons at the greatest distance from the nucleus are ones with the weakest attraction to the nucleus; thus, they are usually the first electrons to be involved in a chemical reaction. These outer electrons then become the **valence** electrons.

As previously mentioned, orbital theory

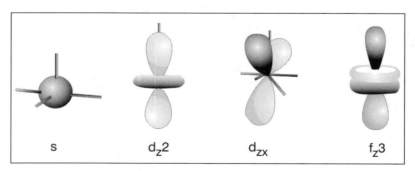

Figure 2.3: Representation of orbital shapes.

Quantum Number Energy Level	Shells	Maximum Number of Electrons	Suborbitals
1	Shell K	2	s2
2	Shell L	8	s2, p6
3	Shell M	18	s2,p6,d10
4	Shell N	32	s2,p6,d10, f14
5	Shell O	32	s2,p6,d10, f14
6	Shell P	18	s2,p6,d10
7	Shell Q	2	s2

Figure 2.4: Table of seven energy levels, shells K through Q, maximum number of electrons in each shell, and the suborbital sequences in each shell.

is based on quantum mechanics, which states that the position of electrons cannot be precisely determined in their orbits, but rather they are somewhat similar to an electron cloud. However, their positions can be predicted by laws of mathematical probability. Orbitals take the shape of indefinite spheres and elliptical-shaped doughnuts. Each orbital is assigned a letter: *s, p, d,* or *f.* The energy levels, shells, orbitals, and their electron numbers are presented along with figures of each individual element.

Following is a list of the seven major shells, including the four orbital energy levels, with the letters assigned for their position extending from the nucleus. The number of electrons required to fill each shell and orbital is shown. This is also the maximum number of possible electrons in the shells of the known elements.

Up to this point, we have been describing single atoms and their electrons. **Chemical reactions** occur when electrons from the outer shells of atoms of two or more different elements interact. **Nuclear reactions** involve interactions of particles in the nucleus (mainly protons and neutrons) of atoms, not the atoms' electrons. This distinction is fundamental. The former is atomic chemistry (or electron chemistry), and the latter is nuclear chemistry (or nuclear physics).

Some confusion exists. For instance, in normal chemical reactions, the electrons of different element's atoms interact to form molecules of new chemical compounds. These compounds have different characteristics from their original elements. For example, the explosion of gunpowder is a chemical reaction involving the interactions of electrons of different elements that release a great deal of energy and gases when their atoms combine to form new molecular compounds. In contrast, the atom bomb or an atomic energy power plant derives its energy from the nucleus, not the electrons of different atoms combining. Therefore, it is inappropriate to refer to the "bomb" as an atomic bomb, because the electrons of the atom are not the source of the energy. The **fission** (splitting) of nuclei or the **fusion** (combining) of nuclei is the source of the energy that causes nuclear explosions. A detonation of a so-called atom bomb or the production of energy in a nuclear power plant is not a chemical reaction but a nuclear reaction. To nonscientists there is little distinction between a chemical weapon and a nuclear weapon—both are deadly, but the distinction is fundamental to chemistry and physics.

The Nucleus and Radiation

Many scientists have contributed to concepts of radiation as produced by the particles making up the nuclei of atoms, and descriptions of some of those scientists and their work follow.

While experimenting with cathode ray tubes (similar to television screens of the 1950s), Wilhelm Konrad Röentgen (1845–1923) discovered an unfamiliar type of radiation. He named these "X-rays", using "X" to signify the "unknown."

Antoine Henri Becquerel (1852–1908) discovered **alpha particles** that are, in essence, the nuclei of the element helium (He, or, as the ion, He^{++}). Compared to electrons, they are heavy, only travel short distances, and may be stopped by a sheet of cardboard.

Ernest Rutherford is one of the discoverers of **beta particles**, which are somewhat like high-speed electrons that can travel at the speed of light over great distances. An example is electrons traveling over wires to our homes. Actually, an individual electron may not make the entire trip over the wire. Rather, electrons are "pushed" along the wire by a potential difference in voltage from their source to their destination. High-energy beta particles (^-e) stripped from the atom can penetrate several sheets of paper. An electron has a mass that is almost two thousand times less than the mass of a proton in the nucleus. Rutherford is also credited with the discovery of the fundamental positive particle of the nucleus known as the proton. Both Rutherford and James Chadwick (1891–1974) conducted experiments that led to the discovery of the neutron, the other fundamental particle of the nucleus. The neutron has approximately the same mass as the proton, but it is neutral with no electrical charge. In 1932, Chadwick was credited with the neutron's discovery.

Marie Curie and Pierre Curie (1859–1906), who discovered radium by extracting it from uranium ore, experimented with **gamma rays.** Gamma rays have a short wavelength, travel at the speed of light, and can penetrate several inches of lead, depending on their source and energy level. There are several types of gamma rays. Some examples are the product of short bursts of protons in a **cyclotron** or "atom smasher," cosmic radiation from outer space, and the natural disintegration of the nuclei of radioactive atoms in the Earth. Most of the heavier elements with an atomic number above 88 are **radioactive** to some degree See Table 2.1 for some information on the major sources and forms of radiation.

Previously mentioned was the classic experiment by Ernest Rutherford in the early 1900s that established the concept that most of the mass of an atom exists as a tiny center of positively charged matter, called the nucleus. This nucleus is only about one millionth of the diameter of the whole atom. Rutherford's experiment also established that the negative electrons surrounding the nucleus are at a relatively great distance from this very dense central mass. In addition, the electrons make up only an extremely small fraction of the atom's weight (mass). Thus, most of the atom is empty space, consisting of a tiny, positively charged, relatively heavy, dense core surrounded by distant, and much lighter, negatively charged electrons.

Many other experiments also verified the concept that the nucleus consists of two major types of particles: **protons,** which carry a positive charge, and **neutrons,** which have a similar mass to the protons, but have no electrical charge. Thus, the total mass of a nucleus consists of the total number of both protons and neutrons. Together, they make up all but a tiny fraction of the weight of an atom. The protons, by themselves, are the source of the positive electrical charge of the nucleus, which is balanced by the negatively charged electrons.

Table 2.1: Major atomic particles and forms of radiation.

Type	Symbol	Mass	Charge
Electron	-e	1/1,835 of proton	-1
Proton	+p	1	+1
Neutron	n	~1	0
Alpha	α	4 (helium nuclei)	+2
Beta	β	~0 (hi energy electrons)	-1
Gamma	γ	0	0

The science of particle physics continues to study electrons, protons, and neutrons, which are considered **subatomic** particles. The quest continues for even smaller subatomic, or rather subnuclear, particles. Most subnuclear particles are fleeting in time of existence, are practically weightless, and are thus very difficult to detect and measure.

The Fermi National Accelerator Laboratory, located near Chicago, operates the Tevatron, which is the second-largest "atom smasher" in the world. (The largest is CERN, the Conseil Européen pour la Necherche Nucléaire located near Geneva, Switzerland.) It is used by particle physicists to advance research in high-energy physics and to determine the nature of matter. In the Tevatron, counter-rotating beams of protons and antiprotons produce collisions, allowing scientists to examine the most basic building blocks of matter and the forces acting on them. Important discoveries have been made relative to the interactions of the particles and forces that determine the nature of matter in the universe. Some examples of research that Fermilab is addressing in the twenty-first century are the following:

- Why do particles have mass?
- Does neutrino mass come from a different source?
- What is the true nature of quarks and leptons? Why are there three generations of elementary particles?
- What are the truly fundamental forces?
- What are the differences between matter and antimatter?
- What are the dark particles that bind the universe together?
- What is the dark energy that drives the universe apart?
- How does the universe work?
  (*Research at Fermilab*, http://www.final.gov/pub/about/experiments/, accessed 18 March 2004)

The Tevatron is a 4.5-mile-long tunnel through which scientists send high-speed protons while sending high-energy **antiprotons** (**antimatter**) from the other direction. When the particles meet in the Tevatron, a tremendous burst of energy results in the production of smaller particles.

Both protons and antiprotons are made of **quarks.** When their quarks collide, there is evidence of smaller particles. Quarks are hypothetical entities that carry very small electrical charges. They are considered the major constituents of the smallest bits of matter. Both quarks and **leptons** (several lighter atomic elementary particles) are the basic building blocks of matter. There are three basic leptons, each with an associated particle known as a **neutrino.** The three major leptons are (1) the negative electron (-e) with its neutrino, (2) the negative **muon** (-µ) with its neutrino, and (3) the negative **tau** (-τ) with its neutrino. The electron also has an **antiparticle** called the **positron** (+p). Of these three types of leptons, chemistry is most interested in the electron and its characteristics and associations with chemical reactions. All of these subatomic particles (and many more) serve the purpose of facilitating calculations. Some smaller, more fleeting bits of matter than the quark exhibit properties of both particles and waves. There are many such wave/particles with odd names that have been detected by more elaborate experimental equipment. By separating and measuring smaller and smaller bits of matter, scientists develop new understanding of the origin and nature of matter.

Chemical Bonding

Chemical bonding is a force exerted between atoms that is strong enough to combine two or more atoms together as a single unit (molecule). If just two elements combine to form a molecule, it is known as a binary compound. All compounds consisting of two or more different elements follow the same rules of bonding. These types of chemical reactions are interactions of electrons that "join" the shells surrounding the atoms of different elements. How these electrons of different atoms combine to form molecules is the essence of the science of chemistry. There are three major energy reactions related to **ionic** and **covalent** types of chemical bonding:

1. *Ionization energy* is the energy required to remove an electron from a neutral atom, changing the atom into a positive **ion.** The alkali metals on the left side of the periodic table exhibit the least ionization energy because they have only one electron in their outer shell. In general, the ionization energy increases from left to right for the elements on the periodic table. This means that the noble gases with a closed shell exhibit the maximum ionization energy.
2. *Electron affinity* occurs when a neutral atom attracts an electron to become a negative ion. Electron affinity is related to the energy changes when an electron is added to a neutral atom, changing it into a negative ion.
3. *Electronegativity* is the ability (in terms of structure and energy level of electrons) of either neutral atoms or molecules to attract bonding electrons to them. In essence, electronegativity is a measurement of how effectively an atom within a molecule is able to attract bonding electrons to itself.

Let's define the difference between an atom and an ion. The atom has a neutral charge because of the equal number of negative electrons and positive protons. Ions that have negative charges are called **anions,** and positive ions are called **cations.** The charge on an atom that transforms the atom into an ion results from the gain or loss of one or more electrons. **Radicals** also act as ions, or rather polyatomic ions, in chemical reactions because of their electrical charge. Some of these types of ions are chemicals known as "**free radicals**" that are fragmented molecules having one or more unpaired electrons, thus "free" electrons. They are short-lived and highly reactive. In 1815, the French scientist Joseph-Louis Gay-Lussac (1778–1850) experimented with gases used to fly balloons. He discovered one gas that turned out to be poisonous, which he named "cyanogen" (C_2N_2). He determined that the CN combination of carbon and nitrogen atoms, which he called the "cyano group," had tight bonds between the C and N atoms, and thus the group acted together as single charged atoms in chemical reactions. This knowledge advanced the field of organic chemistry. These groups or fragments of molecules became known as "organic radicals." There are three types of free radicals:

1. simple single atoms such as O^-, F, Cl;
2. inorganic free radicals, such as OH, SO_4^{--}, NH_4^+; and
3. organic free radicals, such as CH_3^+, $C_6H_6^{++}$.

Generally, free radicals are formed when the stable bonds in molecules are ruptured, resulting in two molecular fragments, each with unpaired electrons. Radicals can be single-charge atoms or just separated parts of a molecule that have one or more "free" electrons. Most free radicals are extremely reactive, but short-lived. These characteristics cause some concern with regard to their negative role in health and aging.

Chemical compounds are composed of two or more atoms formed into molecular arrangements wherein the total energy of the new molecule is less than the total energy of the constituent atoms of the molecule. Atoms have the unique capacity to maintain electrons in their outer shells, each having a complete number of the required electrons (2, 8, 18, 32). For instance, if an atom of an element has a single electron in its outermost shell, it will reactively give that electron up or share it with another atom that requires an electron to complete its outer shell. A simple example: The element sodium is a very reactive pure metal with a single electron in its outermost shell that it wants to give up so that it can have a complete shell at the next lower level. It will react with an element, usually a nonmetal such as chlorine that has only seven electrons in its outer shell. Chlorine "wants" to gain one electron to complete its outer shell at eight. Thus, both elements react to complete their outer shells. Sodium gives up an electron, and chlorine accepts it, creating a new molecule of sodium chloride. When this occurs, sodium has a 1^+ charge, and chlorine has a 1^- charge, and the joining between the two is known as an **ionic bond.** In other words, the ionic bond exists where one or more electrons that form a neutral atom are moved to another atom, which in turn results in both a positive and negative ion in a new molecule. In this example, ionic bonds hold the atoms together to form the new molecule of a compound called sodium chloride, or table salt, which has none of the characteristics of the two original elements: $(2Na + Cl_2 = 2Na^+Cl)$.

Another type of bond is the **covalent bond,** in which one, two, or more pairs of electrons are shared by two or more atoms. Unlike ionic bonds, covalent bonds involve the sharing of electrons by atoms. The simplest example of covalent bonding occurs when two atoms of hydrogen bond to form a diatomic hydrogen molecule ($H\bullet + \bullet H$ yield H:H, or H_2).

There are several other specialized types of bonding as well:

1. *Polar covalent bonds* (in which the electrons spend more time with one atom or another atom in a molecule)
2. *Non-polar covalent bonds* (in which the electrons spend equal amounts of time with all atoms of the molecule)
3. *Metallic bonds* (somewhat similar to ionic bonding)
4. *Proton bridging* (hydrogen bonding where hydrogen acts as either a + or –ion)

The most common element that exhibits covalent bonding is carbon, which has four electrons in its outermost shell, thus permitting carbon to mutually exchange or share electrons with other elements. The molecules and compounds formed by covalent bonding reactions involving carbon are mostly organic compounds (biochemistry) and hydrocarbons (fossil fuels).

Organic Chemistry

Perhaps a more accurate term than "organic chemistry" might be "carbon chemistry," considering that the major element in organic molecules is carbon. Carbon is a unique element with a structure of six protons, six neutrons, and six electrons that results in its being a most versatile element, and it is used to produce a multitude of organic (carbon) compounds. (See the section on carbon for more about this element.) Organic compounds have been used by humans, as well as all other living organisms, since the beginning of life on Earth. However, it was not until the late nineteenth century when scientists begin exploring the nature of all the organic chemicals that we began to use it every day. Organic chemistry is the study of two basic types of carbon compounds. One, *biochemistry* is the chemistry of compounds that make up living organisms and their products, such as sucrose (table sugar) with the formula of $C_{12}H_{22}O_{11;}$ and two, *hydrocarbon* chemistry includes the study of both natural and man-made compounds composed mainly of carbon and hydrogen, such as natural gas (methane) with the formula CH_4 and propane cooking gas with the formula C_3H_8.

The organic carbon compounds identified in the science of biochemistry are composed of some very large molecules. For instance, all carbohydrates (sugars, starches), fats/lipids (lard), proteins (meat), and nucleic acids (DNA) consist of organic compounds composed of many large molecules. Many of these compounds are constituents of living organisms. The products we use, such as fossil fuels (coal, oil, gasoline, natural gas) and some cosmetics, wax, soap, and alcohol, are hydrocarbons that are related to biochemistry only in that they contain carbon. Hydrogen and several other elements combine with carbon to form many interesting and useful compounds. Although this book deals mostly with inorganic (nonliving and mostly noncarbon) chemistry rather than organic chemistry, it also emphasizes everyday use of the chemical elements. This requires some understanding of organic chemistry and how and why carbon is such a major factor in many of our useful chemical products.

Over the years, many people contributed to the development of the field of organic chemistry. To better understand how this science provides so many useful items for our daily use, it is necessary to be familiar with some of the nomenclature of organic chemistry. There are two basic types of hydrocarbon substances, namely, **aliphatic** and **aromatic.** There are three basic types of aliphatic hydrocarbon molecules defined by the number of bonds involved in straight linear-chained molecules. If the basic structure of a hydrocarbon molecule is a ring instead of a straight chain, they are known as aromatic hydrocarbons, typified by the benzene ring.

Aliphatic hydrocarbon molecules, normally called "hydrocarbons," are divided into several major groups. Their molecules are mostly formed as straight or branched chains that form three major subgroups:

1. *Alkanes* (saturated straight or branched chains of carbon atoms)

    For example, H_3C—CH_2—CH_3 (Propane), and

$$CH3$$
$$|$$
$$H3C—CH—CH3 \quad (Butane)$$

The general formula for alkanes is C_nH_{2n+2} (where n = a small whole number such as 1, 2, 3).

(Note: The term *alkyl* is often used to identify a paraffinic hydrocabon group of alkanes by dropping one hydrogen atom from the formula, for example, methyl CH_3, or ethyl C_2H_5.)

2. *Alkenes* (unsaturated and very reactive olefins)
3. *Alkynes* (highly saturated acetylenes with triple bonds)

Hydrocarbon molecules that have only single bonds (C–C) are known as **saturated hydrocarbons,** whereas **unsaturated hydrocarbon** molecules have double or triple bonds (C=C or C≡C). A very logical system that assigns names to the structures of these types of hydrocarbons uses Greek prefixes to identify the number of carbon atoms in a particular type of hydrocarbon molecule (see Table 2.2).

The molecular formulas just shown for 10 alkane hydrocarbon molecules represent the proportions of carbon to hydrogen in each molecule. These formulas do not reveal much about their structures, but rather indicate the proportions of each element in their molecules. Each molecule may have several different structures while still having the same formula. Molecules with different structures but the same formulas are called **isomers.** For example, n-butane is formed in a straight chain, but in an isomer of butane, the CH_3 branches off in the middle of the straight chain. Another example is ethane, whose isomeric structure can be depicted as H_3C H_3C–CH_3. The name for the normal structure sometimes uses "n" in front of the name.

In 1825 Michael Faraday (1791–1867) discovered an unknown substance that was produced from heated whale oil. Later, Eilhardt Mitscherlich (1794–1863) isolated this new compound and named it benzene.

In the mid-19th century Fredrich von Kekule (1829–1886) determined that carbon was a tetravalent atom with a valence of four, capable of forming a number of different compounds including a great variety of organic molecules found in living tissues. It was Michael Faraday who determined that the molecule for the aromatic compound benzene contained six carbon atoms with a total of 24 bonding electrons, but benzene also had six hydrogen atoms but with

Table 2.2: Ten alkane hydrocarbon molecules.

Chemical name	Numerical prefix	Molecular formula	Number of isomers
Methane	meth (1)	CH4	1
Ethane	eth (2)	C2H6	1
Propane	prop (3)	C3H8	1
Butane	but (4)	C4H10	2
Pentane	pent (5)	C5H12	3
Hexane	hex (6)	C6H14	5
Heptane	hept (7)	C7H16	9
Octane	oct (8)	C8H18	18
Nonane	non (9)	C9H20	?
Decane	dec (10)	C10H22	?

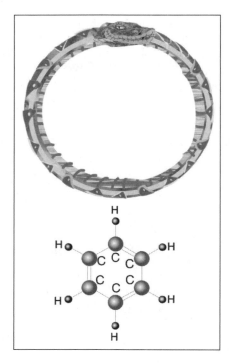

Figure 2.5: Artist's depiction of Friedrich August von Kekule's dream of a snake eating its own tail, a dream that aided von Kekule in solving the problem of the structure for the organic compound benzene, composed of a ring of carbon atoms.

only one bonding electron each. When this combination of 24 carbon bonds and six hydrogen bonds were diagrammed, either as a straight line of carbon atoms with the hydrogen atoms attached or as a branching structure, the bonding arrangement just did not add up. Thus, $6 \times 4 = 24$ for carbon, and $6 \times 1 = 6$ for hydrogen resulted in too few electrons to satisfy the octet rule for a linear or branched structure such as C_6H_6. It is reported that Kekule solved this problem one night in a dream in which he saw a different configuration that resembled a snake eating its own tail. He woke up excited and, by working the rest of the night, came up with the structure of the benzene ring.

The ring consisted of each carbon atom sharing two of its four bonding electrons with another carbon atom, one valence electron with a partner on the other side on the ring, and one valence electron with a hydrogen atom outside the ring, resulting in the classical hexagonal benzene ring. This answered many questions and was a revolution for organic chemistry because it was now possible to substitute other atoms, molecules, or radicals for one or more of the six hydrogen atoms in one or more connected rings.

It was not until the 1930s that the complete picture of the benzene ring became clear. Linus Pauling (1891–1994) determined that the benzene rings are of two alternating structures wherein the double and single bonds between the carbon atoms in the ring alternate positions between the two side-by-side rings. Therefore, the molecule should be thought of as **resonance structure** that is a hybrid of the two joined rings. Because of this hybrid structure, the benzene molecule is very stable and thus requires high temperatures and pressures while using catalysts to substitute atoms of other elements for the hydrogen on the outside of the ring. Many useful products are made by substituting various aliphatic groups for the hydrogen atoms of benzene. Two benzene rings may join to form aromatic compounds such as naphthalene. Through organic synthesis, putting together organic molecules, as with building blocks, millions of new and useful organic compounds are produced.

Fullerenes and Nanotechnology

The 1996 Nobel Prize in chemistry was awarded to three American scientists for their identification of a new allotrope of aromatic carbon molecules called "fullerenes." These unusual carbon molecules form a closed-cage structure of joined carbon atoms. The original soccer ball–shaped carbon molecule called buckminsterfullerene contained 60 carbon atoms and was nicknamed "Bucky Ball" in honor of Buckminster Fuller (1859–1983), who used similar shapes in some of his architectural structures. Since then additional organic pentagon structures beyond the original icosahedral fullerene (C_{60}) have been developed, all with an even number of fully saturated carbon atoms, ranging up to C_{240}. Because they are extremely rare and expensive to create, they have not found many uses outside of the field of chemistry.

Nanotechnology combines the sciences of chemistry, physics, and engineering at the quantum level to build structures the size of one billionth of a meter (10^{-9}) from single atoms and molecules. For comparison, a human blood cell is 10 micrometers in diameter, an electrical path on a computer chip is 1 micrometer (10^{-6}), and a virus is 100 nanometers in diameter. Bucky Balls are 1 nanometer in diameter, and nanotubes are less than 1.5 nanometers in diameter. These nanotubes are stronger than steel and could reduce the weight while increasing the strength of cables and wires as well as airplanes, cars, and other products. Many existing and new companies are investing millions of dollars in this new industrial revolution. The U.S. government is spending millions to assist in developing new and improved products using nanotechnologies. Some possible future products include small lasers, new nano drugs and drug-delivery systems resulting in the cure of many human ailments, faster and smaller laptop computers with longer-lasting batteries, extremely sharp computer and TV screens, improved efficiency for hydrogen fuel-cells used in electric cars, improved crop production, improved scratch resistant products, and much more. The possibilities for future use of nanotechnologies are almost endless. It is speculated that current high manufacturing costs related to nanotechnology will rapidly drop as molecular machines construct more molecular machines that in turn will use individual atoms and molecules to manufacture new or to improve existing products.

Just as modern society will be different with nanotechnology, a similar revolution occurred with the development of both inorganic and organic chemistry. The knowledge of organic molecular compounds and the ability to manipulate them to form a multitude of useful products increased our and the world's standard of living. The development and use of nanotechnology will result in similar advancements.

The development of new technologies, industries, and products can be intimidating. People are often more comfortable with the status quo, and they can be easily misled. At times, the public is bewildered by the term "organic," particularly when purchasing some foods now labeled "organic" that are regulated by the federal Organic Foods Production Act (OFPA). Actually, all food except supplements such as vitamins and some seasonings (e.g., table salt) are organic in the sense that they consist mainly of carbon compounds. The only difference with so-called organic foods is that they are grown using fewer synthetic insecticides and chemical fertilizers, and the crops are grown in specific ways. Nitrogen and other elements are required for plant growth. Plants do not distinguish the source of their required nutrients.

three

The Periodic Table of the Chemical Elements

History

Without a doubt, the periodic table of the chemical elements is the most elegant organizational chart ever devised. Conceptually, many individuals recognized that certain chemical elements have similar characteristics to other chemical elements before the table was developed. This was accepted even though the atoms of a particular element are different from atoms of related elements. How do we determine what the relationships of these similar characteristics are when compared with different elements?

The periodic table of the chemical elements is organized as a matrix of rows of horizontal "periods" that list the elements in their increasing atomic numbers and, generally, according to their atomic weights. For instance, the fourth row in the table is period 4 and starts with group 1 or (1A) element $_{19}K$ (potassium with 19 protons in is nucleus) and continues through element $_{36}Kr$ (krypton with 36 protons in its nucleus) in group 18 or (VIIIA). (There are two different numbering systems used for identifying the groups in the table. This is explained in the section on the periodic table.) Historically, the elements in periods 2 and 4 were identified as repeating their characteristics after eight elements, each increasing in their atomic weights, thus forming a new period. This was known as the "octet rule."

The matrix also includes vertical columns in which elements are arranged somewhat according to similarities between their chemical and physical properties and those properties of the elements located just above and below them in the column, or "group." Thus, the three somewhat similar elements in a group might be thought of as a "triad."

A number of prominent chemists of the late nineteenth and early twentieth centuries contributed to the development of the periodic table of the chemical elements. These scientists arranged, listed, and categorized various chemicals according to observed properties. In 1829, Wolfgang Döbereiner (1780–1849) recognized that similar elements appeared in groups of three, which he called "triads." Two of the triad sets he proposed were chlorine/bromine/iodine and lithium/sodium/potassium. (Notice that the three are in their same vertical group on the periodic table.) Döbereiner realized that the atomic mass (weight) of the middle element of the triad was about halfway between the atomic masses of the other two. Julius Lothar Meyer

(1830–1895) and John Alexander Reina Newlands (1837–1898) both developed charts based on the atomic weights of the elements as known in their time. They recognized the "octal" (eight) nature of repeating characteristics of several elements, which preceded the concept of the horizontal periods for the current periodic table. Newland's table grouped elements in vertical columns of seven elements according to their atomic mass. (The noble elements that would have been the eighth column had not yet been discovered.) Julius Meyer and Dimitri Mendeleev (1834–1907) independently produced periodic charts of the elements. However, Meyer's chart was based on the physical characteristics of elements (e.g., volume versus mass, melting points, and so on), whereas Mendeleev placed data for each element on separate cards and then arranged them in various orders until he noticed gaps in some of the sequences of atomic masses in his card arrangement. He not only corrected the known atomic masses of some elements, but he also left blank spaces in his chart to represent unknown, yet-to-be discovered elements. When the gap was in the middle of a triad of two known elements, he would guess at the **atomic mass number** (the average mass of the other two elements), **atomic proton number,** and other characteristics for the missing element. Then he named these with the prefix "eka," meaning "first" in Sanskrit. For instance, eka-aluminum was later identified as gallium, and eka-silicon proved to be germanium. Within 15 years the "missing" elements were discovered with the basic chemical and physical characteristics based on Mendeleev's predictions. Meyer's chart was published in 1870, and Mendeleev's table appeared the following year.

Several discoveries about the structure of the atom provided information that led to improvements of the older tables and the development of the modern periodic table. For example, in 1897 J. J. Thomson discovered that the tiny negatively charge particle in the atom had a mass of less than 1/2000 of that of the nucleus of hydrogen. In 1920, Ernest Rutherford discovered that during the spontaneous disintegration of radioactive elements, such as uranium, a small but heavy positively charged particle was produced. This was later identified as the alpha particle that was actually the nucleus of helium. When Rutherford shot alpha particles toward a thin gold foil, most particles went through as if the foil did not exist. Just a few of the alpha particles bounced back, and a few were deflected sideways of the foil, giving evidence of the small, heavy positively charged nucleus surrounded by vast amounts of space. In 1932, James Chadwick (1891–1974) discovered the neutron, which had no electrical charge but was of approximately the same mass as the proton. In the late 1800s, Henry Gwyn Moseley (1887–1915) used an X-ray spectrometer to examine the electromagnetic wavelengths of different atoms that he exposed to the X-rays. He observed that each element produced its own specific wavelength, which he then considered as a separate integer that was proportional to the square root of the frequency of its wavelength. This integer is now referred to as the element's atomic number.

However, Mendeleev received credit for devising the modern periodic table of the elements, even though his table was based on atomic mass numbers rather than the atomic proton numbers of the elements. In 1871 he arranged the elements not only by their atomic mass in horizontal rows (periods), but also in vertical columns (groups, also called families) by their valences as well as other chemical and physical characteristics.

The configuration of electrons around the nuclei of atoms is related to the structure of the periodic table. Chemical properties of elements are mainly determined by the arrangement of electrons in the outermost **valence** shells of atoms. (Other factors also influence chemical properties, such as the atoms' size and mass.) An important factor is the "ground state" of an atom, which is the basic energy level in which atoms are usually found. The first electron out from the atom's nucleus is in the lowest energy state. The second shell of electrons represents the next higher level, and so on. Each atom's shell and orbital can hold only so many electrons before additional electrons form new orbitals.

Remember, valence is the combining power, in whole numbers, of one element with another. The most common combining ratio of atoms of different elements to form molecules is 1:1, but other ratios are possible. For instance, hydrogen (H) and chlorine (Cl) both have a valence of one, so they combine to form a molecule of HCl, hydrogen chloride or hydrochloric acid. Nitrogen has a combining power (valence) of three. When hydrogen combines with nitrogen, it takes three H atoms for each N atom to form the resulting compound, ammonia (NH_3).

Mendeleev's chart with the vacant spaces provided him and other scientists a plan to predict where new elements would fit as new discoveries were made. As the chart began to fill in, some problems became evident. In 1913, the British physicist Henry Gwynn Mosely, by using X-rays, identified that the positive charges of nuclei increased as the atomic mass became greater. Thus, the sizes of atoms increased. This discovery led to a correction of the periodic table. Instead of arranging the elements according to their atomic weight, Mosely arranged them by their atomic number—that is, the number of positive protons in the nucleus. This is how the periodic table is constructed today.

Let's take a more detailed look at the periodic table and discover its great symmetry and usefulness. (Refer to the periodic table of chemical elements reproduced on the inside front and back covers of this book.)

Rules for Cataloging the Elements

1. The order in which the elements appear in the periodic table follows specific rules.
2. Changes in properties of the elements repeat in orderly ways.
3. The arrangement of the elements in the table is according to the following factors:
    a. A configuration of **electrons** surrounds the nuclei of all atoms. The electrons are negatively charged particles (units) located at different energy levels, referred to as shells, surrounding the nucleus. For most elements, the number of the electrons in the outermost shells determines the element's "combining power" or valence, or oxidation number.
    b. The number of positively charged particles units (protons) in the nucleus of an atom dictates that atom's atomic number. In other words, the number of protons determines the atomic number, which is also referred to as the proton number.
    c. All of the positive protons plus all of the neutrons, which have no electrical charge, are found in the nucleus. Together, the total number of both the protons and neutrons determines the atomic weight (mass) of each atom as well as the isotopes for that element. For example: atomic number (protons) + number of neutrons = atomic weight. (Note: The atomic weight is actually the average of that element's isotopes' weights as arrived at by their respective proportions found in a particular element.)
4. Periods are the rows that run left to right and follow the octet rule.
5. Both the atomic number (protons) and electrons increase in number from left to right in each period.

6. When a principal energy level (shell) receives its full complement of electrons (e.g., inert noble gases in group 18 (VIII)), a new row begins, which is the start of new period.

7. The vertical columns represent "families" of elements that exhibit similar characteristics. These families of elements are called groups. Elements in the same group, in general, exhibit similar combining powers (valence), but do not exhibit the same degree of reactivity to other elements. Some of these family (group) characteristics are **atomic radius, ionization energy,** and **electron affinity.**

8. Most elements in the same group (family) have the same number of valence electrons in their outermost shell. The outer electrons of an atom involved in chemical reactions are the valence electrons. These electron configurations assist in explaining the recurrence of chemical and physical properties of elements in the same group (family).

9. In general, for groups, when the table is read from left to right, the number of valence electrons in the outer shell of elements increases. For instance, lithium is located at the start of period 2 and has one electron in its outer shell and is found in group 1 (IA), and neon, located at the end of period 2 in group 18 (VIIIA), has eight electrons in its completed outer shell.

   There are several methods for numbering the groups from left to right. *The International Union of Pure and Applied Chemistry* (IUPAC) uses a notation system of Arabic numerals, 1 to 18, for the groups. Roman numerals are used in the old form. Hydrogen is located on the upper left of the table in the first period in group 1. Because hydrogen has properties of both a metal (**electropositive**) and a nonmetal (**electronegative**), some periodic tables place hydrogen at the top of group VIIA, next to helium, as well as in the first period of group 1. Hydrogen is the only element to exhibit both characteristics.

10. **Metals** (on the left side and central area of the table), in general, have fewer electrons in their outer valence shell than do **nonmetals** on the right side of the table.

11. Metals give up valence electrons (thus are electropositive) or share electrons with nonmetals. The most active metals are the ones on the left side of the table that have the least number of valence electrons.

12. Most nonmetals located in the groups on the right side of the table have more electrons in their outer valence shells than do metals.

13. The most reactive nonmetals (electronegative) are those in group 17 (VIIA) on the right side of the table. (See exception noted in rule 19c.) They tend to accept valence electrons from the metals to complete their outer valence shells from seven electrons to form full outer shells of eight electrons.

14. Atomic and ionic sizes of elements, in general, increase from the top of the table to the bottom as the atomic mass and number of electron shells increase.

15. Thus, for both metals and nonmetals, the atomic size increases as the atomic mass (weight) and atomic number increase.

16. The neutron is a fundamental particle of matter found in the nucleus. The neutron has about the same mass as the proton, but, unlike the proton, the neutron has no electrical charge.

17. The total atomic mass (weight) of an atom consists almost entirely of the total mass of both the protons and the neutrons. The negatively charged electron has a mass of less than 1/2000 that of the proton. (The precise figure is 1/1837.) Therefore, the electron's mass is negligible when considering the total atomic mass of an atom.

18. The atomic number is the number of protons (positive charge units) in the nucleus.

19. Exceptions:

   a. Elements in the far right column, group 18 (VIIIA), all have completed outer shells of eight valence electrons. Thus, they do not easily react with other elements. They are known as the **noble elements** or **inert gases.** However, under certain circumstances, they can form some compounds.

   b. By exchanging and sharing outer electrons in chemical reactions, atoms tend to gain a full complement of two or eight electrons in their outer shells and become neutral. If the outer shell has less than four electrons, the element normally gives up electrons in chemical reactions and, thus, is electropositive. On the other hand, if the outer shell has more than four electrons, the atom tends to accept or gain enough electrons to complete the outer shell with a complement of eight electrons. Those atoms that gain electrons are **electronegative.** The result is that atoms that have combined to form a new molecule have now become more stable.

   c. Two elements that do not fit the primary pattern of the table are hydrogen and helium. In some forms of the periodic table, hydrogen and helium are placed together in a separate category because they do not seem to fit any other position in the table.

   Hydrogen, with one valence electron tends to be a diatomic gas (H_2). Although not a metal, hydrogen is sometimes placed in the table to head the alkali metals in the first group. Hydrogen has one electron in the K shell, which is its first and only shell. Because as an atom it has only one electron, it can also collect an electron from another element, including another metal, to form a **hydride**—for example, $2 Na + H_2 \rightarrow 2 NaH$ (sodium hydride). So hydrogen might be thought of as both electropositive and electronegative considering that it can give up its electron, receive an electron, or share its electron. (See the entry for hydrogen in the guide section of this book for more on hydrides.)

   Helium is placed at the top of the far right column of group 18 (VIIIA) because it has two electrons in the K shell (the first shell), which completes its outer valence shell. Helium is inactive. Therefore, it is included in the group with the noble inert gases, even though it has a completed outer "K" shell of two instead of eight electrons in the outer most shell of the other inert gases.

20. The **transition elements** are found in three series in the center of the periodic table, starting in group 3. The first series is in period 4, from scandium ($_{21}Sc$) to zinc ($_{30}Zn$). The second series is in period 5, yttrium ($_{39}Y$) to cadmium ($_{48}Cd$). The third series is a bit different. It is in period 6, following the lanthanide series, and starts at group 4 at the element hafnium ($_{72}Hf$) and continues to include mercury ($_{80}Hg$). The transition elements represent a change in the chemical characteristics of the elements in these series from metals to nonmetals. They show a gradual shift from being strongly electropositive (giving up electrons), as do elements in groups 1 (IA) and 2 (IIA). The shift (transition) continues to the electronegative elements, those gaining electrons, as in groups 15, 16, and 17. Thus, they progress from having some properties and characteristics similar to those of metals to having several properties more like those of **metalloids, semiconductors,** and nonmetals.

21. There are other series of elements that appear separately in the periodic table, namely the **lanthanide** series and the **actinide** series. The superactinide series and the super heavy elements (SHE) are additional series of newly discovered elements. These series of elements are extensions to the normal periodic order of the periodic table.

a. The lanthanide series is also considered metal-like because these elements have two electrons in their outer shells. Because they are difficult to find and identify, they were initially considered scarce. Although it has been determined that they are not scarce, they are still called **rare-earth elements.** There are 15 elements in this series, starting with group 3 in period 6. They include the elements lanthanide ($_{57}$La) to lutetium ($_{71}$Lu).

b. The actinide series is both metal-like and radioactive.

This series also starts at group 3, but in period 7. It includes the element actinium ($_{89}$Ac) and ends with lawrencium ($_{103}$Lr). They are unstable and radioactive.

22. The transuranic elements are a subseries within the actinide series with atomic numbers higher than uranium ($_{92}$U). They include the actinides neptunium ($_{93}$Np) up to lawrencium ($_{103}$Lr). They are man-made, very unstable, and radioactive with very short half-lives. The actinides with higher atomic masses are so unstable that they exist only for microseconds or minutes, and only a few atoms of some have been artificially created.

The transactinide series includes elements that range from rutherfordium ($_{104}$Rf) up to the element with atomic number 113 or 114.

The superactinide series begins where the transactinide series ends, usually with the element with atomic number 114, and continues to the element with atomic number 118.

Super heavy elements (SHE) and possible future elements may someday be discovered beyond element 118.

Isotopes

In chemistry, an isotope is a different form of an element, similar to a different species of plant or animal in biology. All isotopes of an element have the same atomic number (protons in the nucleus), but they have a different number of neutrons and thus different atomic masses. Even though an isotope of an element has a different atomic mass, it essentially maintains the same chemical characteristics of that element. An isotope of an element occupies the same position in the periodic table as does the "regular" form of that element. The atomic weight given for each of the elements that have isotopes is really the average of all the atomic masses for each isotope in proportion to their prevalence (in terms of percentage) within that particular element. When this leads to a fractional average atomic mass, the figure is often rounded off and stated as the closest whole number. There are three kinds of isotopes categorized by their origins. One type is found in nature. It is stable and not radioactive. The second type is naturally radioactive and thus unstable. Isotopes in the third category are produced by artificial bombardment of atoms with neutrons, and they are also unstable and radioactive. Artificially radioactive isotopes are products of nuclear bombardments in nuclear reactors, nuclear particle accelerators (atom smashers), nuclear bombs, and similar sources of high-energy particles. Not all elements have isotopes. In these cases, all the atoms of these elements have the same number of neutrons and atomic mass. The 83 most abundant elements have one or more isotopes. All of these except Be have an odd number of protons (atomic number) in their nuclei. The most common form of an element (highest proportion of atoms with that atomic weight) that makes up the average weight of its isotopes (if any) is considered representative of that element. In addition, this average atomic weight for the most represented form of that element is identified in the periodic table. All other forms of those elements that exist but with different atomic masses are called isotopes.

At one time, the hydrogen atom with one proton and no neutron was used as the standard to define 1 atomic mass unit (1 amu). Today, chemists use carbon-12, the most abundant isotope of carbon for the standard amu, which is defined as 1/12 of the C-12 atom. Therefore, the actual atomic weight for an element is in average mass units (numbers), taking into account all the isotopes (atoms) of that element.

Examples of isotopes are abundant. The major form of hydrogen is represented as ^{1}H (or H-1), with one proton; ^{2}H, known as the isotope deuterium or heavy hydrogen, consists of one proton and one neutron (thus an amu of 2); and ^{3}H is the isotope of hydrogen called tritium with an amu of 3. Carbon-12 (^{12}C or C-12) is the most abundant form of carbon, though carbon has several isotopes. One is the ^{14}C isotope, a radioactive isotope of carbon that is used as a tracer and to determine dates of organic artifacts. Uranium-238 is the radioactive isotope ^{238}U. (Note: The atomic number is placed as a subscript prefix to the element's symbol—for example, $_8$O—and the atomic mass number can be written either as a dash and number following the elements symbol [e.g., O-16] or as superscript prefix to the elements symbol [e.g., ^{16}O]. Because of the difficulty of placing both numbers as prefixes preceding the element's symbol, this text mostly uses the O-16 style to represent the elements' mass number.)

Isotopes of the chemical elements found on Earth are known to exist in space, on the stars, and on other planets. The ratio or proportion of these extraterrestrial isotopes to their main atoms is different from those on Earth.

Radioactivity

Radioactivity is emission of electromagnetic radiation from atoms (or more correctly the nuclei of atoms) resulting from natural or artificial nuclear transformation (transmutation). In 1895 Antoine Henri Becquerel (1852–1908) inadvertently discovered radiation when he experimented with exposing fluorescent crystals to X-rays. He placed his crystals in a drawer on top of a photographic plate that was wrapped in dark paper. When he developed the plate, it showed dark areas where it had been exposed to the crystal through the paper. From this Becquerel hypothesized that the crystals emitted some form of penetrating radiation. Later, other scientists identified this new radiation and other types of emissions as three forms of radiation: **alpha radiation** (α) features the nuclei of helium, the key particles in **beta radiation** (β) are similar to high-energy electrons, and **gamma radiation** (γ) is characterized by highly penetrating, short-wavelength electromagnetic radiation (ranging from 10^{-14} to 10^{-11} meters). All three forms of radiation result from both natural and man-made sources. The radiation results from a form of transmutation or decay process of the nuclei of some heavy elements. An important point is that when alpha particles (and to some extent beta particles) are emitted from the nuclei of a particular element's atoms, the atomic number is changed, thus resulting in nuclei of a different element. For example, the decay of radium-226 emits natural alpha particles, and a new product, radon-222, results. Radon-222, in turn, decays by emitting a series of alpha and beta particles to form other unstable isotopes until the end product is the stable isotope of the element lead-206. In 1899 Ernest Rutherford discovered this nuclear reaction and named the resulting radiation "radioactivity." He referred to this nuclear reaction as "transmutation" and coined the terms for the two types of radiation. He called one "alpha" ($_2$He-4) and the other "beta" ($_{-1}$e-0). Transmutation is defined as the natural or artificial change of atoms of one element into atoms of a different element, resulting from a nuclear interaction (reaction). The previously noted example is a type of transmutation in which the nuclei of larger atoms disintegrate (transmutate) into nuclei of smaller atoms. Another type of transmutation occurs when nuclei of smaller elements combine to form nuclei of larger elements (i.e., fusion). An example of this form of transmutation is the nitrogen nuclei capturing helium nuclei (alpha) to form oxygen nuclei plus hydrogen ($_7$N-14 + $_2$He-4 >nuclei combine> $_8$O-16 + $_1$H-1).

The time it takes for radioactive decay to occur is the **half-life** for that isotope. The half-life is the time required for the activity to decrease to one-half of the radioactivity of the original isotope, followed by a decrease of one-half of that remaining half, and half of that remaining half of the nuclei decaying, and so on. This means that the nuclei of radioisotopes do not decay all at once, but rather undergo decay to more stable forms of atoms—giving off radiation in each of the sequences of decay until stable nuclei of an element result. Interestingly, scientists cannot measure or predict the half-life of a particular single nucleus. They do not have the computing power for this analysis, so they work with averages and mathematical probabilities.

The half-lives of some radioisotopes are measured in billions of years; for others, the half-life is measured in fractions of seconds. Following are some examples of the half-lives of a few isotopes: uranium-238 = 4.6 billion years; carbon-14 = 5730 years; strontium-90 = 38 years; phosphorus-32 = 14.3 days; radon-222 = 3.8 days; uranium-239 = 23.5 minutes.

Radiation is one area of science not well understood by the lay public, and often the media information relating to radioactivity is misleading and misunderstood. To some extent, the topics of radioactivity and radiation have become a political issue. The public is somewhat scientifically illiterate about radiation, and many people do not have a very clear understanding of the physical nature, sources, uses, benefits, and dangers of radiation and radioactivity. We can all learn more about radioactivity so that it can be used for the benefit of mankind without undue fear. After all, it is very natural and universal. Radioactivity takes place both inside and on the surface of our Earth. Not only does it exist in space, but it also is penetrating our bodies at all times from natural sources, and small amounts of radiation exist in our tissues and organs. It is part of all life.

Some of the sources of radiation that affect us follow:

1. Potassium is essential to our diets and is found in many foods. Our bodies cannot distinguish between potassium-39 (nonradioactive) and the smaller quantities of radioactive potassium-40 found in our foods. Radioactive K-40 makes up almost one-fourth of all the atomic radiation we normally receive.

2. Radon is a radioactive gas that seeps into our homes, schools, and offices. It is produced by the natural decay of radium in the ground. Radon gas is thought to be a cause of some cancers, particularly lung cancer, as it seeps into the ground levels of buildings. Kits are available for testing the levels of radon that may exist in your home—particularly the basement or ground-level areas.

3. There are other sources of radiation from the decay of radioactive elements in the Earth's crust.

4. We receive radiation from outer space as cosmic rays, solar radiation, and upper-atmosphere radiation. The higher the altitude at which you live, the greater will be your exposure to cosmic radiation from space. Since nuclear radiation accumulates in our bodies

over time, people living in Denver, Colorado, receive more radiation in their lifetimes than do people living in areas at sea level.

5.   Humans are exposed to radiation from the testing and explosion of nuclear weapons and the wastes of nuclear reactors and power plants. Strontium-90 is a fission product from nuclear reactors. It is of particular concern because it has a long half-life of 38 years and becomes concentrated in the food chain, particularly plants-to-milk. The ban on atmospheric testing of nuclear weapons has reduced this hazard. Strontium-90 does have some industrial uses.

6.   Most people in developed countries receive minor exposure to radiation through medical procedures such as X-ray and various treatments for some diseases.

The greatest hazard from strong radiation is the ionization of atoms in cells and tissues of humans that may result in changes in somatic (body) cells. In addition, reproductive sperm and egg cells may also be affected by strong radiation as well as by cosmic rays. Radioactive potassium is thought to be a source of genetic mutations (genetic changes) in the DNA of plant and animal cells. This natural process of radiation and mutation has a relationship to the evolution of plant and animal species because all plants and animals require potassium to survive and live with constant cosmic radiation. As mentioned earlier, living organisms cannot tell stable potassium from the radioactive form in their diets.

Our exposure to man-made radioactive sources, such as from nuclear power plants, is negligible when compared to the total radiation we receive. Man-made radiation accounts for less than 3% of the total radiation we receive in the United States, but in some countries, this figure is higher. The vast majority of the 3% of man-made doses of radiation we receive in our lifetime results from medical uses, and the vast majority of the 97% of the total exposure to all radiation we receive comes from natural sources.

In summary, the structure and recurring characteristics of elements are represented in the catalog-like periodic table of chemical elements. A review of the material covered thus far follows:

1.   The negatively charged electrons surround the positively charged nucleus. Electrons can exist in shells or energy levels orbiting around the nucleus, or they can easily be stripped from the atom and exist as **free electrons** as electricity or beta particles. An electron is 1/1837 the mass of a proton. Even so, an electron's electrical charge balances the positive charge of a proton. In addition, the distance between the electrons and the nucleus of an atom is about one million times greater than the diameter of the nucleus. In a neutral atom, the number of electrons equals the number of protons.

2.   The positively charged protons are compacted in a tiny, dense center of the atom called the nucleus. The number of protons in the nucleus determines the atomic number for each element. The periodic table lists the number of protons in progression from the first number, hydrogen ($_1$H, with one proton), to the most recently discovered superactinide elements and yet-to-be-discovered elements with the highest atomic numbers.

3.   The neutron was initially more difficult to identify because it has no electrical charge. Neutrons are found in the nucleus with the protons. They have approximately the same mass as the protons, and together they make up the atomic mass of the atoms for each of the elements. To determine the number of neutrons in the atoms of an element, one can subtract the atomic number (protons) from the total atomic weight.

4. Isotopes are atoms with more than the usual number of neutrons in their nuclei. Their atomic number does not change, but their atomic mass does. Many isotopes of atoms with high atomic masses are radioactive.

5. Ions are atoms that have gained or lost one or more electrons in a chemical reaction to form a molecule of a new compound. Thus, one might think of ions as atoms with electrical charges.

The background presented here will aid you in learning how to use the periodic table of the chemical elements. Once you study and understand the organization of this remarkable chart, you can glean a great deal of information about the Earth's chemical elements. The structure and characteristics of elements may be identified by their position in the table. How a given element will react with other elements can be predicted by the element's placement in relation to other elements in the table. The periodic table contains an abundance of intriguing and useful information. The periodic table of the chemical elements is the Rosetta Stone for decoding the nature of chemistry.

As mentioned, a number of different forms of tables, charts, and diagrams for the arrangements of the chemical elements have been proposed and designed. One of the newest and unique designs is the Chemical Galaxy version of the periodic table devised by Philip Stewart. Copies of this unique periodic table are available at the Chemical Galaxy Web site (http://www.chemicalgalaxy.co.uk/).

Guide to the Elements

Introduction

This section provides an outline of how this book is organized, following how the elements are listed in the periodic table of the chemical elements. Elements are not only arranged according to the number of protons in their nuclei, but are also classified as metals or nonmetals, with a group of transition elements between the two categories. The major section of this book is the guide that is organized as follows.

The guide starts with the alkali metals and continues with the alkali earth metals, followed by the three series of transition elements. These series are followed by the metallics, which leads to the metalloids of group 14 and then to the metalloids of group 15. This is the end of the common elements that are considered "metals" and the beginning of the nonmetals, halogens, and noble gases. The nonmetals are then followed by groups of less common metals know as the lanthanide series, the actinide series (including the transuranic series), and finally the transactinide series, which includes the newer artificially man-made radioactive superactinides (or super heavy elements) and possible future elements.

Most of the elements listed on the periodic table are considered metals or metal-like, which is related to the oxidation states of metals. Metallic elements have oxidation states that form positive ions (cations) when they combine with nonmetals by giving up or sharing one or more electrons. Most metals have less than four electrons in their outer valence shell. Some of the heavier metals are unique in that they can also add electrons to the shells inside the outer shell. Since they "give up" or share electrons in chemical reactions, they are classified as having low or weak electronegativity, and thus they become positive ions. If they are in an electrolytic solution, they are called cations because they collect at the negative pole or cathode. Cations have more protons in their nuclei than electrons in their shells. In addition, their remaining electrons are more strongly attracted to their positive nuclei; thus, metallic ions are generally smaller than atoms of the same elements. This ionization process can be depicted for the metallic elements lithium as $Li \rightarrow Li^+ + e^-$, and for calcium as $Ca \rightarrow Ca^{++} + 2e^-$.

Anions are just the opposite. They are atoms that have gained electrons and thus are ions with a negative charge; they are "electronegative" because they have a tendency to be negatively charged. Anions collect at the positive anode in an electrolytic solution. The electron attractions between nuclei are weaker for anions; thus, in general, negative ions are larger than are the atoms of the same elements from which they were formed. When nonmetal atoms gain electrons, they become anions. For example, oxygen gains two electrons in most chemical reactions: $O + 2e^- \rightarrow O^{2-}$. The halogens and sulfur, in addition to oxygen, are also highly electronegative.

The metals are grouped on the left side of the periodic table and, to some extent, to the center. As you move from the far left side of the table to the right side, the very reactive metals (alkali metals and alkali-earth metals) are followed by less active metals, or transition metals, and the table then moves to metalloids and semiconducting elements before proceeding to nonmetals, whose outer valence shells tend to gain electrons when reacting with metals. As the number of electrons in the outer shells increases, the distance between the nucleus and electrons also increases. The further the electrons are from the nucleus, the weaker the attraction between the electrons and the positive nucleus. Thus, in general, the ionization energy increases from left to right in the periodic table. Therefore, the larger the molecules of metallic elements, the greater their volume, as is true with all three-dimensional objects. In addition, their densities become greater as do their atomic weights, and in general, the higher their atomic mass, the lower will be their melting and boiling points.

Ionization is the process whereby a chemical reaction forms ions (atoms with a negative or positive charge) from the breakup of neutral molecules of some inorganic compounds. A common example is the neutral molecule of sodium chloride (NaCl, salt). When it dissociates (breaks apart) into positive metallic ions of Na^+ by the loss of an electron, the nonmetal chlorine ion Cl^- gains the negative charge given up by the sodium atom.

Approximately 75% of all elements found on and in the Earth are metals. They are **crystalline** solids that at room temperature range from hard to butter-like soft to liquid (mercury). They are generally good conductors of heat and electricity as a result of the swarm of relatively "free" electrons in their outer shell that move without much resistance to other elements, particularly those with a dearth of electrons in their outer shells. In pure states, most metals have a shiny luster when cut. Those located at the far left of the table have only one electron in their outer shell. Therefore, they are very reactive and are not usually found in pure form. Instead, they are found in compounds, **minerals,** or **ores** that must be processed to extract the pure metal from the other elements in the compounds.

Metals may be classified by a variety of categories. Some metals fit more than one of the following categories:

1. *Alkali metals*—Group 1 (IA), soft, silvery, and very active (electropositive). They are all solids and have one electron in their outer valence shell. The alkali metals begin each period, 2 to 7, on the periodic table. They are very reactive and, in the pure metallic state, must be stored in oil and not exposed to air or water. Even though hydrogen may be considered a nonmetal, it is included in group 1 with the alkali metals because it has a single proton in its nucleus and just one electron in its valence shell and, thus, usually acts like a metal. Under certain situations, hydrogen can also act as a nonmetal.

2. *Alkali earth metals*—Group 2 (IIA), shades of white to subtle colors, **malleable, machinable,** and less active than alkali metals. They are all solids and have two electrons in their outer valence shell.

3. *Transition metals*—found in the groups located in the center of the periodic table, plus the lanthanide and actinide series. They are all solids, except mercury, and are the only elements whose shells other than their outer shells give up or share electrons in chemical reactions. Transition metals include the 38 elements from groups 3 through 12. They exhibit several **oxidation** states (oxidation numbers) and various levels of electronegativity, depending on their size and valence.

4. *Other metals*—a classification given to seven metals that do not fit the characteristics of transition metals. They do not exhibit variable oxidation states, and their valence electrons are found only on the outer shell. They are aluminum, gallium, indium, tin, thallium, lead, and bismuth.

5. *Metalloids*—metals found in the region of the periodic table between groups of metals and nonmetals. Thus, they have some characteristics of both metals and nonmetals. Some are **semiconductors.** They are boron, silicon, germanium, arsenic, antimony, tellurium, and polonium.

6. **Noble metals**—refers to several unreactive metals that do not easily dissolve in **acids** or **oxidize** in air (e.g., platinum, gold, and mercury). They include the platinum group of metals (see next item). They are called "noble" because of their resistance to reacting with other elements.

7. *Platinum metals*—includes unreactive transition elements located in groups 8, 9, and 10 of periods 5 and 6. They have similar chemical properties. They are ruthenium, rhodium, palladium, osmium, iridium, and platinum.

8. *Rare-earth metals*—a loose term for less well-known metallic elements. They include the so-called rare-earths. The rare-earths are not actually rare (scarce); historically, some of them were just difficult to find, isolate, and identify.

9. *Lanthanide metals*—also rare-earth elements with atomic numbers ranging from 57 through 71.

10. *Actinide metals*—includes elements with atomic numbers from 89 to 111. Also includes the transuranic elements (e.g., beyond uranium [$_{92}$U to $_{103}$Lr]) and the superactinides (elements with atomic numbers 104 to 118 that are artificial, radioactive, and unstable with very short half-lives).

11. *Light metals*—a general term for elements relatively light in weight but strong enough for construction. Some examples are aluminum, magnesium, titanium, and beryllium.

12. *Heavy metals*—a general term for metals with an atomic mass number greater than 200. Several heavy metals are extremely toxic.

Metals are extremely important not only for chemical reactions but also for the health and welfare of plants and animals. Some examples of metals required for good nutrition, even in trace amounts, are iron, copper, cobalt, potassium, sodium, and zinc. Other metals—for example, mercury, lead, cadmium, barium, beryllium, radium, and uranium—are very toxic. Some metals at the atomic and ionic levels are crucial for the oxidation process that **metabolizes** carbohydrates for all living cells.

PERIODIC TABLE OF THE ELEMENTS

— TRANSITION ELEMENTS —

GROUPS / PERIODS	1 IA	2 IIA	3 IIIB	4 IVB	5 VB	6 VIB	7 VIIB	8 VIII	9 VIII	10 VIII	11 IB	12 IIB	13 IIIA	14 IVA	15 VA	16 VIA	17 VIIA	18 VIIIA
1	1 H 1.0079																	2 He 4.00260
2	3 Li 6.941	4 Be 9.01218											5 B 10.81	6 C 12.011	7 N 14.0067	8 O 15.9994	9 F 18.9984	10 Ne 20.179
3	11 Na 22.9898	12 Mg 24.305											13 Al 26.9815	14 Si 28.0855	15 P 30.9738	16 S 32.066(6)	17 Cl 35.453	18 Ar 39.948
4	19 K 39.0983	20 Ca 40.08	21 Sc 44.9559	22 Ti 47.88	23 V 50.9415	24 Cr 51.996	25 Mn 54.9380	26 Fe 55.847	27 Co 58.9332	28 Ni 58.69	29 Cu 63.546	30 Zn 65.39	31 Ga 69.72	32 Ge 72.59	33 As 74.9216	34 Se 78.96	35 Br 79.904	36 Kr 83.80
5	37 Rb 85.4678	38 Sr 87.62	39 Y 88.9059	40 Zr 91.224	41 Nb 92.9064	42 Mo 95.94	43 Tc (98)	44 Ru 101.07	45 Rh 102.906	46 Pd 106.42	47 Ag 107.868	48 Cd 112.41	49 In 114.82	50 Sn 118.71	51 Sb 121.75	52 Te 127.60	53 I 126.905	54 Xe 131.29
6	55 Cs 132.905	56 Ba 137.33	★	72 Hf 178.49	73 Ta 180.948	74 W 183.85	75 Re 186.207	76 Os 190.2	77 Ir 192.22	78 Pt 195.08	79 Au 196.967	80 Hg 200.59	81 Tl 204.383	82 Pb 207.2	83 Bi 208.980	84 Po (209)	85 At (210)	86 Rn (222)
7	87 Fr (223)	88 Ra 226.025	▲	104 Unq (261)	105 Unp (262)	106 Unh (263)	107 Uns (264)	108 Uno (265)	109 Une (266)	110 Uun (267)	111 Uuu (272)	112 Uub	113 Uut	114 Uuq	115 Uup	116 Uuh	117 Uus	118 Uuo

★ 6 Lanthanide Series (RARE EARTH)

▲ 7 Actinide Series (RARE EARTH)

★ 6	57 La 138.906	58 Ce 140.12	59 Pr 140.908	60 Nd 144.24	61 Pm (145)	62 Sm 150.36	63 Eu 151.96	64 Gd 157.25	65 Tb 158.925	66 Dy 162.50	67 Ho 164.930	68 Er 167.26	69 Tm 168.934	70 Yb 173.04	71 Lu 174.967
▲ 7	89 Ac 227.028	90 Th 232.038	91 Pa 231.036	92 U 238.029	93 Np 237.048	94 Pu (244)	95 Am (243)	96 Cm (247)	97 Bk (247)	98 Cf (251)	99 Es (252)	100 Fm (257)	101 Md (258)	102 No (259)	103 Lr (260)

Alkali Metals: Periods 1 to 7, Group 1 (IA)

Introduction

The alkali metals are the elements in group 1 (IA), the first vertical grouping of elements on the left side of the periodic table. They begin periods 2 through period 7 (horizontal rows). Hydrogen ($_1$H) is not really an alkali metal, but it best fits in group 1 because it gives up its single electron found in its outer valence shell to form a positive ion, as do all the other alkali metals. Therefore, it is assigned to group 1 in period 1.

All of the alkali metals are electropositive and have an oxidation state of 1^+ and form cations (positively charged ions) by either giving up or sharing their single valence electron. The other elements of group 1 are lithium ($_3$Li), sodium ($_{11}$Na), potassium ($_{19}$K), rubidium ($_{37}$Rb), cesium ($_{55}$Cs), and francium ($_{87}$Fr). Following are some characteristics of the group 1 alkali metals:

1. They are relatively soft metals in purified forms. Francium is the exception.
2. When freshly cut, they are silvery colored but soon oxidize in air, turning darker in color.
3. Their **melting** and **boiling points** become lower as their atomic weights and size increase, (i.e., proceeding down group 1, from period 2 through period 7). For example, lithium melts at about 180 degrees Celsius whereas francium melts at an estimated temperature of 27 degrees Celsius, which is just above room temperature.
4. Their atomic volumes (and radii) become larger as their atomic mass increases, and thus their **densities** also increase.
5. They produce distinctive colored flames when burned: lithium = crimson; sodium = yellow; potassium = violet; rubidium = purple; cesium = blue; and the color of francium's flame is not known. Many of francium's characteristics have not been determined owing to the fact that it is rare and all of its many radioactive isotopes have short half-lives.
6. Alkali metals react with **water** and weak acids. Those with higher atomic weight react explosively.

7. Their alkalinity is above **pH** 7 (7 is neutral in the pH acid/base scale). They become more alkaline (higher than 7) as their atomic mass becomes greater—with the exception of francium.

Following is detailed information about the seven alkali metals and hydrogen.

HYDROGEN

SYMBOL: H **PERIOD:** 2 **GROUP:** 1 (IA) **ATOMIC NO:** 1
ATOMIC MASS: 1.0079 amu **VALENCE:** 1 **OXIDATION STATE:** +1, −1
NATURAL STATE: Gas
ORIGIN OF NAME: Hydrogen was named after the Greek term *hydro genes,* which means "water former."
ISOTOPES: The major isotope of hydrogen has just one proton and no neutrons in its nucleus ($_1$H-1). It is by far the most abundant form of hydrogen on Earth and in the universe. Deuterium (^{2}D or H-2) has a nucleus consisting of one proton plus one neutron. It is rare and found in "heavy" water (DOD), and it is present as only one part in almost 7,000 parts of regular water. Tritium (^{3}T or H-3), another variety of heavy water (TOT), has nuclei consisting of one proton and two neutrons. It is man-made by nuclear reactions and is radioactive.

ELECTRON CONFIGURATION

Energy Levels/Shells/Electrons Orbitals/Electrons

    1-K = 1   s1

Properties

Hydrogen's atom is the simplest of all the elements, and the major isotope (H-1) consists of only one proton in its nucleus and one electron in its K shell. The density of atomic hydrogen is 0.08988 g/l, and air's density is 1.0 g/l (grams per liter). Its melting point is −255.34°C, and its boiling point is −252.87°C (**absolute zero** = −273.13°C or −459.4°F). Hydrogen has two oxidation states, +1 and −1.

Characteristics

H_2 is a diatomic gas molecule composed of two tightly joined atoms that strongly share their outer electrons. It is an odorless, tasteless, and colorless gas lighter than air. Hydrogen is included in group 1 with the alkali metals because it has an oxidation state of +1 as do the other alkali metals. Experiments during the 1990s at the Lawrence Livermore National Laboratory (LLNL), in Livermore, California, lowered the temperature of H_2 to almost absolute zero. By exploding gunpowder in a long tube that contained gaseous hydrogen, the gas that was under pressure of over one million times the normal atmospheric pressure was compressed into a liquid. This extreme pressure on the very cold gas converted it to liquid hydrogen (almost to the point of solid metallic hydrogen), in which state it did act as a metal and conduct electricity.

Hydrogen gas is slightly soluble in water, alcohol, and ether. Although it is noncorrosive, it can permeate solids better than air. Hydrogen has excellent **adsorption** capabilities in the way it attaches and holds to the surface of some substances. (Adsorption is not the same as **absorption** with a "b," in which one substance intersperses another.)

Abundance and Source

Hydrogen is the most abundant element in the universe. For every carbon atom in the universe, there are about 12,000 hydrogen atoms. On Earth, hydrogen is only the 10th most abundant element and occurs mainly in a combined form as water, and oxygen is the most abundant element on Earth. Hydrogen is also found in organic and hydrocarbon compounds. Only traces of hydrogen are found in the atmosphere. Because hydrogen gas is lighter than air, it escapes from Earth's gravity into outer space.

Hydrogen may have been formed by the "big bang" at the beginning of the universe about 13 to 15 billion years ago, accounting for over 98% of all the atoms in existence at this explosive event. Today hydrogen makes up over 90% of all the atoms and about 75% of the entire weight (mass) of the universe. The "big bang" is only one theory for the origin of the universe, but it is the one with the most current and persuasive evidence, and it argues that billions of years ago, an incredibly small, but dense, ball or point of matter, or energy, exploded, dispersing matter and energy in an ever-expanding three-dimensional pattern. Astronomical observations indicate that the universe continues to expand. The energy and hydrogen produced by the big bang are thought to be the building blocks of matter and sources of all energy in the current universe, including the solar system and Earth.

There is some evidence that the larger planets such as Jupiter and Saturn are composed of several gases, mainly hydrogen. Because of their size, the internal pressure is about 100 million times that of Earth's normal air pressure, which is strong enough to compress hydrogen gas to a **liquid** phase, under which conditions it then takes on characteristics of a metallic conductor. The rotation of the planets around their liquid hydrogen cores may be responsible for the large magnetic fields and radio emissions of these large planets.

History

Hydrogen was most likely produced accidentally by ancient alchemists who knew it as a gas that burned. The first recorded event of its discovery was by Theophrastus Bombastus von Hohenheim (known as the alchemist and physician Paracelsus; 1493–1541). It was a well-known explosive gas produced by pouring acids over zinc metal. Paracelsus was unaware that the gas produced by this chemical reaction was hydrogen—the name had not yet been designated.

In 1671 Robert Boyle (1627–1691), famous for Boyle's Law, which stated there is an inverse relationship between the pressure and volume of a gas if the temperature remains constant ($P \times V = c$), described the generation of hydrogen gas by adding iron filings to a dilute acid in a glass beaker. He explained that, when ignited by a candle, the resulting gas "belched hot flames and stinking fumes."

Scientists throughout the world used this simple process of acid + iron filings = exploding gas ($Fe_2 + 4HCl \rightarrow 2FeCl_2 + 2H_2$) to justify their theory of **phlogiston,** which states that

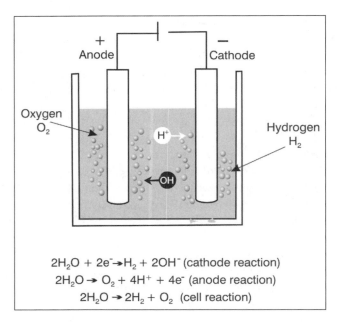

$$2H_2O + 2e^- \rightarrow H_2 + 2OH^- \text{ (cathode reaction)}$$
$$2H_2O \rightarrow O_2 + 4H^+ + 4e^- \text{ (anode reaction)}$$
$$2H_2O \rightarrow 2H_2 + O_2 \text{ (cell reaction)}$$

Figure 4.1: In electrolysis, the positive ions (cations) are attracted to the negative cathode, and the negative ions (anions) are attracted to the positive anode. The electrolyte is the solution that conducts the current between the poles.

things that burn contain a substance called phlogiston, which is consumed during combustion.

In 1766 Henry Cavendish (1731–1810) identified the gas from the iron filings-plus-acid reaction as "fire air" and determined that the gas, when burned in air, produced water. He is given credit for the discovery of hydrogen even though the chemical reaction that produced the gas had been known for hundreds of years. In 1783 Antoine Lavoisier (1743–1794) gave the gas its name: hydrogen.

Hydrogen gas can be produced by several types of gasification processes using coal. The **hydrogasification** process involves adding hydrogen and steam to hot pulverized coal, peat, and lignite under extreme pressure to form methane gas or methanol that can then be used as pipeline fuel. Another process is known as the Fischer-Tropsch method, which involves passing steam at a temperature of about 800°C over hot coals in the presence of air. This reaction produces methane that is broken down by the hot steam into carbon monoxide and hydrogen gas. Additional hydrogen gas is produced when the temperature is increased, resulting in carbon monoxide reacting with hot steam (water) to produce carbon dioxide and hydrogen.

$C + \Delta H_2O \rightarrow CO + H_2\uparrow$ (Note: Methane CH_4 can be substituted for the coal or charcoal.) The carbon monoxide reaction is $CO + \Delta H_2O \rightarrow CO_2 + H_2\uparrow$. This synthetic gas, called "water gas," was produced in the "old gasworks" and was used for home lighting and cooking before natural gas became widely available. Both methods of producing non–petroleum-based fuels were used by some countries, including Germany during World War II. It is still used to produce limited amounts of methane and methanol in South Africa and some other countries. None of these processes will be commercially economical to produce methane gas or hydrogen on a large worldwide scale until the price of petroleum drastically increases.

Natural gas is a hydrocarbon of low molecular weight consisting of about 85% methane (CH_4), which is recovered as an underground gas in areas where petroleum is found. It is also economically produced by conversion of crude oil to gas. Possible future sources of a hydrogen-based fuel are large frozen methane deposits recently found on ocean bottoms.

The most efficient method of producing hydrogen (but not the most economical) is **electrolysis** of water (H_2O), where the hydrogen gas is separated from the oxygen gas by the passing of an electrical charge through the water. During electrolysis positive hydrogen ions (cations) collect in the water at the negative cathode pole to form H_2 gas bubbles, while negative oxygen ions (anions) collect at the positive anode pole to form O_2 gas. Other methods of producing hydrogen-based fuels are being explored.

Common Uses

Hydrogen is an excellent **reducing agent,** meaning it removes oxygen from compounds or adds hydrogen to substances, resulting in the production of many useful products.

Production of ammonia (NH_3): Anhydrous (dry) ammonia is the fifth most produced industrial compound. The Haber-Bosch process uses steam on hot coke, which is mostly used in South Africa. In the United States, it is mostly produced from partial combustion of natural gas (methane) or by combining several gases using steam. Other methods use coke-oven gas, refinery gas (mostly methane), or even solar energy. Ammonia is toxic if inhaled and has a high pH value when mixed with water (**hydration**) to form ammonium hydroxide (NH_4OH), which has many uses, including as a household cleaner. Ammonia forms many compounds, including ammonium nitrate in fertilizer, rocket fuel, and explosives. Ammonia is also explosive when mixed with mercury or silver or when mixed as part of nitrocellulose.

Ethanol (ethyl alcohol made from grains): Ethanol (C_2H_5OH) is one of many types of alcohol. Grain alcohol can be produced by fermentation of agricultural waste, corn, or other grains. Another method is the hydration of ethylene: the reaction of water with ethylene (C_2H_4), a reaction in which the alcohol molecule is formed; the water is then split off by heat.

Pure 100% ethyl alcohol is a colorless, volatile liquid with a pungent taste. One hundred–proof alcoholic drinks are about 50% ethanol. When consumed, ethanol is a depressant and may be habit-forming. It rapidly oxidizes in the body, but even small amounts cause dizziness, nausea, headaches, and loss of motor control. "Proof" was how whiskey salesmen of the Old West demonstrated that their product was potent. They would place some gunpowder in a tin dish and then pour on some of their whiskey. If a match would ignite the mixture, this was claimed to serve as "100% proof" of the whiskey's quality. It just happens that 100-proof whiskey is about 50% ethanol.

Hydrogenation of vegetable oils: Hydrogenation is a two-step process.

1. Hydrogen is added to the carbon double bonds of unsaturated or polyunsaturated molecules where not all these double carbon bonds are used. This process changes unsaturated fats (liquids) to saturated fats (solids).
2. The added hydrogen forms **saturated** molecules where all single-carbon valence bonds become satisfied or attached to other atoms, thus producing solid saturated or trans fatty acids that may increase LDL cholesterol (the bad cholesterol) attributed to numerous health risks.

Hydrogenation is an important process in the production of and in the extension of the shelf life of many of our modern foods. Hydrogenation occurs when high heat and pressure are used to force hydrogen atoms to be added to each double bond in the chain of carbon atoms of alkane or alkene molecules. In other words, it is an addition-type reaction wherein hydrogen atoms use one of the double bonds between the carbon atoms. This process is used to convert unsaturated liquid vegetable oils, such as corn, coconut, or soybean, into partially saturated or fully saturated **solid** fatty acids, such as margarine and other fast-food products. This process requires a metallic catalyst, such as platinum or nickel. Converting unsaturated vegetable oils into saturated solid results in a variety of different food products with much longer shelf lives. At the same time, saturated fatty acids, also known as trans fatty acids, increase LDL and lower HDL (the good cholesterol), therefore increasing the risk of heart disease and type 2 diabetes. The healthiest oils and fats are polyunsaturated or monounsaturated fats, which include olive oil, canola oil, soybean oil, and corn oil. These have not been saturated with hydrogen to produce solid fats.

Halogenation: Another common substitution reaction in which hydrogen atoms are replaced by halogen atoms. (See the group VIIA of the periodic table for more on this group of active halogen elements.) This is an example of a substitution-type reaction that forms a group of compounds known as chlorofluorocarbons or CFCs. CFCs are chemicals formerly used as refrigerants in air conditioners and to create pressure in aerosol cans that some scientists claim are partially responsible for the hole in the ozone layer of the atmosphere. The two most destructive halogenation compounds to the ozone layer are trichlorofluoromethane (known as Freon-11) and dichlorodifluoromethane (known as Freon-12). New refrigerants have been developed since the chlorofluorocarbons harmful to the ozone have been banned by most countries.

Polymers: Formed when multiple addition reactions occur that link many smaller hydrocarbon molecules into long chains. Ethylene (C_2H_4) and propylene (C_3H_6) are alkynes with the ability to string their molecules together in what is known as "addition polymerization" to form long chains of repeating molecules. The repeating units in the chain are known as **monomers**; for example, each of the individual C_2H_4 monomer is formed into a straight polymer chain. Either ethylene or propylene (as well as other hydrocarbons) is utilized in the production of valuable synthetic products such as polypropylenes, polyvinyl chloride (PVC), and Teflon. Many other products, such as plastics, are manufactured from the millions of tons of polymer types of hydrocarbon molecules produced in the United States each year. A process of polymerization is also used to produce very high-octane gasoline.

Aromatic hydrocarbons: Unsaturated cyclic (ring) molecular compounds that resist addition reactions. These ring-shaped hydrocarbons are highly unsaturated. They are called "aromatic" because of their distinctive aroma that aided in their discovery, as well as the unique arrangements of one or more of their carbon rings. The hexagonal-shaped benzene ring is the most commonly shaped aromatic molecule. All other aromatic hydrocarbons are derived from one or more benzene (C_6H_6) groups. They are very reactive and versatile compounds made from petroleum or coal tars from which many useful products are manufactured.

Hydrocracking *of petroleum and coal:* Hydrogen is used for a process called destructive hydrogenation, which breaks down large hydrocarbon molecules to form more useful liquid and gaseous fuels. Gasoline is a mixture of branched-chain alkyne paraffins, cycloparafins, and some aromatics with an octane number of at least 60 (or higher) that can be ignited by a spark plug inside an internal combustion engine. There are several processes used to produce different qualities of gasoline, but by far, most is produced by the catalytic decomposition of crude oil and unrefined petroleum products at high temperatures. The catalytic cracking process produces gasoline with an octane rating of 80 to 100.

Reducing agent: Hydrogen is electropositive when it provides electrons to other substances in chemical reactions and, thus, is a reducing agent. Reduction is the opposite of oxidation.

Lighter-than-air balloons: Hydrogen gas that is lighter than air was used to inflate some of the early "free flight" balloons and was used later in some dirigibles until the danger of explosions led to the use of helium as a substitute.

As a rocket fuel: Large amounts of liquid hydrogen, along with liquid oxygen, are used in the U.S. Space Program.

As an "ion" fuel for nuclear rocket engines: By stripping the single electron from a hydrogen atom, you end up with a light positive particle (proton) that can be accelerated by a negatively charged source. A great mass of protons reach a very high velocity in an ion engine as they are ejected, thus producing thrust for the engine.

As a possible fuel for "fuel cells": Hydrogen and oxygen gases can be combined in a fuel cell to produce electricity with water as the only by-products. This reaction is the opposite of the electrolysis of water, in which electricity separates the water molecules to form H_2 and O_2. On May 14, 1996, the Daimler-Benz German automobile maker, in cooperation with Ballard Power Systems of Canada, announced development of the first fuel cell–powered passenger car. A similar fuel-cell power system developed by the Canadian company was tested on several buses in cities in the United States. The source of the fuels, at least for the near future, will be natural gas for the hydrogen and air for the oxygen. The fuel-cell power system produces very little pollution and has proven to be a reliable source of both electricity and drinking water for astronauts. However, both the U.S. government and energy companies have reduced their research support for fuel cells designed as a long-term replacement of petroleum for transportation. It is expected that research to improve the practicality of fuel cells for transportation use will increase as the cost of petroleum increases.

Examples of Compounds
Acids and Gases

Anhydrous ammonia (NH_3) is a colorless **gas** with a sharp, irritating odor, lighter than air, easily liquefied. An important commercial compound. It was the first complex molecule identified in outer space.

Boric acid (H_3BO_3) is a solid, soft, smooth, solid weak acid that is used in pharmaceutical and cosmetic industries.

Carbonic acid (H_2CO_3) is produced by dissolving carbon dioxide in water. When formed under pressure, it is the gas used in carbonated drinks. In nature, it dissolves the limestone in caves, resulting in the formation of stalactites and stalagmites. It is corrosive as are other acids, although it is considered a rather weak acid.

Hydrochloric acid (HCL) is also known as hydrogen chloride, and in a less than pure form it is commonly called muriatic acid. It is used by many industries and is mainly obtained as a by-product of the organic chloride chemicals used in the manufacturing of plastics. It can be produced in pure form by exploding a mixture of hydrogen and chlorine gases. The stomach's digestive juice is a form of hydrochloric acid.

Hydrogen peroxide (H_2O_2) in a purified form is explosive. In a dilute form in water, it is used as an antiseptic and oxidizing agent.

Hydrosulfuric acid or *hydrogen sulfide* (H_2S) is known for its "rotten egg" odor and is a deadly poisonous gas even at 100 ppm of air. The laboratory reaction is $FeS + 2HCl \rightarrow H_2S\uparrow + FeCl_2$.

Nitric acid (HNO_3) is an important industrial acid used to alter or produce many products such as fertilizers and explosives. It reacts with ammonia to produce ammonium nitrate, an important commercial chemical.

Sulfuric acid (H_2SO_4) is the most used acid in many industries and chemical laboratories. Produced in large batches, it is stable and safe for storage and shipment in vast quantities in various grades of purity. A relatively low grade of sulfuric acid is used in the manufacture of fertilizers.

Sulfurous acid (H_2SO_3) can be produced by burning sulfur to form sulfur dioxide (SO_2) gas and by then dissolving the gas in water to form sulfurous acid. This is the acid produced by burning coal that has a high sulfur content; the gaseous sulfur dioxide by-product of combustion then combines with atmospheric water to form "acid rain."

Hydrides

Hydrides are inorganic compounds in which at least one of the elements is hydrogen; therefore, there are thousands of carbon-hydrogen hydride compounds. One of the main building blocks of hydrocarbon chemistry is methane (CH_4), which is ordinary natural gas, sometimes called "swamp gas" because it is one of the by-products of decaying organic matter. The bonding of hydrides can be either ionic or covalent, and the molecules may be binary or complex. Common metallic forms are hydrides of lithium (LiH), sodium (NaH), potassium (KH), and aluminum (AlH_3). Hydrides react violently in both air and water as well as with some nonmetals, resulting in an explosion of hydrogen gas.

Isotopes of Hydrogen

There are three isotopes of the element hydrogen; the first and most common isotope has just one proton in its nucleus (^1H). This isotope is also called protium. The second isotope contains one proton and one neutron and is called deuterium (^2D). The third has an atomic mass of three and is called tritium (^3T) and is very rare. In water, only one tritium atom exists for every 10^{18} atoms of normal hydrogen. The oxide of deuterium (D_2O or DOD) molecule is almost twice as heavy as a regular water molecule and thus is referred to as deuterium oxide or "heavy water." There is only about one molecule of deuterium (D_2O) for every 6,500 molecules of regular H_2O. Deuterium is used as heavy water for a moderator in nuclear accelerators. There are several processes used to produce and collect heavy water. One is fractional distillation; another is by electrolysis, during which about 100,000 gallons of regular water is required to produce just one gallon of pure heavy water.

Hydrocarbons and Organic Compounds

We have already discussed the role of carbon in these types of molecules. Hydrogen also plays a major role in the formation of both organic foods and hydrocarbons.

Hazards

Hydrogen gas is very explosive when mixed with oxygen gas and touched off by a spark or flame. On May 6, 1937, the German rigid-frame dirigible or zeppelin, *Hindenburg,* which was inflated with hydrogen, exploded while landing at Lakehurst, New Jersey. The explosion killed thirty-six people and injured many more when the dirigible attempted to land during a thunder and lightning storm. Many hydrides of hydrogen are dangerous and can become explosive if not stored and handled correctly.

Many organic and hydrocarbon compounds are essential for life to exist, but just as many are poisonous, carcinogenic, or toxic to living organisms.

LITHIUM

SYMBOL: Li   **PERIOD:** 2   **GROUP:** 1 (IA)   **ATOMIC NO:** 3
ATOMIC MASS: 6.941 amu   **VALENCE:** 1   **OXIDATION STATE:** +1   **NATURAL STATE:**
  Solid
ORIGIN OF NAME: The name lithium comes from the Greek word *lithos,* meaning "stone"
  because it was found in rocks on Earth.
ISOTOPES: There are two stable lithium isotopes: Li-6.015, which makes up 7.5% of all lithium atoms, and Li-7.016, which makes up 92.5% of lithium atoms found in the Earth's
  crust. Less prevalent isotopes of lithium are Li-4, Li-5, Li-8, Li-9, Li-10, and Li-11. They
  are unstable with short half-lives and make up only a very small fraction of Lithium's
  total averaged atomic weight.

ELECTRON CONFIGURATION

Energy Levels/Shells/Electrons	Orbitals/Electrons
1-K = 2	s2
2-L = 1	s1

Properties

In the metallic state, lithium is a very soft metal with a density of 0.534 g/cm³. When a small piece is placed on water, it will float as it reacts with the water, releasing hydrogen gas. Lithium's melting point is 179°C, and it has about the same heat capacity as water, with a boiling point of 1,342°C. It is electropositive with an oxidation state of +1, and it is an excellent conductor of heat and electricity. Its atom is the smallest of the alkali earth metals and thus is the least reactive because its valence electron is in the K shell, which is held closest to its nuclei.

Characteristics

While classified as an alkali metal, lithium also exhibits some properties of the alkali earth metals found in group 2 (IIA). Lithium is the lightest in weight and softest of all the metals and is the third lightest of all substances listed on the periodic table, with an average atomic weight of about 7. (The other two are hydrogen and helium.) Although it will float on water, it reacts with water, liberating explosive hydrogen gas and lithium hydroxide ($2Li + 2H_2O \rightarrow 2LiOH + H_2\Delta$). It will also ignite when exposed to oxygen in moist air ($4Li + O_2 \rightarrow 2Li_2O$). It is electropositive and thus an excellent reducing agent because it readily gives up electrons in chemical reactions. Lithium is the only metal that reacts with nitrogen at room temperature. When a small piece of the metal, which is usually stored in oil or kerosene, is cut, the new surface has a bright, shiny, silvery surface that soon turns gray from oxidation.

Abundance and Source

Lithium ranks 33rd among the most abundant elements found on Earth. It does not exist in pure metallic form in nature because it reacts with water and air. It is always combined with other elements in compound forms. These lithium mineral ores make up only about 0.0007%, or about 65 ppm, of the Earth's crust.

Lithium is contained in minute amounts in the mineral ores of spodumene, lepidolite, and amblygonite, which are found in the United States and several countries in Europe, Africa, and South America. High temperatures are required to extract lithium from its compounds and by electrolysis of lithium chloride. It is also concentrated by **solar evaporation** of salt brine in lakes.

Metallic lithium is produced on a commercial scale by electrolysis of molten lithium chloride (LiCl) that is heated as a mixture with potassium chloride (KCl). Both have a rather high melting point, but when mixed, the temperature required to melt them (400°C) is several hundred degrees lower than their individual melting points. This liquid mixture of LiCl and KCl becomes the electrolyte. The **anode** is graphite (carbon) and the **cathode** is steel. The molten liquid positive lithium cations collect at the cathode while negative chlorine anions collect at the anode, and the potassium chloride remains in the electrolyte. Each positive ion of lithium that collects at the cathode gains an electron, thus producing neutral atoms of molten lithium metal, which is then further purified.

History

The mineral petalite was mined as an ore in Sweden. In 1817 Johan August Arfwedson (1792–1841) analyzed this new mineral. After identifying several compounds in the ore, he realized there was a small percentage of the ore that could not be identified. After applying more analytical procedures, he determined it was a new alkali. It turned out that petalite contains lithium aluminum silicate, $LiAl(Si_2O_5)_2$. In 1818 the first lithium metal was prepared independently by two scientists, Sir Humphry Davy (1778–1892) and W. T. Brande (1788–1866). Lithium was discovered at a time in the early nineteenth century when numerous "new" elements were discovered and identified by other scientists. Many of these newly named elements were predicted by the use of the periodic table of the chemical elements.

Common Uses

Lithium has many uses in today's industrial society. It is used as a **flux** to promote the fusing of metals during welding and soldering. It also eliminates the formation of oxides during welding by absorbing impurities. This fusing quality is also important as a flux for producing ceramics, enamels, and glass.

A major use is as lithium stearate for lubricating greases. It makes a solid grease that can withstand hard use and high temperatures.

Lithium is used to manufacture electric storage cells (batteries) that have a long shelf life for use in heart pacemakers, cameras, and so forth.

Some lithium compounds are used as rocket propellants, nuclear reactor coolants, alloy hardeners, and deoxidizers and to make special ceramics.

Lithium is a source of alpha particles when bombarded in a nuclear accelerator. The following occurs: lithium nuclei (3 protons + 4 neutrons) are targeted by high-speed protons (hydrogen nuclei), resulting in lithium absorbing a proton to form 4 protons + 4 neutrons, which become two alpha particles (helium nuclei). For example, see the following equation:

$_1$H + $_3$Li = $_2$He + $_2$He. This is an example of the first man-made nuclear reaction produced by Sir John Cockcroft (1897–1967) and Ernest Walton (1903–1995) in 1929.

Several compounds of lithium are used as **pharmaceuticals** to treat severe psychotic depression (as antidepressant agents). And lithium carbonate is also used as a sedative or mild tranquilizer to treat less severe anxiety, which is a general feeling of uneasiness or distress about present condition or future uncertainties. Lithium is also used in the production of vitamin A.

Examples of Compounds

There are numerous compounds of lithium. Its atoms combine with many other elements to form a variety of compound molecules. Some form as single oxidation states, with one lithium cation combining with one anion (+1 combines with –1), and the more complex compounds involve two positive lithium cations combining with two negative anions (+2 combines with –2). Some examples follow:

Lithium chloride (Li$^+$ + Cl$^-$ → LICl) is used as an antidepressant, especially in the treatment of manic depression and bipolar disorders.

Lithium hydroxide (Li$^+$ + OH$^-$ → LiOH) is used in storage batteries and soaps and as CO_2 absorber in spacecrafts.

Lithium hydride (Li$^+$ + H$^-$ → LiH) is a bluish-white crystal that is flammable in moisture. Used as a source of hydrogen gas that is liberated when LiH becomes wet. LiH is an excellent **desiccant** and reducing agent as well as a shield that protects from radiation created by nuclear reactions.

Lithium Fluoride (Li$^+$ + F$^-$ → LiF) is used to produce ceramics and rocket fuel and is used as welding and soldering flux and in light-sensitive scientific instruments (e.g., X-ray diffraction, which is the scattering of X-rays by crystals that produce a specific pattern of that crystal's atoms, thus producing a technique for identifying different elements).

Lithium oxide (2Li^{++} + O$^-$ → Li$_2$O) is a strong alkali that absorbs carbon dioxide and water from the atmosphere. It is used in manufacturing ceramics and special types of glass.

Lithium carbonate (2Li^{++} + (CO$_3$)$^-$ → Li$_2$CO$_3$) is used as a compound for producing metallic lithium. Lithium carbonate is the result of treating the mineral spodumene with sulfuric acid and then adding calcium carbonate. It is used as an antidepressant.

Lithium aluminum deuteride (LiAlD$_4$) is used as a source of deuterium atoms (heavy hydrogen, $_1$D-2) to produce tritium ($_1$T-3) (super heavy hydrogen) for cooling nuclear reactors.

Hazards

Lithium metal is highly flammable, explosive, and **toxic.** It will ignite when exposed to water, acids, and even damp air. Metallic lithium is a reducing agent that readily gives up an electron to active oxidizing agents that require an electron to complete their outer valence shell—thus the violent chemical reaction that follows. Lithium will even burn in nitrogen gas, which is relatively stable. In addition, many of its compounds also react violently when exposed to water.

As an element (metal), it must be stored in oil or in some type of air and moisture-free container, given that many of its compounds will also burn when exposed to air or water. Lithium fires are difficult to extinguish. If water is poured on the fire, lithium will just burn faster or explode. A supply of special chemicals or even dry sand is required to extinguish such fires.

Solutions and powders of several lithium salts are very toxic to the human nervous system, thus requiring close observation by a physician when used as antidepressant drugs.

SODIUM

SYMBOL: Na   **PERIOD:** 3   **GROUP:** 1 (IA)   **ATOMIC NO:** 11
ATOMIC MASS: 22.9898 amu   **VALENCE:** 1   **OXIDATION STATE:** +1
  NATURAL STATE: Solid
ORIGIN OF NAME: The Latin name for the symbol for "sodium" (Na) is *natrium,* and the name "sodium" in Latin is *sodanum,* which was known as an ancient headache remedy and was called "soda" in English.
ISOTOPES: Sodium has 14 isotopes. The only stable isotope of sodium has an average atomic weight of 23 (^{23}Na) and makes up about 100% of all the isotopes of the element sodium found on Earth. All the other 13 isotopes (from ^{19}Na to ^{31}Na) are radioactive with relatively short half-lives and thus are unstable.

ELECTRON CONFIGURATION

Energy Levels/Shells/Electrons	Orbitals/Electrons
1-K = 2	s2
2-L = 8	s2, p6
3-M = 1	s1

Properties

Sodium is a soft, wax-like silver metal that oxidizes in air. Its density is 0.9674 g/cm³, and therefore it floats on water as it reacts with the water releasing hydrogen. It has a rather low melting point (97.6°C) and a boiling point of 883°C. Sodium is an excellent conductor of heat and electricity. It looks much like aluminum but is much softer and can be cut with a knife like butter. Its oxidation state is +1.

Characteristics

On the periodic table sodium is located between lithium and potassium. A fresh cut into sodium looks silvery but turns gray as sodium oxidizes rapidly in air, forming sodium oxide on its surface.

Sodium is extremely reactive. It reacts explosively in water as it releases hydrogen from the water with enough heat to ignite the hydrogen. The resulting compound of this reaction is sodium hydroxide ($2Na + 2H_2O \rightarrow 2NaOH + H_2\uparrow$). Due to its extremely electropositive reactivity, there are few uses for the pure metallic form of sodium. Because of its reactivity, hundreds of sodium compounds are found on the Earth's surface.

An unusual characteristic of several alkali metals is that a mixture of two or more has a lower melting point than the melting point of the separate metals. This is referred to as a **eutectic system** of metallic alloys. For instance, sodium has a melting point of 97.6°C, and potassium's melting point is 63.25°C, but when the two are mixed, the eutectic melting point (turning into a liquid phase) of the combined Na-K system is below zero degrees Celsius (–10°C). If cesium metal (melting point of 38.89°C) is added to the Na and K mixture, the melting point of this eutectic alloy (Na-K-Cs) is the lowest of any eutectic alloy at –78°C.

Abundance and Source

Sodium is the sixth most abundant of the Earth's elements. Since it is a highly electropositive metal and so reactive with nonmetals, it is not found in its pure elemental form on Earth. Rather, it is found in numerous compounds in relatively abundant quantities. About 2.83% of the Earth's crust consists of sodium in compounds.

Sodium is produced by an electrolytic process, similar to the other alkali earth metals. (See figure 4.1). The difference is the electrolyte, which is molten sodium chloride (NaCl, common table salt). A high temperature is required to melt the salt, allowing the sodium cations to collect at the cathode as liquid metallic sodium, while the chlorine anions are liberated as chlorine gas at the anode: 2NaCl (salt) + electrolysis → $Cl_2\uparrow$ (gas) + 2Na (sodium metal). The commercial electrolytic process is referred to as a Downs cell, and at temperatures over 800°C, the liquid sodium metal is drained off as it is produced at the cathode. After chlorine, sodium is the most abundant element found in **solution** in seawater.

History

Before the year 1700, chemists were unable to distinguish the differences among the various alkali metals. Sodium was often confused with potassium, which was artificially produced by slowly pouring water over wood ashes and then drying the resulting alkaline crystal deposits. Some natural alkali metals were also found at the edges of dried lakebeds and mines and even on exposed surfaces of the Earth.

In the early 1700s, Henri-Louis Duhamel du Monceau (1700–1782) was the first to realize that many minerals exhibited similar alkaline (basic) characteristics. He studied samples of salts both derived artificially and found in nature, including saltpeter (potassium nitrate used in gunpowder), table salt, Glauber's salt, sea salt, and borax.

In 1807 Sir Humphry Davy (1778–1829) devised an electrolysis apparatus that used electrodes immersed in a bath of melted sodium hydroxide. When he passed an electric current through the system, metallic sodium formed at the negative (cathode) electrode. He first performed this experiment with molten potassium carbonate to liberate the metal potassium, and he soon followed up with the sodium experiment. Today, sodium and some of the other alkali metals are still produced by electrolysis. The types of electrolytes may vary using a mixture of sodium chloride and calcium chloride and then further purifying the sodium metal.

Common Uses

Because sodium is such a reactive element and is not found in its elemental form, it is responsible for the formation of many compounds on the Earth's surface. Sodium oxide (Na_2O), also known as sodium monoxide, is the most abundant and caustic salt of sodium in the Earth's crust, but sodium chloride (NaCl) is probably the most common and useful. Other useful sodium salts are sodium bicarbonate (baking soda), sodium carbonate (soda), sodium chloride (rock salt), sodium borate (borax), and sodium sulfate (used in paper and photo industries). Mineral springs have a variety of sodium salts, as well as other trace elements, that give the water its "fresh" taste.

Sodium is used in both low-pressure and high-pressure sodium vapor lamps. The low-pressure arc uses just a small amount of Na along with some neon for a starter. The lamp is economical and bright. The illumination with its single yellow color (electromagnetic frequency) makes it difficult for us to recognize other colors. In addition to sodium, the high-pressure lamp uses mercury, which provides a more natural color rendition of light. The very bright light of sodium-mercury lamps makes them ideal for use in sports stadiums and highways.

Because the melting point of sodium metal is about 98° C (a bit lower than the boiling point of water), it is heated into a liquid phase and then transported in rail tank cars, where it cools and solidifies. When it arrives at its destination, heating coils in the tanks warm it back to the liquid stage, and it is then stored for use. Because sodium has a high **specific heat** rating, a major use is as a liquid coolant for nuclear reactors. Even though sodium (both solid and liquid) is extremely reactive with water, it has proven safe as a coolant for nuclear reactors in submarines.

Soda niter or sodium nitrate ($NaNO_3$) is the most abundant of the nitrate minerals. It is used for fertilizer, explosives, and preservatives. The natural deposits are located in northern Chile, which was the original source for many years. More recently, nitrogen fixation, which extracts nitrogen from air, has been used for producing sodium nitrate. This synthetic process has greatly increased the availability of this useful sodium salt by eliminating the need for the natural source. It is used to preserve and cure meats and is used in photography, in pharmaceuticals, and as a color fixative in fabrics.

Of course, the most common use is everyday table salt, sodium chloride ($NaCl$). Salt is vital for health; the body must have a small amount to survive. In the past, wars were fought over salt mines and salt deposits by nations that did not have any natural sources. Excessive sodium chloride in the diet can also be harmful to one's health.

Following are some of the more useful of the hundreds of existing sodium compounds.

Examples of Compounds

Sodium carbonate (Na_2CO_3) is the eleventh most used industrial chemical in the United States. It is commonly used as a bleaching agent and is manufactured in a two-step process. First, ammonia is combined with carbon dioxide to form sodium chloride and water, which reacts to form sodium bicarbonate and ammonium chloride ($NH_3 + CO_2 + NaCl + H_2O \rightarrow NaHCO_3 + NH_4Cl$). Sodium bicarbonate, commonly known as baking soda, is used as a leavening agent in baking, as an antacid to relieve stomach acid, and as a component for fire extinguishers. The second step is known as the Solvay process, wherein the sodium bicarbonate is heated and converted into sodium carbonate ($NaHCO_3 \Delta \rightarrow Na_2CO_3 + H_2O + CO_2$).

There are two forms of *sodium sulfate* (Na_2SO_4). One is a called salt cake and contains no water and, thus, is called anhydrous sodium sulfate. When one molecule of the dry form combines with 10 molecules of water (**decahydrate**), it is known as Glauber's salt ($Na_2SO_4 \cdot 10H_2O$). Sodium sulfate is produced by several processes, one of which is $2NaCl + H_2SO_4 \rightarrow Na_2SO_4 + 2HCl$. A large number of sodium salts can exist in both the **anhydrous** (dry) form and the **hydrous** (attached to water) form.

There are several other compounds of sodium and sulfur, including the following:

Sodium sulfite (Na_2SO_3) is an **antioxidant,** used as a preservative except with meats. It is also used for water treatment and in photography and textile bleaching.

Sodium sulfide (Na_2S) is used in the dye industry, in the oxidation process of gold, lead, and cooper metal ores, as a sheep dip, and to process paper.

Sodium thiosulfate ($Na_2S_2O_3$) is known as photographers' "hypo." It dissolves the unexposed silver salts from photographic negatives and prints during the process of "fixing" the image so that the film or print will no longer be light-sensitive.

Sodium permanganate ($NaMnO_4$) is a purple crystal that is soluble in water and is used as an **oxidizing agent,** disinfectant, and bactericide and as an antidote for morphine poisoning.

Sodium silicate (Na_2O) is known as "water glass" and is used in water treatment and in making soaps, detergents, adhesives, drilling fluids, bleaches.

Sodium hydroxide (NaOH) is one of the most useful industrial sodium compounds. It is also known as lye or caustic soda and is one of the strongest base alkalis (high pH value) on the household market. It is used as a drain and oven cleaner, and it saponifies fats in the manufacture of soap. It must be used with care because it is also capable of producing serious skin burns.

Rock salt (halite, native NaCl) is an unpurified, coarse common salt that is spread on highways to melt snow and ice.

Hazards

Sodium as the elemental metal is very dangerous because of its extreme electropositive nature, particularly when it comes in contact with moist air, water, snow, or ice or other oxidizing agents. It readily gives up electrons to electronegative atoms (nonmetals). In these reactions, it releases hydrogen gas with enough heat to explosively ignite the hydrogen

Sodium perchlorate ($NaClO_4$) is extremely dangerous. It is easily detonated and thus used to set off explosives. It is also used as a jet fuel.

Sodium peroxide (Na_2O_2) is explosive when in contact with water. It is a strong oxidizing agent that is very irritating.

Numerous sodium compounds are hazardous as **carcinogens** (cancer-causing) and as toxins (poisons) in plants and animals. On the other hand, we benefit greatly from the many compounds containing the element sodium. We could not live without it.

POTASSIUM

SYMBOL: K **PERIOD:** 3 **GROUP:** 1 (IA) **ATOMIC NO:** 19
ATOMIC MASS: 39.0983 amu **VALENCE:** 1 **OXIDATION STATE:** +1
  NATURAL STATE: Solid
ORIGIN OF NAME: Its symbol "K" is derived from the Latin word for alkali, *kalium,* but it is commonly called "potash" in English.
ISOTOPES: A total of 18 isotopes of potassium have been discovered so far. Just two of them are stable: K-39 makes up 93.2581% of potassium found in the Earth's crust, and K-41 makes up 6.7301% of the remainder of potassium found on Earth. All the other 16 potassium isotopes are unstable and radioactive with relatively short half-lives, and as they decay, they produce beta particles. The exception is K-40, which has a half-life of 1.25×10^9 years.

ELECTRON CONFIGURATION

Energy Levels/Shells/Electrons	Orbitals/Electrons
1-K = 2	s2
2-L = 8	s2, p6
3-M = 8	s2, p6
4-N = 1	s1

Properties

Elemental potassium is a soft, butter-like silvery metal whose cut surface oxidizes in dry air to form a dark gray potassium superoxide (KO_2) coating. KO_2 is an unusual compound, in that it reacts with both water and carbon dioxide to produce oxygen gas. It appears more like a hard wax than a metal. Its density (specific gravity) is 0.862 g/cm^3, its melting point is 63.25°C, and its boiling point is 760°C. It has an oxidation state of +1 and reacts explosively with room temperature air or water to form potassium hydroxide as follows: $2K + 2 H_2O \rightarrow \Delta\ 2KOH + H_2$. This is an endothermic reaction, which means the heat generated is great enough to ignite the liberated hydrogen gas. Potassium metal must be stored in a non-oxygen, non-aqueous environment such as kerosene or naphtha.

Characteristics

Because its outer valence electrons are at a greater distance from its nuclei, potassium is more reactive than sodium or lithium. Even so, potassium and sodium are very similar in their chemical reactions. Due to potassium's high reactivity, it combines with many elements, particularly nonmetals. Like the other alkali metals in group 1, potassium is highly alkaline (caustic) with a relatively high pH value. When given the flame test, it produces a violet color.

Abundance and Source

Potassium is the eighth most abundant element in the Earth's crust, which contains about 2.6% potassium, but not in natural elemental form. Potassium is slightly less abundant than sodium. It is found in almost all solids on Earth, in soil, and in seawater, which contains 380 ppm of potassium in solution. Some of the potassium ores are sylvite, carnallite, and polyhalite. Ore deposits are found in New Mexico, California, Salt Lake in Utah, Germany, Russia, and Israel. Potassium metal is produced commercially by two processes. One is **thermochemical** distillation, which uses hot vapors of gaseous NaCl (sodium chloride) and KCl (potassium chloride); the potassium is cooled and drained off as molten potassium, and the sodium chloride is discharged as a slag. The other procedure is an electrolytic process similar to that used to produce lithium and sodium, with the exception that molten potassium chloride (which melts at about 770°C) is used to produce potassium metal at the cathode (see figure 4.1).

History

In 1807 Sir Humphry Davy was the first to isolate potassium metal. He used his famous electrolysis method of passing an electric current through melted potassium chloride. He noticed small globs of silvery metallic potassium formed at the cathode that reacted strongly with water releasing hydrogen gas. The reaction was strong enough to raise the temperature to the degree that the hydrogen was ignited. Through this reaction, Davy recognized that he had produced elemental potassium.

For years, several potassium compounds were collected and identified by leeching water through wood ashes, drying the solutions, and then examining their crystals.

Common Uses

As with other alkali metals, potassium compounds have many uses. For example, almost all of the compound potassium chloride is used in fertilizers. Currently potassium chloride is mined or derived from seawater. Many years ago, potassium was secured for human use by burning wood and plant matter in pots to produce an ash called potash, which was mostly potassium carbonate and used as a caustic, mainly for making soap when mixed with fats.

Several explosives are mixtures of compounds of potassium and other substances. It is an important raw material for making explosives, matches, and fireworks.

Liquid potassium, when mixed with liquid sodium (NaK), is an alloy used as a heat-exchange substance to cool nuclear reactors.

Potassium is an important **reagent** (something that is used in chemical reactions to analyze other substances) that forms many compounds used in chemical and industrial laboratories.

It is used to manufacture both hard and soft soaps, as a bleaching agent, and where a highly caustic chemical is required.

Potassium is essential to all living organisms. It is a trace element required for a healthy diet and is found in many foods. One natural source is bananas.

Examples of Compounds

Potassium nitrate (KNO_3) has been known for hundreds of years and was originally known as "saltpeter," a name derived from the Greek word *petra,* which stands for "rock." Potassium nitrate was thought to be an element until the 1700s, when it was discovered that several of the alkali metals (including potassium) were combined with nitrogen. All of the early sources of potassium nitrate were found as residues around salty lakes, in decayed matter, in guano (bat dung), or in natural deposits. In the tenth century the Chinese were using natural saltpeter mixed with charcoal and sulfur to make fireworks and an early form of gunpowder. It was not until the thirteenth century that this technology arrived in Europe, where the production of gunpowder and firearms soon became common. Finding sources of the scarce saltpeter was a considerable problem that was not solved until the 1700s, when chemists realized that nitrogen was also an element in this compound (sometimes called "niter"). Potassium nitrate can be prepared by treating sodium hydroxide (from leeched wood ashes) with nitric acid: $HNO_3 + KOH \rightarrow KNO_3 + H_2O$. Other processes are possible as well. In addition to fireworks and fertilizers, potassium nitrate is also used for preserving foods and in match heads, and at one time, it was used as a "sexual depressant." However, the evidence for effectiveness of the treatment is scanty.

Potassium bromide (KBr) is used in photography and as a medical sedative.

Potassium chloride (KCl) is used in drug preparations and as a food additive and chemical reagent. It is possible to reduce the sodium in your diet by substituting potassium chloride for table salt (sodium chloride), which may be healthier. Molten potassium chloride is also used in the electrolytic production of metallic potassium. KCl is also found in seawater brine and can be extracted from the mineral carnallite.

Potassium carbonate (K_2CO_3), also known as potash or pearl ash, is the 33rd most produced chemical in the United States. It is a white, granular powder that becomes alkaline when dissolved in water. It is produced by a chemical process involving magnesium oxide, potassium chloride, and carbon dioxide, or by a laboratory process of bubbling CO_2 through a solution of KOH. Commercially, both the Haber-Bosch and the Solvay processes are employed in its manufacture. Potassium carbonate is used in the production of color TV tubes and other optical devices, as a food additive, for making paint and ink pigments, in the manufacture of some glass and soaps, and as an agent for **dehydration** (a material that will "dry out" or remove moisture from substances).

Potassium hydroxide (KOH) is commonly referred to as caustic potash or lye because of its extreme alkalinity (high pH value). It is produced in an electrolytic cell reaction ($2KCl + 2H_2O \rightarrow H_2 + Cl_2 + 2KOH$). It is used to manufacture soaps and as a caustic drain cleaner because it is strong enough to dissolve animal fats that clog household and restaurant drains.

Potassium chromate (K_2CrO_4) is soluble in water and is used to make bright yellow inks and paint pigments. It is also used as a reagent in chemical laboratories and as a **mordant** to "fix" dyes in colored textiles.

Potassium cyanide (KCN) is a white crystalline substance with a slight odor of bitter almonds. It is produced when hydrogen cyanide is absorbed in potassium hydroxide. It is used to extract gold and silver from their ores, in electroplating computer boards, and as an insecticide. Potassium cyanide is very toxic to the skin or when ingested or inhaled, and it is used as a source of cyanide (CN) gas in gas chambers.

Potassium permanganate ($KMnO_4$) is a dark purple-bluish sheen crystal with a slightly sweet taste. It is produced by oxidizing manganate in an electrolytic cell or by passing carbon dioxide through a hot solution of manganate and then cooling until permanganate crystals form. It is a strong oxidizing agent, particularly with organic matter, which makes it a good disinfectant, deodorizer, bleach, and antiseptic.

Potassium iodide (KI) is added to table salt (known as iodized salt) to help prevent enlargement of the thyroid gland, a condition known as a goiter.

Hazards

Elemental potassium as a metal is not found in its pure form in nature, but is derived from its numerous compounds. The metal is very dangerous to handle. It can ignite while you are holding it with your hands or as you cut it. The metal must be stored in an **inert** gas atmosphere or in oil. Potassium fires cannot be extinguished with water—it only makes matters worse because it results in the formation of potassium hydroxide and hydrogen gas with enough heat to ignite the hydrogen. Dry chemicals such as soda ash, graphite, or dry sand can be used.

A particular hazard, which has been with humans since the beginning of time, is the radioactive isotope potassium-40 (K-40). Less than 1% of all potassium atoms on Earth are in the form of this radioactive isotope. It has a half-life of 1.25 billion years. Its decay process ends with the formation of the noble gas argon, which can then be analyzed to determine the age of rocks. This system (K-40 → argon) has been used to establish that the oldest rocks on Earth were formed about 3.8 billion years ago. Every living thing needs some potassium in its diet, including humans, who cannot escape this source of radiation, given that the human body cannot distinguish the radioactive potassium from the nonradioactive form. Along with cosmic rays and other naturally radioactive elements in the Earth's crust, potassium-40 contributes to the normal lifetime accumulation of radiation. It makes up almost one-fourth of the total radiation the human body receives during a normal life span.

Many of the potassium "salts" mentioned are hazardous because they are explosive when either heated or shocked. Some are also toxic to the skin and poisonous when **ingested.** On the other hand, numerous compounds of potassium make our lives much more livable.

RUBIDIUM

SYMBOL: Rb **PERIOD:** 5 **GROUP:** 1 (IA) **ATOMIC NO:** 37
ATOMIC MASS: 85.4678 amu **VALENCE:** 1 **OXIDATION STATE:** +1
  NATURAL STATE: Solid
ORIGIN OF NAME: Rubidium is named for the Latin word *rubidus,* meaning "reddish."
ISOTOPES: There are 30 isotopes of rubidium, ranging from Rb-75 to Rb-98. Rb-85 is the only stable form of rubidium and constitutes 72.17% of all rubidium isotopes found in the Earth's crust. Rb-87 is radioactive (a half-life of 4.9×10^{10} years) and makes up about 27.83% of the remainder of rubidium found in the Earth's crust. All the other 28 isotopes make up a tiny fraction of all the rubidium found on Earth and are radioactive with very short half-lives.

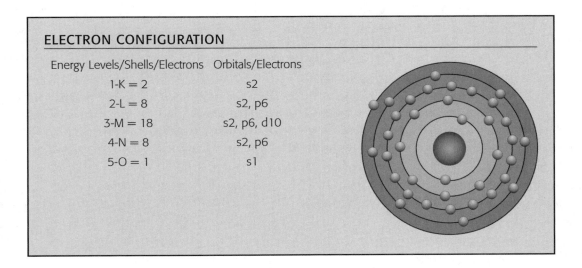

ELECTRON CONFIGURATION

Energy Levels/Shells/Electrons	Orbitals/Electrons
1-K = 2	s2
2-L = 8	s2, p6
3-M = 18	s2, p6, d10
4-N = 8	s2, p6
5-O = 1	s1

Properties

Rubidium is a silvery-white lightweight solid at room temperature, but it melts at just 38.89°C (102°F), which is just over the human body's normal temperature. Its boiling point is 686°C, its density is 1.532 g/cm³, and it has an oxidation state of +1.

Characteristics

Rubidium is located between potassium and cesium in the first group in the periodic table. It is the second most electropositive alkali element and reacts vigorously and explosively in air or water. If placed on concrete on a sunny day, it would melt and then react violently with moist air to release hydrogen with enough heat to burn the hydrogen. If a chunk of rubidium metal is left on a table exposed to the air, it combusts spontaneously. Rubidium must be stored in oil, such as kerosene.

Abundance and Source

Rubidium does not exist in its elemental metallic form in nature. However, in compound forms it is the 22nd most abundant element on Earth and, widespread over most land areas in mineral forms, is found in 310 ppm. Seawater contains only about 0.2 ppm of rubidium, which is a similar concentration to lithium. Rubidium is found in complex minerals and until recently was thought to be a rare metal. Rubidium is usually found combined with other Earth metals in several ores. The lepidolite (an ore of potassium-lithium-aluminum, with traces of rubidium) is treated with hydrochloric acid (HCl) at a high temperature, resulting in lithium chloride that is removed, leaving a residue containing about 25% rubidium. Another process uses thermochemical reductions of lithium and cesium ores that contain small amounts of rubidium chloride and then separate the metals by fractional distillation.

History

Two nineteenth-century chemists, Robert Bunsen (1811–1899) and Gustav Kirchhoff (1824–1887), collaborated in a search of alkali metals in the mineral lepidolite. Using Bunsen's new gas burner flame and the new instrument called a **spectroscope,** they noticed that, when heated or burned, different elements produce distinctive colors of light when transmitted through a **prism.** When the spectrum of elements in the sun are viewed on Earth via a spectroscope, the element's light passes through the sun's light, which (by absorbsion of that element's light) produces a dark line in that element's spectrum. In other words, instead of viewing bright lines from the burning element, one sees dark lines because the sun's light absorbs the element's bright line. Using the spectroscope, the scientists noticed spectral lines for several colors when they burned lepidolite. The spectral lines for sodium, potassium, lithium, and iron were identified. They also noticed a new distinctive spectral line with a deep red color and in 1861 named its source "rubidium" because it resembled the color of a ruby.

Common Uses

Because rubidium is a much larger atom than lithium or sodium, it gives up its outer valence electron easily, thus becoming a positive ion (oxidation state = Ru^+).

Rubidium forms numerous compounds, but only a few are useful. One of the main uses for rubidium is as a **getter** in **vacuum** tubes used in early radios, TVs, and cathode-ray tubes. These kinds of tubes work best if all the air is removed, so a getter absorbs the remaining few atoms of air that cannot be removed mechanically by vacuum pumps, thus extending the life of the vacuum tube.

Rubidium chloride is used for the production of rubidium metal, which, in the liquid form, has a high heat transfer coefficient, making it useful as a coolant (along with other alkali metals) for nuclear reactors.

When rubidium gas is placed in sealed glass cells along with an inert gas, it becomes a rubidium-gas cell clock. Because of the consistent and exact frequency (vibrations) of it atoms, it is a very accurate timekeeper.

Rubidium and selenium are used in the manufacture of photoelectric cells, sometimes called electric eyes. When light strikes these elements, electrons are knocked loose from the outer shells of their atoms. These free electrons have the ability to carry an electric current. **Photoelectric cells** are used to send a beam across a driveway or garage door. When the beam is broken, an alarm sounds or the door opens.

Rubidium is a very caustic alkali (base) with a high pH value that makes it an excellent reducing agent (highly electropositive) in industry and chemical laboratories.

A unique use is its ability to locate brain tumors. It is a weak radioisotope able to attach itself to diseased tissue rather than healthy tissue, thus making detection possible.

Examples of Compounds

Because rubidium is a highly reactive metal, it forms many compounds. A few examples follow.

Rubidium carbonate (Rb_2O_3) is used to make special types of glass.

Rubidium chlorides (RbCl) is a source of rubidium metal and is used as a chemical reagent.

Rubidium hydroxide (RbOH) is very **hygroscopic** (absorbs large amounts of water for its weight). It is also an excellent absorber of carbon dioxide. Rubidium hydroxide can be used to etch glass and as an electrolyte in low-temperature electric storage batteries for use in vehicles in the subarctic.

Rubidium also has the ability to form what are called double sulfates.

Rubidium cobalt sulfate ($Rb_2SO_4 \bullet CoSO_4 \bullet 6H_2O$) is an example of several double sulfates that rubidium has the ability to form. Rubidium cobalt sulfate is a combined rubidium-cobalt compound in the form of ruby-red crystals. Other rubidium sulfate crystal compounds and their colors are rubidium + copper = white; rubidium + iron = dark green; and rubidium + magnesium = colorless.

Hazards

The major hazard is from fire and explosions of the elemental metallic form of rubidium. It must be stored in an inert atmosphere or in kerosene. When rubidium contacts skin, it ignites and keeps burning and produces a deep, serious wound. Water and blood just make it react more vigorously.

Many of the compounds of rubidium are toxic and strong irritants to the skin and lungs. It is one of the elements best left to experienced handlers.

Very small traces of rubidium are found in the leaves of tobacco, tea, and coffee, as well as in several edible plants, but these radiation traces are harmless when used in moderation.

CESIUM

SYMBOL: Cs   **PERIOD:** 6   **GROUP:** 1 (IA)   **ATOMIC NO:** 55
ATOMIC MASS: 132.90546 amu   **VALENCE:** 1   **OXIDATION STATE:** +1
   NATURAL STATE: Solid
ORIGIN OF NAME: In 1860 Gustav Kirchhoff and Robert Bunsen named the element "Cesium," using the Latin word *caesius,* which means bluish-gray.

ISOTOPES: Cs-133 is the only stable isotope of cesium, and it makes up all of the naturally occurring cesium found in the Earth's crust. In addition to Cs-133 there are about 36 radioactive isotopes of Cs, most of which are artificially formed in nuclear reactors. All are produced in small numbers of atoms with relatively short half-lives. The range of Cs isotopes is from Cs-113 (amu = 112.94451) to Cs-148 (amu = 147.94900). Most of these radioisotopes produce **beta** radiation as they rapidly decay, with the exception of Cs-135, which has a half-life of 3×10^6 yr, which makes it a useful research tool. Cs-137, with a half-life of 33 years, produces both beta and gamma radiation.

ELECTRON CONFIGURATION

Energy Levels/Shells/Electrons	Orbitals/Electrons
1-K = 2	s2
2-L = 8	s2, p6
3-M = 18	s2, p6, d10
4-N = 18	s2, p6, d10
5-O = 8	s2, p6
6-P = 1	s1

Properties

Like the other alkali metals, cesium is a soft-solid silvery metal, but much softer than the others. It is the least electronegative and most reactive of the Earth metals. Cesium has an oxidation state of +1, and because its atoms are larger than Li, Na, and K atoms, it readily gives up its single outer valence electron. The single electron in the P shell is weakly attached to its nucleus and thus available to combine with many other elements. It is much too reactive to be found in its metallic state on Earth.

Cs has a melting point of 29°C, which is lower than the body temperature of humans (37°C), and thus a chunk of cesium will melt in a person's hand with disastrous results. Since it reacts with moisture on skin as well as with the air to release hydrogen, it will burn vigorously through the palm of one's hand.

Cesium's boiling point is 669.3°C and its density is 1.837 g/cm^3. Mercury is the only metal with a lower melting point than cesium. It is extremely dangerous when exposed to air, water, and organic compounds or to sulfur, phosphorus, and any other electronegative elements. It must be stored in a glass container containing an inert atmosphere or in kerosene.

Cesium reacts with water in ways similar to potassium and rubidium metals. In addition to hydrogen, it forms what is known as superoxides, which are identified with the general formula CsO_2. When these superoxides react with carbon dioxide, they release oxygen gas, which makes this reaction useful for self-contained breathing devices used by firemen and others exposed to toxic environments.

Characteristics

Cesium is located between rubidium and francium in group 1 of the periodic table. It is the heaviest of the stable alkali metals and has the lowest melting point. It is also the most reactive of the alkali metals.

Cesium will decompose water, producing hydrogen, which will burn as it is liberated from H_2O. Cesium is extremely dangerous to handle and will burn spontaneously or explode when exposed to air, water, and many organic compounds.

Abundance and Source

The stable form of Cs-133 is the 48th most abundant element on Earth, but because it is so reactive, it is always in compound form. The Earth's crust contains only about 7 ppm of Cs-133. Like the other alkali metals, it is found in mixtures of complex minerals. Its main source is the mineral pollucite ($CsAlSi_2O_6$). It is also found in lepidolite, a potassium ore. Pollucite is found in Maine, South Dakota, Manitoba, and Elba and primarily in Rhodesia, South Africa.

One problem in refining cesium is that it is usually found along with rubidium; therefore, the two elements must be separated after they are extracted from their sources. The main process to produce cesium is to finely grind its ores and then heat the mix to about 600°C along with liquid sodium, which produces an alloy of Na, Cs, and Ru, which are separated by fractional distillation. Cesium can also be produced by the thermochemical reduction of a mixture of cesium chloride (CsCl) and calcium (Ca).

History

In 1860 Gustav Kirchhoff and Robert Bunsen discovered cesium while experimenting with their spectroscope. While analyzing mineral water, they identified diffraction lines for sodium, lithium, calcium, and strontium. As they removed these elements from their samples, they noticed two bright blue lines indicating the discovery of a new element, and they named cesium for the blue color of the lines seen through the prism in their **spectroscopic analysis** of the heated sample. Kirchhoff and Bunsen were the first to publish the technology of **spectroscopy,** in a book titled *Chemical Analysis through Observation of the Spectrum.* Their system became a valuable asset in the identification of many other elements.

Common Uses

Because of some of its longer-lived isotopes, cesium has become valuable for its ability to produce a steady stream of beta particles (β) as electrons.

Light is strong enough to "knock off" electrons from cesium, which makes this phenomenon useful as a coating for photoelectric cells and electric eye devices. Cesium iodide (CsI) is used in scintillation counters (**Geiger counters**) to measure levels of external radiation. It is also useful as a "getter" to remove air molecules remaining in vacuum tubes.

In 1960 the International Committee of Weights and Measures selected radioactive cesium-137 (with a half-life of about 33 years) as the standard for measuring time. They equated the second with the radiation emitted by a Cs-137 atom that is excited by a small energy source. Thus, the second is now defined as 9,192,631,770 vibrations of the radiation emitted by an atom of Cs-137. There are about 200 atomic clocks around the world that collaborate their efforts to maintain this extremely accurate clock that never needs winding or batteries.

Cesium is used as a **hydrogenation** catalyst to enhance and assist the reaction in the conversion of liquid oils to solids forms (e.g., in the production of margarine).

In a molten state, it is used as a heat-transfer fluid in plants generating electric power.

Cesium is used experimentally as a **plasma** to produce a source of ions to power outer space vehicles using ion engines.

Cesium is used in military **infrared** devices and signal lamps as well as in other optical devices.

Cesium is used as a chemical reagent and reducing agent in industry and the laboratory. It can also be used as an antidote for arsenic poisoning.

Examples of Compounds

Cesium bromide (CsBr) crystals are used in **scintillation counters** to detect radiation. The compound is also used to coat the inside of **fluorescent** screens.

Cesium carbonate (Cs_2CO_3) is used in the beer-brewing industry to make the "head" of beer foamier. It is also used in glassmaking and to enhance the taste of mineral water.

Cesium chloride (CsCl) is produced by the reaction of cesium metal with chlorine gas (Ca^+ + Cl^- → CsCl). It is also used in the beer brewing industry, to coat fluorescent screens, and to improve the taste of mineral water.

Cesium hydroxide (CsOH) is the strongest base (alkali) with the highest pH value of any chemical yet found. It is easy to produce by just placing cesium metal in water (which is very reactive). After the hydrogen escapes, cesium hydroxide remains in the water ($2Cs + H_2$ → $2CsOH + H_2$).

Cesium-137 is a highly useful radioisotope that emits its radiation at a very steady and controllable rate. This makes it useful as an atomic clock because it is extremely accurate and never needs winding or a new battery. It is also useful as a radiation source for treatment of malignant cancers. Cs-137 has replaced the much more dangerous cobalt-60 as a source of radiation in industry and medicine.

Hazards

Although cesium has many of the properties and characteristics of the other alkali metals, because of the large size of its atoms, cesium metal is much more reactive and dangerous to handle. Special precautions need to be taken to keep it away from air, water, and organic substances with which it can vigorously react. Its use should be restricted to laboratories and industries capable of using it safely.

Cesium-137, with a half-life of about 30 years, produces dangerous radiation and can cause radiation poisoning if mishandled. It is used to sterilize wheat, potatoes, and other foods to protect them from insect damage and rotting. It is also used to kill bacteria in the treatment of sewage sludge.

FRANCIUM

SYMBOL: Fr   PERIOD: 7   GROUP: 1 (IA)   ATOMIC NO: 87
ATOMIC MASS: 223 amu   VALENCE: 1   OXIDATION STATE: +1 (?).   NATURAL STATE:
   Solid, unstable, radioactive.

ORIGIN OF NAME: In 1939 Marguerite Perey (1909–1975), a French physicist who worked for the Curie institute in Paris, named the newly discovered element after her country—France.

ISOTOPES: There are no stable isotopes of francium found on Earth. All of its 33 isotopes (ranging from Fr-201 to Fr-232) are radioactive; therefore, the one with the longest half-life of about 20 minutes (Fr-223) is the one used to determine its atomic weight. Fr-223 is the only radioisotope of francium that is found naturally as a decay product from other unstable elements.

ELECTRON CONFIGURATION

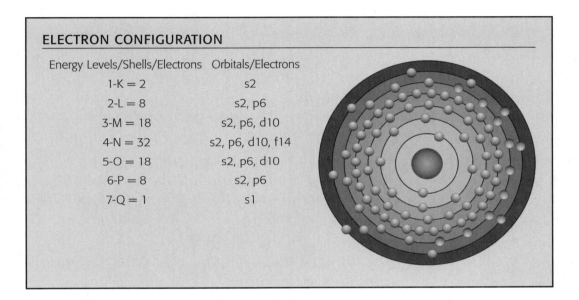

Energy Levels/Shells/Electrons	Orbitals/Electrons
1-K = 2	s2
2-L = 8	s2, p6
3-M = 18	s2, p6, d10
4-N = 32	s2, p6, d10, f14
5-O = 18	s2, p6, d10
6-P = 8	s2, p6
7-Q = 1	s1

Properties

Not a great deal is known about francium's properties, but some measurements of its most stable isotope have been made. Its melting point is 27°C and its boiling point is 677°C, but its density is unknown. It is assumed to have a +1 oxidation state (similar to all the other alkali metals)

Francium's atoms are the largest and heaviest of the alkali metals in group 1 (IA). It is located just below cesium on the periodic table, and thus it is assumed to be an extremely reactive reducing agent even though it is the most scarce of the alkali metals. Its most stable isotope (Fr-223) exists for about 21 or 22 minutes. No one has figured out how to refine francium from natural minerals (ores) because the atoms of the most stable isotope found in nature (Fr-223) are scattered very thinly over the Earth's crust. All of the other 30 isotopes are produced for study by nuclear decay of other radioactive elements.

Characteristics

Francium is the only element from atomic numbers 1 to 92 that has a half-life of less than 30 minutes. It is the only known element with more identified isotopes than identified compounds. All samples of francium available for study are produced artificially.

Abundance and Source

Only about one ounce of natural francium exists in the Earth's crust. All the other isotopes of francium are artificially produced in very small amounts (just a few atoms at a time) that exist for a few seconds to minutes.

Two methods to secure very small samples of francium for examination use the decay processes of other radioactive elements. One is to bombard thorium with protons. The second is to start with radium in an accelerator, where, through a series of decay processes, the radium is converted to actinium, which in turn rapidly decays into thorium, and finally, thorium decays naturally into francium. Following is a schematic of the decay process used for the production of small amounts of Fr-223 which, in turn, after several more decay processes ends up as stable lead (Pb):

$$_{90}\text{Th-231} \rightarrow {}_{91}\text{Pa-231} \rightarrow {}_{89}\text{Ac-227} \rightarrow {}_{87}\text{Fr-223} \rightarrow \text{through more decays} \rightarrow {}_{82}\text{Pb-207}.$$

History

The placement of an unknown element with an atomic number of 87 in group 1, period 7 of the periodic table was one of Dimitri Mendeleev's ideas based on the chemical properties and physical characteristics of the other alkali metals. In the late nineteenth century, Mendeleev named this unknown element "eka-cesium" and predicted its properties based on what was known of cesium's placement on the periodic table. This led to worldwide searches for element number 87, which were not all successful but which did result in proposed names for eka-cesium (moldavium, virginium, russium).

Marguerite Catherine Perey, an assistant to Marie Curie, is credited with the discovery of francium-223 in 1939. Perey discovered the sequence of radioactive decay of radium to actinium and then to several other unknown radioisotopes, one of which she identified as francium-223. Since half of her sample disappeared every 21 minutes, she did not have enough to continue her work, but a new element was discovered.

Common Uses

Given that all isotopes of francium are radioactive with relatively short half-lives, there are few practical uses for it—except as a source of radiation to study the radioactive decay process.

Examples of Compounds

Francium has many more isotopes (33) than compounds. However, knowing how the other alkali metals form compounds, one may speculate on several possibilities. Its metal ion most likely is Fr^+, which means it has a very low level of electronegativity and would combine vigorously with anions of nonmetals that have a very high electronegativity. For example, if it reacted with chlorine, it would form FrCl; and if a chunk of metallic francium (which would be hard to find) were dropped in water, it would explode and form francium hydroxide ($2Fr + 2H_2O \rightarrow 2FrOH + H_2\uparrow$).

Hazards

Because all of francium's isotopes are radioactive, they are a potential hazard to humans and should be handled only by experienced laboratory personnel. On the other hand, the scarcity of these isotopes makes it almost impossible to consider them harmful in our everyday lives.

Alkali Earth Metals: Periods 2 to 7, Group 2 (IIA)

Introduction

The alkali earth metals are the elements in group 2 (IIA) from periods 2 to 7 in the periodic table. They are beryllium ($_4$Be), magnesium ($_{12}$Mg), calcium ($_{20}$Ca), strontium ($_{38}$Sr), barium ($_{56}$Ba), and radium ($_{88}$Ra).

During the Middle Ages and Renaissance, any element that could not be easily recognized by its outward physical properties, that was not considered a metal, and that was not soluble in water was considered an earth element. This classification was a holdover from the ancient Greek concept that all elements fit into one of the three categories of earth, fire, and air. In addition, any substance that acted in ways similar to potash and soda (alkaline caustics) were classed as alkali earth metals.

Alkali earth metals are not found as free elements in nature, but rather in compounds of minerals and ores. They all have an oxidation number of +2 (as well as additional oxidation numbers) that make them electropositive, resulting in strong reactions with electronegative nonmetals. However, they are not as reactive as the alkali metals. Alkali earth metals do not burn at room temperatures. They are less volatile, denser, and not as soft as the alkali metals. In general, they have higher melting and boiling points than do group 1 metals. Each of the alkali earth metals produces its own distinctive brilliant colored flame when burned, which makes the metals useful for fireworks. Most are malleable, which means, as metals, they can be worked into different shapes as rods, wires, or sheets.

BERYLLIUM

SYMBOL: Be   **PERIOD:** 2   **GROUP:** 2 (IIA)   **ATOMIC NO:** 4
ATOMIC MASS: 9.012182 amu   **VALENCE:** 2   **OXIDATION STATE:** +2   **NATURAL STATE:** Solid
ORIGIN OF NAME: Beryllium was originally known as "glucina" (glucose) from the Greek word *glukos,* meaning "sugar," because of the sweet taste of a few of its salt compounds. Later, beryllium was given the Greek name beryllos after the greenish-blue gemstone beryl (emeralds) that was later found to contain the element beryllium.

PERIODIC TABLE OF THE ELEMENTS

GROUPS	1 IA	2 IIA	3 IIIB	4 IVB	5 VB	6 VIB	7 VIIB	8 VIII	9 VIII	10 VIII	11 IB	12 IIB	13 IIIA	14 IVA	15 VA	16 VIA	17 VIIA	18 VIIIA
PERIODS																		
1	1 H 1.0079																	2 He 4.00260
2	3 Li 6.941	4 Be 9.01218											5 B 10.81	6 C 12.011	7 N 14.0067	8 O 15.9994	9 F 18.9984	10 Ne 20.179
3	11 Na 22.9898	12 Mg 24.305											13 Al 26.9815	14 Si 28.0855	15 P 30.9738	16 S 32.066(6)	17 Cl 35.453	18 Ar 39.948
4	19 K 39.0983	20 Ca 40.08	21 Sc 44.9559	22 Ti 47.88	23 V 50.9415	24 Cr 51.996	25 Mn 54.9380	26 Fe 55.847	27 Co 58.9332	28 Ni 58.69	29 Cu 63.546	30 Zn 65.39	31 Ga 69.72	32 Ge 72.59	33 As 74.9216	34 Se 78.96	35 Br 79.904	36 Kr 83.80
5	37 Rb 85.4678	38 Sr 87.62	39 Y 88.9059	40 Zr 91.224	41 Nb 92.9064	42 Mo 95.94	43 Tc (98)	44 Ru 101.07	45 Rh 102.906	46 Pd 106.42	47 Ag 107.868	48 Cd 112.41	49 In 114.82	50 Sn 118.71	51 Sb 121.75	52 Te 127.60	53 I 126.905	54 Xe 131.29
6	55 Cs 132.905	56 Ba 137.33	★	72 Hf 178.49	73 Ta 180.948	74 W 183.85	75 Re 186.207	76 Os 190.2	77 Ir 192.22	78 Pt 195.08	79 Au 196.967	80 Hg 200.59	81 Tl 204.383	82 Pb 207.2	83 Bi 208.980	84 Po (209)	85 At (210)	86 Rn (222)
7	87 Fr (223)	88 Ra 226.025	▲	104 Unq (261)	105 Unp (262)	106 Unh (263)	107 Uns (264)	108 Uno (265)	109 Une (266)	110 Uun (267)	111 Uuu (272)	112 Uub	113 Uut	114 Uuq	115 Uup	116 Uuh	117. Uus	118 Uuo

— TRANSITION ELEMENTS —

6 ★ Lanthanide Series (RARE EARTH)	57 La 138.906	58 Ce 140.12	59 Pr 140.908	60 Nd 144.24	61 Pm (145)	62 Sm 150.36	63 Eu 151.96	64 Gd 157.25	65 Tb 158.925	66 Dy 162.50	67 Ho 164.930	68 Er 167.26	69 Tm 168.934	70 Yb 173.04	71 Lu 174.967
7 ▲ Actinide Series (RARE EARTH)	89 Ac 227.028	90 Th 232.038	91 Pa 231.036	92 U 238.029	93 Np 237.048	94 Pu (244)	95 Am (243)	96 Cm (247)	97 Bk (247)	98 Cf (251)	99 Es (252)	100 Fm (257)	101 Md (258)	102 No (259)	103 Lr (260)

©1996 R.E. KREBS

ISOTOPES: There are nine isotopes of beryllium, ranging from Be-6 to Be-14. Only Be-9 is found in nature and makes up about 100% of the element found in the Earth's crust. All the other eight isotopes are radioactive and artificially produced with half-lives ranging from Be-8 = 0.067 seconds to Be-14 = 1.6×10^6 years.

ELECTRON CONFIGURATION

Energy Levels/Shells/Electrons   Orbitals/Electrons

1-K = 2	s2
2-L = 2	s2

Properties

As the first element in group 2 (IIA), beryllium has the smallest, lightest, and most stable atoms of the alkali earth metals. Its melting point is 1278° C, its boiling point is 2970°C, and its density is 1.8477 g/cm³. Its color is whitish-gray.

Characteristics

Beryllium is one-third as dense as aluminum. Fresh-cut surfaces of the metal oxidize, thus resisting further oxidation, as does aluminum. It is a lightweight, hard, brittle metal. It can be machined (rolled, stretched, and pounded) into many shapes and is used to produce lightweight alloys.

Abundance and Source

Since its discovery, beryllium has been classed as the 36th most abundant of the elements found in the Earth's crust. Beryllium's principle source is a mineral composed of a complex of beryllium, silicon, and oxygen. It is usually found in deposits as hexagonal crystalline forms in Brazil, Argentina, South Africa, and India as well as in Colorado, Maine, New Hampshire, and South Dakota in the United States. Some deposits have been found in Canada. Many crystals of the mineral may be very large One chunk that measured 27 feet long length and weighed a 25 tons was found in Albany, Maine in 1969.

One method of obtaining beryllium metal is by **chemical reduction,** whereby beryllium oxide is treated with ammonium fluoride and some other heavy metals to remove impurities while yielding beryllium fluoride. This beryllium fluoride is then reduced at high temperatures using magnesium as a **catalyst,** which results in deposits of "pebbles" of metallic beryllium.

Another method for obtaining beryllium metal is by electrolysis of a solution of beryllium chloride ($BeCl_2$) along with NaCl as an electrolyte in solution that is kept molten but below the melting point of beryllium. ($_4$Be has a relatively high melting point of 2,332.4°F.) The beryllium metal does not collect at the negative cathode as do metals in other electrolytic cells, but rather beryllium metal pieces are found at the bottom of the cell at the end of the process.

History

From the days of the Egyptians, when emeralds were a particular favorite of kings, beryl has also been a favored gemstone. It was not until the late eighteenth century that Abbe René Just Haüy (1743–1822), the father of crystallography, studied the crystalline structures and densities of emeralds and beryl and determined that they were the same mineral. At about the same time, in 1798, Louis-Nicolas Vauquelin (1763–1829) discovered that both emeralds and beryl were composed of a new element with four protons in its nucleus. The element was named "glucina" because of its sweet taste. It was not until the nineteenth century that the metal beryllium was extracted from beryllium chloride ($BeCl_2$) by chemical reactions. Late in the nineteenth century, P. Lebeau (dates unknown) separated the metal by the electrolytic process.

Common Uses

In the mid-twentieth century, the determination that beryllium has a number of unique properties led to the production of beryllium metal by electrolysis on a commercial scale. It proved valuable as an alloy metal to produce specialized, strong—but light—structural metals for use in satellites, aircraft, and spacecraft.

A 2% beryllium mixture with copper produces a unique alloy of bronze that is six times stronger than copper metal. This alloy does not give off sparks when struck with a hammer—a valuable characteristic when metals must be used in explosive gaseous environments. This alloy sometimes contains small amounts of other metals such as nickel or cobalt, which makes for excellent electrical conductivity for switching equipment, given the alloy's simultaneous hardness and nonsparking qualities. Beryllium is also "transparent" to X-rays, which makes it ideal for windows for X-ray tubes.

In 1932 James Chadwick (1891–1974) bombarded beryllium with alpha particles (helium nuclei) that produced free neutrons. Since then, this nuclear process has made beryllium a reliable neutron emitter for laboratory nuclear research. Beryllium is not only an excellent **moderator** to slow down high-speed neutrons in nuclear reactors, but it also can act as a reflector of neutrons as well.

Beryllium is an excellent source of alpha particles, which are the nuclei of helium atoms. Alpha particles (radiation) are not very penetrating. These particles travel only a few inches in air and can be stopped by a sheet of cardboard. Alpha particles are produced in cyclotrons (atom smashers) and are used to bombard the nuclei of other elements to study their characteristics.

In the first part of the twentieth century, beryllium was used as coating inside **fluorescent** electric light tubes, but proved **carcinogenic** (causes cancer) when broken tubes produced beryllium dust that was inhaled. Because of this potential to cause cancer, since 1949 beryllium has no longer been used as the inside coating of fluorescent tubes. Beryllium is also used for computer parts, electrical instrument components, and solid propellant rocket fuels. Because it is one of the few metals that is transparent to X-rays, it is used to make special glass for X-ray equipment.

Examples of Compounds

Beryllium carbide (Be_2C) is used for the cores in nuclear reactors.

Beryllium chloride ($BeCl_2$) is used as a catalyst to accelerate many organic reactions, and beryllium chloride is the electrolyte used along with NaCl in the electrolytic process to produce beryllium metal.

Beryllium copper is not really a compound, but a very useful alloy that often contains other metals such as cobalt or nickel in small amounts. It is a hard, strong alloy with excellent electrical conductivity, which makes it very useful in electrical switching equipment owing to its nonsparking qualities. It makes excellent spot-welding electrodes, springs, and metal bushings, cams, and diaphragms.

Beryllium fluoride (BeF_2) is an example of beryllium that has an oxidation state of +2, combining with a negative anion element with an oxidation state of –1. Beryllium fluoride is also used along with magnesium metal in the chemical reduction process to produce beryllium metal.

Beryllium hydride (BeH_2) liberates hydrogen gas when mixed with water. It is used as a source of hydrogen in experimental rockets and fuel cells.

Beryllium oxide (BeO) is a beryllium compound produced in significant commercial quantities. The chemical process starts with minerals containing aluminum silicate and silicon dioxide and undergoes a number of chemical reactions, some at high temperatures, to end up with BeO.

Hazards

The elemental metallic form of beryllium is highly toxic, as are most of its compounds. When inhaled, the fumes, dust, or particles of beryllium are highly carcinogenic. Some beryllium compounds are toxic when they penetrate cuts in the skin (e.g., when an old fluorescent tube breaks). Beryllium oxide when inhaled can result in a fatal disease known as berylliosis (similar to, but more toxic than, silicosis).

As with many other chemicals, beryllium has its positives and negatives. Although it is an important industrial chemical, the handling of beryllium is best left to experienced workers and laboratory personnel in proper facilities.

MAGNESIUM

SYMBOL: Mg   **PERIOD:** 3   **GROUP:** 2 (IIA)   **ATOMIC NO:** 12
ATOMIC MASS: 24.305 amu   **VALENCE:** 2   **OXIDATION STATE:** +2   **NATURAL STATE:** Solid

ORIGIN OF NAME: Magnesium is named after Magnesia, an ancient region of Thessaly, Greece, where it was mined. Magnesium is often confused with another element, manganese. One way to eliminate the confusion is to think of magnesium (Mg) as "12" and manganese (Mn) as "25" and to use the mental trick of remembering that "g" comes before "n" in the alphabet, so magnesium is the one with lower atomic number.

ISOTOPES: There are 15 isotopes of magnesium, ranging from Mg-20 to Mg-34. Three of these isotopes are stable: Mg-24 makes up 78.99% of all magnesium found in the Earth's crust. Mg-25 makes up 10%, and Mg-26 constitutes most of the rest at 11%. The other 12 isotopes are radioactive and are produced artificially with half-lives ranging from microseconds to a few hours.

ELECTRON CONFIGURATION

Energy Levels/Shells/Electrons	Orbitals/Electrons
1-K = 2	s2
2-L = 8	s2, p6
3-M = 2	s2

Properties

Magnesium is a lightweight, silvery-white, malleable alkali earth metal that is flammable. It has a weak electronegativity (–1.31), which means it is highly reactive as it combines with some nonmetals. As with other alkali earth metals, magnesium is a good conductor of heat and electricity. Its melting point is 648.8°C, its boiling point is 1090°C, and its density is 1.74 g/cm^3, making it about one-fifth the density of iron and only two-thirds as dense as aluminum.

Characteristics

While in a thin solid form, magnesium ignites at 650°C, and it is more easily ignited in a fine powder form. Burning magnesium produces a brilliant white light. It is also used as an oxidizer to displace several other metals from their compound minerals, salts, and ores. It is alloyed with other metals to make them lighter and more machinable, so that they can be rolled, pounded, formed into wires, and worked on a lathe.

The ground water in many regions of the United States contains relatively high percentages of magnesium, as well as some other minerals. A small amount improves the taste of water, but larger amounts result in "hard" water, which interferes with the chemical and physical action of soaps and detergents. The result is a scum-like precipitate that interferes with the cleansing action. The solution is the use of water softeners that treat hard water with either sodium chloride or potassium chloride, which displace the magnesium—making the water "soft," resulting in a more effective cleansing action.

Abundance and Source

Magnesium is the eighth most abundant of the elements found in the entire universe, and the seventh most abundant found in the Earth's crust. Its oxide (MgO) is second in abundance to oxide of silicon (SiO_2), which is the most abundant oxide found in the Earth's crust. Magnesium is found in great quantities in seawater and brines, which provide an endless supply. Each cubic mile of seawater contains about 12 billion pounds of magnesium. Although magnesium metal cannot be extracted from seawater directly, it can be extracted by several chemical processes through which magnesium chloride ($MgCl_2$) is produced. Electrolysis is then used with the magnesium chloride as the electrolyte at 714°C to produce metallic magnesium and chlorine gas. Another method of securing magnesium is known as the Pigeon process. This procedure uses the magnesium minerals dolomite or ferrosilicon. Dolomite ($CaCO_3$), which also contains $MgCO_3$, is crushed and then heated to produce oxides of Ca and Mg. The oxides are heated to about 1200°C along with the ferrosilicon (an alloy of iron and silicon), and the silicon reduces the magnesium, producing a vapor of metallic magnesium that, as it cools, condenses to pure magnesium metal.

History

Magnesium was known in ancient times, but it was not identified as an element until much later. As the story goes, a seventeenth-century farmer's cows refused to drink water from a mineral well in Epsom, England. The farmer tasted the water and found it to be bitter but somewhat refreshing. He also noted that it seemed to heal a rash on his skin. After several years, the demand for his Epsom water was more than he could manage, so other sources were located and exploited. To no one's surprise, the product became known as Epsom salt.

In the mid-eighteenth century, hundreds of chemists determined that the substance in Epsom salt (also known as magnesia alba) was a salty mixture of a sulfate and an oxide of what was believed to be a new element. In 1755 Joseph Black (1728–1799) separated magnesium oxide from lime. However, he was unable to separate the magnesium from the oxygen. Magnesium was not identified as an element until 1808, by Sir Humphry Davy (1778–1829). Davy used the electrolysis process with an electrolyte of a mixture of magnesium oxide (MgO = magnesia) and mercury oxide (HgO). A mixture of magnesium and mercury (an alloy-like **amalgam**) collected at the negative cathode. He then heated this amalgam, vaporizing the mercury with its lower melting point and leaving the magnesium metal behind.

Davy also discovered several other elements (potassium, barium, calcium, and strontium) by isolating the metals from their compounds through electrolysis. His work led to the development of **electrochemistry,** which is the use of electricity as the energy source to break up the oxides of these alkali and alkali earth elements.

Common Uses

Small particles of powdered magnesium metal burn with a bright white flame that makes the magnesium ideal for aerial flares dropped from airplanes that will light up ground areas. It is has also been used in aerial firebombs during wars to devastate a city by fire because water will not extinguish the flames—sand must be used. In the past decades, thin magnesium wire or foil was placed inside glass bulbs containing pure oxygen to form flash bulbs for photographic purposes. When an electric charge ignites the magnesium, a brilliant light is produced. Today most flash cameras use a strobe light instead of flash bulbs.

Pure magnesium metal is lighter in weight than aluminum and, thus, would make an excellent construction metal were it not for its high reactivity and flammability at a rather low temperature when compared to other metals. It is an excellent metal to alloy with other metals for use in the aircraft, space, and automobile industries.

It is used for the production (thermal reduction) of other metals, such as zinc, iron, titanium, zirconium, and nickel. For instance, because of its strong electropositive nature, magnesium can "desulfurize" molten iron when it combines with the sulfur impurities in the iron to produce high-grade metallic iron plus MgS.

Milk of Magnesia is an alkaline (basic) water suspension and "creamy-like" suspended form of magnesium hydroxide, $Mg(OH)_2$. It is used as an antacid to neutralize excess stomach acid. Magnesium can also be used in the form of Epsom salts as a treatment for rashes and as a laxative. A more important commercial use of Epsom salts is in the tanning of leather, as well as in the dyeing of fabrics.

Magnesium is essential for proper nutrition in humans as well as other living organisms. It plays an important role in the process of **photosynthesis** in plant chlorophyll and is thus essential to green plants, which are, in turn, essential for most living organisms. Magnesium is also used as a dietary supplement for both humans and animals for maintaining proper enzyme levels.

Magnesium is an important element that acts as a catalyst in many life processes. In addition to photosynthesis, it is also required for the oxidation in animal cells that produce energy and for the production of healthy red blood cells. Humans cannot live without magnesium—which we acquire mainly from various foods.

Examples of Compounds

Magnesium acetate [$Mg(C_2H_3O_2)_2 \bullet 4H_2O$] is used in the textile industry as a mordant ("fixes" dyes so that they will not run). It is also used as a deodorant and antiseptic.

Magnesium chloride (($MgCl_2$) is mostly obtained from saltwater and has many uses, including as the source of magnesium metal during electrolysis, as a catalyst, and in the making of ceramics, lubricants, paper and textiles, and disinfectants.

Magnesium fluoride (MgF_2) is used to polarize corrective lenses of eyeglasses to reduce the glare of sunlight by selecting the orientation of the light waves passing through the lenses. MgF_2 is also used to polarize windows, sunglasses, and similar optical items.

Magnesium chromate ($MgCrO_4$) has several medical uses, including as a dietary supplement and laxative.

Magnesium oxide (MgO) is used as a lining for steel furnaces, as a component in ceramics, as food additives and pharmaceuticals, and to make strong window glass, fertilizers, paper, and rubber manufacturing.

Magnesium hydroxide [$Mg(OH)_2$] is a whitish solid and when suspended in water is used as an antacid known as milk of magnesia.

Magnesium carbonate ($MgCO_3$) is found in a mixture of natural minerals. It can also be produced in several ways, including pumping carbon dioxide through magnesium oxide or magnesium hydroxide. It is used in pharmaceuticals such as magnesium citrate and as a desiccant to keep hydroscopic products from caking (table salt) and to strengthen rubber and produce dyes, inks, and cosmetics.

Hazards

Magnesium metal, particularly in the form of powder or small particles, can be ignited at relatively low temperatures. The resulting fires are difficult to extinguish, requiring dry sand or dirt. Water will just accelerate the fire as hydrogen that will intensify the fire is released from the water.

Some magnesium compounds, whose molecules contain several atoms of oxygen—$Mg(ClO_4)$, for example—are extremely explosive when in contact with moist organic substance, such as your hands.

Although traces of magnesium are required for good nutrition and health, some compounds of magnesium are poisonous when ingested.

CALCIUM

SYMBOL: Ca   **PERIOD:** 4   **GROUP:** 2 (IIA)   **ATOMIC NO:** 20
ATOMIC MASS: 40.078 amu   **VALENCE:** 2   **OXIDATION STATE:** +2   **NATURAL STATE:** Solid

ORIGIN OF NAME: Its modern name was derived from the word *calcis* or calx, which is Latin for "lime."

ISOTOPES: There are 20 isotopes of calcium, ranging from Ca-35 to Ca-54. Of the six stable isotopes, Ca-40 makes up 96.941% of the calcium found in the Earth's crust, and Ca-42 = 0.647%, Ca-43 = 0.135%, Ca-44 = 2.086%, Ca-46 = 0.004%, and Ca-48 = 0.187% found on Earth. All the other isotopes of calcium are radioactive and are artificially produced with half-lives ranging from a few microseconds to 1×10^5 years.

Radioactive Ca-45 emits beta particles (high-speed electrons) and has a half-life of about 163 days. It is used to determine the calcium levels in bones and in soils.

ELECTRON CONFIGURATION

Energy Levels/Shells/Electrons	Orbitals/Electrons
1-K = 2	s2
2-L = 8	s2, p6
3-M = 8	s2, p6
4-N = 2	s2

Properties

Before the nineteenth century, calcium and some other alkali earth metals were considered minerals rather than metals because they readily formed hydroxides and, thus, were classed as alkaline substances (bases). Calcium and other alkali earth metals were mined as ores. Therefore, early chemists considered them earth elements.

Calcium metal is moderately hard and has a lustrous, silvery color when freshly cut, but it soon oxidizes to a dull gray. It is a good conductor of heat and electricity, but has few uses in the electronics industries. Calcium is an important element for the nutrition of living organisms, particularly vertebrates. Vitamin D from foods, sunlight, and milk aids in the deposition of calcium in bones and teeth. The elemental metal is very reactive in water as it releases hydrogen from the water. It is somewhat less reactive in air. Its melting point is 839°C, its boiling point is 1484°C, its density is 1.54 g/cm³, and it has a cubic crystal structure.

Characteristics

Finely powdered calcium metal is flammable in air because it liberates hydrogen from the moisture. It can be extremely reactive in water but can be dissolved in acids. Calcium is harder than sodium metal, but softer than aluminum. In its elemental form it can be machined (cut on a lathe), extruded (pushed through a die), and drawn (stretched into rods or wires).

Calcium is present in a number of products used in our everyday lives. It is found in classroom chalk, teeth, and bones. About 2% of the human body consists of various forms of calcium compounds. Calcium is an essential inorganic element (usually in compound form) for plant and animal life.

Abundance and Source

Calcium is the fifth most abundant element found in the Earth's crust. It is not found as a free element, but as calcium compounds (mostly salts and oxides), which are found on all landmasses of the world as limestone, marble, and chalk. Calcium, particularly as the compound calcium chloride ($CaCl_2$), is found in the oceans to the extent of 0.15%.

Calcium is produced by two methods. One method is the electrolysis of calcium chloride ($Ca^{++} + 2Cl^- \rightarrow CaCl_2$) as the electrolyte at a temperature of ⁻800°C, during which process metallic calcium cations (Ca^{++}) are deposited at the cathode as elemental calcium metal. Calcium can also be produced through a thermal process under very low pressure (vacuum) in which lime is reduced by using aluminum.

In addition to limestone, calcium is also found in other rocks, coral, shells, eggshells, bones, teeth, and stalactites and stalagmites.

History

Several calcium compounds were used by ancient civilizations for thousands of years. The Egyptians used gypsum to form plasters. Gypsum ($CaSO_4 \bullet 2H_2O$) is the common name for calcium sulfate dihydrate and is a mineral produced from seawater. Gypsum plasters are found in Egyptian tombs. The ancient Greeks and Romans also made mixtures of gypsum plaster somewhat similar to the modern drywall boards used extensively in today's construction industry. A desiccated form of gypsum, known as plaster of Paris, when soaked into cloth bandages, was used to support broken bones as far back as the tenth century. The same procedure is still employed to form casts to support broken bones.

By the early 1800s several chemists had separated potassium and sodium as elements from compounds. It was believed that metallic calcium could be obtained by similar methods. In 1808 Sir Humphry Davy finally produced the metallic element calcium from a mixture of lime and mercuric oxide by his experimental electrolysis apparatus. This was the same process he had previously used to discover several other alkali earth metals.

Common Uses

Calcium oxide was used in ancient times to make mortar for building with stone. Both the metal and calcium compounds have many industrial as well as biological uses. Metallic calcium is used as an alloy agent for copper and aluminum. It is also used to purify lead and is a reducing agent for beryllium.

It is used to remove carbon and sulfur impurities during the processing of iron, producing a higher-grade iron for use in the manufacture of **steel.** It is also used as a reducing agent in the preparation of several other important metals.

Calcium is an important ingredient in the diets of all plants and animals. It is found in the soft tissues and fluids of animals (e.g., blood) as well as in bones and teeth. Calcium makes up about 2% of human body weight.

Calcium is the main ingredient of Portland cement and is used to reduce the acid content of soils.

Examples of Compounds

Many calcium compounds are found on both land and sea.

Calcium acetate [$Ca(CH_3COO)_2 \bullet H_2O$] is used as a food additive and a mordant to fix dyes in the textile industry. It is used as an alkali (base) in the manufacture of soaps, to improve some lubricants, and as an antimold to preserve baked goods for a longer shelf life.

Calcium bromide ($CaBr_2$) is used as a developer for photographic film and paper and as a dehydrating agent (drying agent), food preservative, and fire retardant.

Calcium carbide (CaC_2) has a garlic-like odor and reacts with water to form acetylene gas plus calcium hydroxide and heat. In the past, it was used in miners' lamps to continuously produce a small acetylene flame to provide some illumination in coal mines.

Calcium chloride ($CaCl_2$) has many uses. It is used as a drying agent and to melt ice and snow on highways, to control dust, to thaw building materials (sand, gravel, concrete, and so on). It is also used in various food and pharmaceutical industries and as a fungicide.

Calcium carbonate ($CaCO_3$) can be in the form of an odorless crystal or powder and is one of calcium's most stable compounds, better known in its natural state as limestone, marble, chalk, calcite, oyster shells, and the minerals marl and travertine. Calcium carbonate is the source of lime and is used as a "filler" for many products, including paints, plastics, and foods (bread), and as an antacid.

Calcium oxide (CaO) is formed when calcium metal is exposed to atmospheric oxygen ($2Ca + O_2 \rightarrow 2CaO$). It is known as quick lime, unslaked lime, calx, or just lime. It is used as a flux in steel industries and as a food additive and is also used to make paper pulp, insecticides, and fungicides, to remove hair from animal hides, and to produce calcium carbide (CaC_2). Calcium oxide becomes incandescent when heated to a high temperature. This attribute made it useful in bright spotlights for stage lighting. It became known as "limelight" and thus its use as a term for public attention.

Calcium hydroxide [$Ca(OH)_2$] is known as slaked or hydrated lime and is formed by exposing calcium oxide to water. Slaked lime is less caustic than quick lime. Therefore, it is used to line football fields. (Unslaked lime, CaO, is very caustic when wet, and if it is used on playing fields, players may receive caustic burns.) Calcium hydroxide has many uses, including as an ingredient for stonemasons' mortar, cements, whitewash, and soil conditioner (high pH), as a food additive, and as a human **depilatory.**

Calcium nitrate [$Ca(NO_3)_2$] is known as Norwegian saltpeter. It is a strong oxidizer (because of the NO_3) that is flammable in the presence of organic materials (such as hands). It explodes when given a hard shock. It is used in fireworks, matches, and fertilizers.

Hazards

The metallic form of calcium, particularly the powdered form, combines with water or oxidizing agents to release hydrogen that may explode, as do the other alkali metals. There are many useful calcium compounds; some are excellent reducing agents, some are explosive, and others are essential for life.

The radioactive isotope calcium-45 is deposited in bones and teeth as well as other plant and animal tissues. Because our bodies cannot distinguish between Ca-45 and the stable Ca-40, the radioactive isotope Ca-45 is used as a tracer to study diseased bone and tissue. At the same time, a massive overexposure to Ca-45 can displace the stable form of Ca-40 in animals and can cause radiation sickness or even death.

A few calcium compounds, when in powder or vapor form, are toxic when ingested or **inhaled.**

STRONTIUM

SYMBOL: Sr   **PERIOD:** 5   **GROUP:** 2 (IIA)   **ATOMIC NO:** 38
ATOMIC MASS: 87.62 amu   **VALENCE:** 2   **OXIDATION STATE:** +2
NATURAL STATE: Solid
ORIGIN OF NAME: Strontium was named after the town Strontian, located in Scotland in the British Isles.
ISOTOPES: There are 29 isotopes of strontium, ranging from Sr-75 to Sr-102. The four natural forms of strontium are stable and not radioactive. These stable isotopes are Sr-84, which constitutes 0.56% of the element's existence on Earth; Sr-86, which makes up 9.86%; Sr-87, which accounts for 7.00% of the total; and Sr-88, which makes up 82.58% of strontium found on Earth. The remaining isotopes are radioactive with half-lives ranging from a few microseconds to minutes, hours, days, or years. Most, but not all, are produced in nuclear reactors or nuclear explosions. Two important radioisotopes are Sr-89 and Sr-90.

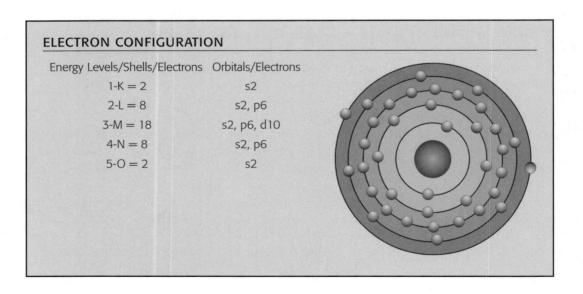

ELECTRON CONFIGURATION

Energy Levels/Shells/Electrons	Orbitals/Electrons
1-K = 2	s2
2-L = 8	s2, p6
3-M = 18	s2, p6, d10
4-N = 8	s2, p6
5-O = 2	s2

Properties

In its elemental state, strontium is a relatively soft, pale yellow metal somewhat similar to elemental calcium. When freshly cut, strontium has a silvery shine to its surface that soon turns grayish as it is oxidized by atmospheric oxygen ($2Sr + O_2 \rightarrow 2SrO$) and nitrogen (3Sr +

$N_2 \rightarrow Sr_3N_2$), which prevents further oxidation. Strontium's melting point is 769°C, its boiling point is 1348°C, and its density is 2.54 g/cm³.

Characteristics

When strontium metal is exposed to water, it releases hydrogen, as do the other earth metals ($Sr + 2H_2O \rightarrow Sr(OH)_2 + H_2\uparrow$). Strontium can ignite when heated above its melting point. When in a fine powder form, it will burn spontaneously in air. It must be stored in an inert atmosphere or in naphtha. Several of its salts burn with a bright red flame, making it useful in signal flares and fireworks.

Abundance and Source

Strontium metal is not found in its elemental state in nature. Its salts and oxide compounds constitute only 0.025% of the Earth's crust. Strontium is found in Mexico and Spain in the mineral ores of strontianite ($SrCO_3$) and celestite ($SrSO_4$). As these ores are treated with hydrochloric acid (HCl), they produce strontium chloride ($SrCl_2$) that is then used, along with potassium chloride (KCl), to form a **eutectic** mixture to reduce the melting point of the $SrCl_2$, as a molten electrolyte in a graphite dish-shaped electrolysis apparatus. This process produces Sr cations collected at the cathode, where they acquire electrons to form strontium metal. At the same time, Cl⁻ anions give up electrons at the anode and are released as chlorine gas $Cl_2\uparrow$.

Two other methods of producing strontium are by thermal reduction of strontium oxide and by the distillation of strontium in a vacuum.

History

In 1787 William Cruikshank (1745–1795) isolated, but did not identify, strontium from the mineral strontianite he examined. In 1790 Dr. Adair Crawford (1748–1794), an Irish chemist, discovered strontium by accident as he was examining barium chloride. He found a substance other than what he expected and considered it a new mineral. He named the new element "strontium" and its mineral "strontianite" after a village in Scotland. In 1808 Sir Humphry Davy treated the ore with hydrochloric acid, which produced strontium chloride. He then mixed mercury oxide with the strontium chloride to form an amalgam alloy of the two metals that collected at the cathode of his electrolysis apparatus. He heated the resulting substance to vaporize the mercury, leaving the strontium metal as a deposit.

Common Uses

Strontium does not have as many practical uses as do some of the other alkali earth metals.

Strontium nitrate [$Sr(NO_3)_2$], when burned, produces a bright red flame, and it is used in fireworks. During military combat, it is used to make "tracer bullets" so that their paths can be tracked at night. Strontium is also used in making specialty metals when alloyed with other metals and in the manufacture of soaps, greases, and similar materials that are resistant to extreme high or low temperatures.

Examples of Compounds

Strontium-90, a radioactive strontium isotope with a half-life of 29 years, is a dangerous fallout source of radiation from atmospheric nuclear bombs. If a person is exposed to it, it will rapidly accumulate in bone tissue and interfere with the production of new red blood cells and may cause death. Strontium-90 does have some utilitarian uses as the radiation source for instruments that measure the thickness of materials during their production (e.g., cigarettes, building materials, and textiles).

Strontium hydroxide [$Sr(OH)_2$] is used to extract sugar from sugar beet molasses. It is also used in the manufacture of soaps, adhesives, plastics, glass, and lubricants that can be used in very high or low temperature environments.

Strontium iodide (SrI_2) is made by treating strontium carbonate with hydrochloric acid. It is used as a medicinal source of iodine.

Strontium nitrate [$Sr(NO_3)_2$], because of the bright red flame it produces when burned, is used in fireworks, matches, marine signals, and so forth.

Strontium carbonate ($SrCO_3$) is used to make radiation-resistant glass and TV picture tubes, as well as pyrotechnics.

Strontium peroxide or strontium dioxide (SrO_2) can cause fires or explode when heated and in contact with organic substances. It is used as both a reducing agent and an oxidizing agent.

Strontium sulfide (SrS) smells like rotten eggs. It is used as a depilatory to remove hair from skin and hides.

Hazards

As a powder, strontium metal may spontaneously burst into flames. Both its metal and some of its compounds will explode when heated. Some of the compounds will explode if struck with a hammer.

Both the metal and some compounds will react with water to produce strontium hydroxide [$Sr(OH)_2$] and release hydrogen gas. The heat from the exothermic reaction may cause the hydrogen to either burn or explode [$Sr + 2H_2O \rightarrow Sr(OH)_2 + H_2\uparrow$].

Some compounds, such as strontium chromate and strontium fluoride, are carcinogens and toxic if ingested. Strontium-90 is particularly dangerous because it is a radioactive bone-seeker that replaces the calcium in bone tissue. Radiation poisoning and death may occur in people exposed to excessive doses of Sr-90. Strontium-90, as well as some other radioisotopes that are produced by explosions of nuclear weapons and then transported atmospherically, may be inhaled by plants and animals many miles from the source of the detonation. This and other factors led to the ban on atmospheric testing of nuclear and thermonuclear weapons.

BARIUM

SYMBOL: Ba   **PERIOD:** 6   **GROUP:** 2 (IIA)   **ATOMIC NO:** 56
ATOMIC MASS: 137.347 amu   **VALENCE:** 2   **OXIDATION STATE:** +2
   NATURAL STATE: Solid
ORIGIN OF NAME: The name barium is derived from the Latin word *barys,* which means "heavy."
ISOTOPES: There are currently 35 known isotopes of barium, ranging from Ba-120 to Ba-148. Seven of these isotopes are stable. The percentages of each as found in nature are as follows: Ba-130 = 0.106%, Ba-132 = 0.101%, Ba-134 = 2.147%, Ba-135 = 6.592%, Ba-136 = 7.854%, Ba-137 = 11.23%, and Ba-138 = 71.7%.

ELECTRON CONFIGURATION

Energy Levels/Shells/Electrons	Orbitals/Electrons
1-K = 2	s2
2-L = 8	s2, p6
3-M = 18	s2, p6, d10
4-N = 18	s2, p6, d10
5-O = 8	s2, p6
6-P = 2	s2

Properties

Barium is the fifth element in group 2 (IIA) of the alkali earth metals and has most of the properties and characteristics of the other alkali earth metals in this group. For example, they all are called alkaline earths because, when first discovered, they exhibited both characteristics of alkaline (basic) substances and characteristics of the earth from which they came. Ancient humans did not know they were metals because their metallic forms do not exist in nature. Barium is a silvery metal that is somewhat malleable and machineable (can be worked on a lathe, stretched and pounded). Its melting point is 725°C, its boiling point is about 1640°C, and its density is 3.51 g/cm^3. (The accurate figures for its properties are difficult to determine because of barium's extreme activity—the pure metal will ignite when exposed to air, water, ammonia, oxygen, and the halogens.)

Characteristics

When barium burns in air, it produces barium oxide ($2Ba + O_2 \rightarrow 2BaO$). When metallic barium burns in water, it forms barium hydroxide [$Ba + 2H_2O \rightarrow Ba(OH)_2 + H_2\uparrow$]. Several barium compounds burn with a bright green flame, which make them useful for fireworks.

Barium is more reactive with water than are calcium and strontium. This is a result of the valence electrons' being further from the positive nucleus. Therefore, barium is more electronegative than the alkali earth metals with smaller nuclei.

In powdered form, it will burst into a bright green flame at room temperature.

Abundance and Source

Barium is the 17th most abundant element in the Earth's crust, making up about 0.05% of the crust. It is found in the minerals witherite, which is barium carbonate ($BaCO_3$), and barite, known as barium sulfate ($BaSO_4$). Pure barium metal does not exist on Earth—only as compounds or in minerals and ores. Barium ores are found in Missouri, Arkansas, Georgia, Kentucky, Nevada, California, Canada, and Mexico.

It is produced by the reduction of barium oxide (BaO), using aluminum or silicon in a high-temperature vacuum. It is also commercially produced by the electrolysis of molten barium chloride ($BaCl_2$) at about 950°C, wherein the barium metal is collected at the cathode and chlorine gas is emitted at the anode.

History

Alchemists in the early Middle Ages knew about some barium minerals. Smooth round pebble-like stones found in Bologna, Italy, were known as "Bologna stones." When these odd stones were exposed to sunlight, or even a primitive reading lamp, they would continue to glow for several years. This characteristic made them attractive to witches as well as the alchemists. These stones are actually the mineral barite, barium sulfide ($BaSO_4$), which today is a major source of barium metal.

Chemists did not discover the mineral witherite ($BaCO_3$) until the eighteenth century. Carl Wilhelm Scheele (1742–1786) discovered barium oxide in 1774, but he did not isolate or identify the element barium. It was not until 1808 that Sir Humphry Davy used molten barium compounds (baryta) as an electrolyte to separate, by electrolysis, the barium cations, which were deposited at the negative cathode as metallic barium. Therefore, Davy received the credit for barium's discovery.

Common Uses

Pure barium metal has few commercial uses because of it reactivity with air and water. Nevertheless, this property makes it useful as a "getter" or scavenger to remove the last traces of gas from vacuum tubes. Barium metal is used to form alloys with other metals. One alloy is used to make sparkplugs that easily emit electrons when heated, thus improving the efficiency of internal combustion engines.

Its compounds have many practical uses. For example, when the mineral barite is ground up into a fine powder, it can be used as a filler and brightener for writing and computer paper. It is also used (along with zinc sulfide) as a pigment, called lithopone, for white paint. Barium compounds are also used in the manufacture of plastics, rubber, resins, ceramics, rocket fuel, fireworks, insecticides, and fungicides and to refine vegetable oils.

A major medical use is a solution of barium sulfide (with flavoring) that is ingested by patients undergoing stomach and intestinal X-ray and CT scan examinations. Barium sulfide is opaque to X-rays, and thus it blocks the transmission of the rays. The organs appear in contrast against a background, which highlights any problems with the digestive system.

Examples of Compounds

Barium acetate [$Ba(C_2H_3O_2)_2 \cdot H_2O$], a white crystal, is used as a dryer for paints and varnishes. It is produced by adding acetic acid to barium sulfate and recovering the crystals by evaporation. It is also used as a textile mordant and catalyst.

Barium bromate [$Ba(BrO_3)_2$] is used as a corrosion inhibitor to prevent rust and as an oxidizing agent and chemical reagent.

Barium nitrate [$Ba(NO_3)_2$] burns with a bright green flame and is used in signal flares and pyrotechnics. It can be produced by treating barium carbonate with nitric acid.

Barium chloride ($BaCl_2$) is used in the manufacture of paint pigments and dyeing textiles and as an additive in oils. It is also used as a water softener.

Barium hydroxide (hydrate) [$Ba(OH)_2$] exists in several forms and has many uses in oil and grease additives, water treatment, vulcanization of rubber, and the manufacture of soaps, beet sugar, glass, and steel.

Barium peroxide (BaO_2) is a grayish-white dry powder that makes an excellent bleaching agent that can be stored in paper packages. Its bleaching qualities are released when mixed with water.

Barium sulfate ($BaSO_4$), as mentioned, is used in medicine as an opaque liquid medium to block X-rays when ingested, thus providing an image of ulcers and intestinal problems. It is also used in the manufacture of paints, rubber, and plastics.

Barium thiosulfate (BaS_2O_3) is mainly used in explosives, matches, varnishes, and photography.

Hazards

Barium metal, in powder form, is **flammable** at room temperature. It must be stored in an oxygen-free atmosphere or in petroleum.

Many of barium's compounds are toxic, especially barium chloride, which affects the functioning of the heart, causing ventricular fibrillation, an erratic heartbeat that can lead to death.

Several of barium's compounds are explosive as well as toxic if ingested or inhaled. Care should be used when working with barium and other alkali metals in the laboratory or in industry.

RADIUM

SYMBOL: Ra   **PERIOD:** 7   **GROUP:** 2 (IIA)   **ATOMIC NO:** 88
ATOMIC MASS: 226.03 amu   **VALENCE:** 2   **OXIDATION STATE:** +2   **NATURAL STATE:** Solid
ORIGIN OF NAME: Radium's name is derived from the Latin word *radius,* which means "ray."
ISOTOPES: There are no stable isotopes of radium. Radium has 25 known radioisotopes, ranging from Ra-206 to Ra-230. Their half-lives range from a fraction of a second to hundreds of years. Radium-226 was discovered by the Curies and has a half-life of about 1630 years. Ra-226 is the most abundant isotope, and thus, Ra-226 is used to determine radium's atomic mass.

Various radium isotopes are derived through a series of radioactive decay processes. For example, Ra-223 is derived from the decay of actinium. Ra-228 and Ra-224 are the result of the series of thorium decays, and Ra-226 is a result of the decay of the uranium series.

ELECTRON CONFIGURATION

Energy Levels/Shells/Electrons	Orbitals/Electrons
1-K = 2	s2
2-L = 8	s2, p6
3-M = 18	s2, p6, d10
4-N = 32	s2, p6, d10. f14
5-O = 18	s2, p6, d10
6-P = 8	s2, p6
7-Q = 2	s2

Properties

Radium is the last element in group 2 and is very similar to the other alkali earth metals, which makes it the largest and heaviest element in the group. It particularly resembles barium, which is just above it in group 2 of the periodic table. Radium is a bright white radioactive luminescent alkali earth metal that turns black when exposed to air. Its melting point is 700°C, its boiling point is 1,140°C, and its density is approximately 5.0 g/cm^3.

Characteristics

Radium is extremely radioactive. It glows in the dark with a faint bluish light. Radium's radioisotopes undergo a series of four decay processes; each decay process ends with a stable isotope of lead. Radium-223 decays to Pb-207; radium-224 and radium-228decay to Pb-208; radium-226 decays to Pb-206; and radium-225 decays to Pb-209. During the decay processes three types of radiation—alpha (α), beta (β), and gamma (γ)—are emitted.

In addition to being radioactive, radium is extremely chemically reactive and forms many compounds. These radium compounds are not only radioactive but also toxic and should be handled by experienced personnel.

Abundance and Source

Radium is the 85th most abundant element found in the Earth's crust. Radium is found in the uranium ores **pitchblende** and chalcolite, which are both very radioactive. Radium metal exists to the extent of only one part to every three million parts of the uranium ore (pitch-blende). Only about one gram of radium is found in every seven or eight tons of uranium ore. This scarcity seems to be the reason that only about five pounds of uranium are produced each year in the entire world. Uranium ores are found in the states of Utah, New Mexico, and Colorado in the United States and in Canada, the Czech Republic, Slovakia, Russia, Zaire, and France.

History

The French chemist Marie Sklodowska Curie (1867–1934) and her husband, Pierre Curie (1859–1906), a physicist and chemist, are credited with the discovery of the element radium at the end of the nineteenth century in 1898. Marie imported tons of pitchblende (uranium ore), and by using small amounts of the ore for each of many procedures, she was able to extract the uranium. She found that the remaining ore was still radioactive after the uranium was removed. Over a period of four years of hard work, she further refined the ores by dissolving the residues in acids, filtering the remainder, and then crystallizing the results. At the end, she had produced just 1/10 of a gram of radium. They used a spectroscope to identify several radioactive elements, including thorium and barium. During the spectroanalysis of a refined specimen, they saw a new spectral line, which she named radium after the Latin word *radius*. The same technique was used to identify another new element. Madam Curie named polonium after her native country, Poland. Marie Curie was one of only four people to receive two Nobel Prizes. In 1903 she and Pierre Curie and Henri Becquerel were jointly awarded the Nobel Prize in Physics for the discovery of spontaneous radioactivity. In 1911 she was awarded her second prize, this one in chemistry, for the discovery of two new elements, radium and polonium. Pierre Curie developed several techniques for measuring the strength of radiation, including his experiments with piezoelectricity that enabled him to measure the effects of radiation emitted by radioactive crystals. He determined that each gram of radium gives off 140 calories of heat per hour and that the element has a half-life of over 1,600 years.

Common Uses

Radium's most important use is as a source of radiation in industry, medicine, and laboratories. The isotope radium-226, which is the most abundant of all the 25 isotopes and has a half-life of 1630 years, is the only useful form of the element. It is used in the medical treatment of malignant cancer growth. It kills cancer cells that have spread throughout the body.

Other uses are to produce phosphorescence and fluorescence in organic compounds and for scintillation screens on instruments used to detect radiation. Radium salts were used in the past to paint the dials of luminous clock faces that glow in the dark.

Examples of Compounds

Radium, like the other alkali earth metals, readily combines with halogens such as chlorine and bromine (which are electronegative). $RaCl_2$ and $RaBr_2$ are the forms in which radium is usually stored and shipped for a variety of uses.

Radium bromide ($RaBr_2$) is a source of radiation for treating cancer and for research in the field of physics.

Radium chloride ($RaCl_2$) has the same use as radium bromide.

Radium hydroxide (RaOH) is the most soluble of all the hydroxides of the alkali earth metals. It is formed when radium reacts with water ($2Ra + 2H_2O \rightarrow 2RaOH + H_2\uparrow$).

Radium sulfate ($RaSO_4$) is the most insoluble sulfate known in chemistry and is created when radium combines with a sulfate ion ($Ra + H_2SO_4 \rightarrow RaSO_4 + H_2\uparrow$).

Hazards

Because radium energetically emits three types of radiation, it poses great danger to anyone handling it. In addition, it is toxic. If it is ingested in even small amounts, it replaces bone tissue, which can result in radiation sickness and death.

The Curies were both exposed to radiation and were the first persons to suffer from radiation sickness, resulting in an early death for Marie. Pierre was killed in an accident with a horse and carriage. Marie's notebooks are still radioactive.

One of the decay products of radium is the gas radon, which can seep up through the Earth's crust into basements and "slab level" homes. Good ventilation assures that the radon does not accumulate to the extent that would be harmful.

At one time, women painted clock and watch dials with luminous radium paint that was a mixture of radium salts and zinc sulfide. They would place the small brushes between their lips and tongue to make the bristles more pointed, in order to paint fine lines with the radium paint. Over the years, they developed cancers that resulted in badly eaten-away and disfigured lips and jaws. Once the danger was known, luminous radium paint was banned for this use. Today, promethium (Pm-147), with a half-life of 2.4 years, is used for this purpose.

Transition Elements: Metals to Nonmetals

Introduction

The transition elements form three series of metals that progress from elements that give up or lose electrons (metals) to elements that gain or accept electrons (nonmetals). The elements that are in transition from metals to nonmetals are located in the center of the periodic table in periods 4, 5, and 6 and are found in groups 3 through 12.

The transition metals are unique in that they represent a gradual shift in electronegativity. Electronegativity is characterized by the neutral atom's ability to acquire electrons from outside itself, and thus, neutral atoms have a tendency to become negatively charged. Electronegativity can be thought of as a gradually shifting scale from very weak electronegativity of metallic atoms to the stronger electronegativity of nonmetallic elements. The positive nuclei of neutral atoms that have less than eight electrons in their valence shells attract more electrons, thus the tendency for some elements to gain electrons. Nuclei of metals in groups 1 and 2 have a very weak attraction for more electrons and exhibit the least electronegativity, whereas transition elements range from elements with weak electronegativity to nonmetallic elements whose atoms are more receptive to gaining electrons and, thus, exhibit greater electronegativity.

←Weak Electronegativity ←——————————————→Stronger Electronegativity→
Groups 1 &and 2 < Transition Elements >   Groups 13, 14, 15, 16, 17

The transition metals are also unique in that their outermost valence shell is not the main energy level being completed by sharing electrons. Generally, the shell next to the outermost shell (energy level or orbit) provides the electrons that make the atoms of the transition elements less electronegative (or they might be thought of as more electropositive). These elements are the only ones that use electrons in the next to the outermost shell in bonding with nonmetals. This is why transition elements have more than one oxidation state (valence). All other metals and nonmetals in the major groups use their outermost shells as their valence shell in bonding. The transition metals are malleable and ductile (can be worked into shapes and pulled into wires), and they conduct electricity and heat. The three transition elements that exhibit the unique property of magnetism are iron, cobalt, and nickel.

Electronegativity is an important concept because the attractive force created by the positive nuclei of atoms makes it possible for both the ionic and covalent bonding of atoms to form molecular compounds.

Many of the compounds formed by transition elements appear in various colors. Several are very toxic. Chromium, zinc, cobalt, nickel, and titanium are carcinogenic.

There are several ways to present the transition elements. We present them as three series found in periods 4, 5, and 6.

The first series starts at period 4, group 3 (IIIB) with the element scandium ($_{21}$Sc). The second series starts at period 5, group 3 (IIIB), with the element yttrium ($_{39}$Y). The third series starts at period 6, group 3 (IIIB) following a number of special metals. We start the third series with the element hafnium ($_{72}$Hf).

Some authorities begin and end the transition elements at different groups in the periodic table. The transition elements sometimes continue beyond group 12 to include some metallic and semiconducting (metalloids) elements in groups 14, 15, and 16. These elements are presented in different sections.

We include transition elements in periods 4, 5, and 6 within groups 3 through 12 as separate sections.

Transition Elements: First Series—Period 4, Groups 3 to 12

SCANDIUM

SYMBOL: Sc   **PERIOD:** 4   **GROUP:** 3 (IIIB)   **ATOMIC NO:** 21
ATOMIC MASS: 44.956 amu   **VALENCE:** 2 and 3   **OXIDATION STATE:** +3
   NATURAL STATE: Solid
ORIGIN OF NAME: From the Latin word *Scandia,* for "Scandinavia."
ISOTOPES: There are 28 isotopes of scandium, ranging from scandium-36 to scandium-57. Scandium-45 is the only stable isotope and contains about 100% of the natural scandium found in the Earth's crust. The remaining isotopes are radioactive with half-lives ranging from nanoseconds to a few minutes to a few hours to a few days, and therefore, they are not found naturally in the Earth's crust. The radioactive isotopes of scandium are produced in nuclear reactors.

ELECTRON CONFIGURATION

Energy Levels/Shells/Electrons	Orbitals/Electrons
1-K = 2	s2
2-L = 8	s2, p6
3-M = 9	s2, p6, d1
4-N = 2	s2

PERIODIC TABLE OF THE ELEMENTS

GROUPS / PERIODS

	1	2	3	4	5	6	7	8	9	10	11	12	13	14	15	16	17	18
	IA																	VIIIA
1	1 H 1.0079	IIA																2 He 4.00260
2	3 Li 6.941	4 Be 9.01218				TRANSITION ELEMENTS							IIIA 5 B 10.81	IVA 6 C 12.011	VA 7 N 14.0067	VIA 8 O 15.9994	VIIA 9 F 18.9984	10 Ne 20.179
3	11 Na 22.9898	12 Mg 24.305	IIIB	IVB	VB	VIB	VIIB	—VIII—			IB	IIB	13 Al 26.9815	14 Si 28.0855	15 P 30.9738	16 S 32.066(6)	17 Cl 35.453	18 Ar 39.948
4	19 K 39.0983	20 Ca 40.08	21 Sc 44.9559	22 Ti 47.88	23 V 50.9415	24 Cr 51.996	25 Mn 54.9380	26 Fe 55.847	27 Co 58.9332	28 Ni 58.69	29 Cu 63.546	30 Zn 65.39	31 Ga 69.72	32 Ge 72.59	33 As 74.9216	34 Se 78.96	35 Br 79.904	36 Kr 83.80
5	37 Rb 85.4678	38 Sr 87.62	39 Y 88.9059	40 Zr 91.224	41 Nb 92.9064	42 Mo 95.94	43 Tc (98)	44 Ru 101.07	45 Rh 102.906	46 Pd 106.42	47 Ag 107.868	48 Cd 112.41	49 In 114.82	50 Sn 118.71	51 Sb 121.75	52 Te 127.60	53 I 126.905	54 Xe 131.29
6	55 Cs 132.905	56 Ba 137.33	★	72 Hf 178.49	73 Ta 180.948	74 W 183.85	75 Re 186.207	76 Os 190.2	77 Ir 192.22	78 Pt 195.08	79 Au 196.967	80 Hg 200.59	81 Tl 204.383	82 Pb 207.2	83 Bi 208.980	84 Po (209)	85 At (210)	86 Rn (222)
7	87 Fr (223)	88 Ra 226.025	▲	104 Unq (261)	105 Unp (262)	106 Unh (263)	107 Uns (264)	108 Uno (265)	109 Une (266)	110 Uun (267)	111 Uuu (272)	112 Uub	113 Uut	114 Uuq	115 Uup	116 Uuh	117 Uus	118 Uuo

6 ★ Lanthanide Series (RARE EARTH)

57 La 138.906	58 Ce 140.12	59 Pr 140.908	60 Nd 144.24	61 Pm (145)	62 Sm 150.36	63 Eu 151.96	64 Gd 157.25	65 Tb 158.925	66 Dy 162.50	67 Ho 164.930	68 Er 167.26	69 Tm 168.934	70 Yb 173.04	71 Lu 174.967

7 ▲ Actinide Series (RARE EARTH)

89 Ac 227.028	90 Th 232.038	91 Pa 231.036	92 U 238.029	93 Np 237.048	94 Pu (244)	95 Am (243)	96 Cm (247)	97 Bk (247)	98 Cf (251)	99 Es (252)	100 Fm (257)	101 Md (258)	102 No (259)	103 Lr (260)

Properties

Scandium is a soft, lightweight, silvery-white metal that does not **tarnish** in air, but over time, it turns yellowish-pink. It resists corrosion. Scandium reacts vigorously with acids, but not water. Scandium has some properties similar to the rare-earth elements. Although its position in group 3 places it at the head of the 17 elements of the lanthanide series of rare-earth metals, scandium, as a metal, is not usually considered a rare-earth. Scandium's melting point is 1,541°C, its boiling point is 2836°C, and its density is 2.989 g/cm^3.

Characteristics

Scandium is the first element in the fourth period of the transition elements, which means that the number of protons in their nuclei increases across the period. As with all the transition elements, electrons in scandium are added to an incomplete inner shell rather than to the outer valence shell as with most other elements. This characteristic of using electrons in an inner shell results in the number of valence electrons being similar for these transition elements although the transition elements may have different oxidation states. This is also why all the transition elements exhibit similar chemical activity.

Abundance and Source

Although scandium is chemically similar to rare-earths, it no longer is considered to be one of them. Scandium is the 42nd most abundant element found in the Earth's crust, making up about 0.0025% of the Earth's crust. It is widely distributed at 5 ppm on the Earth. (It is about as abundant as lithium, as listed in group 1.) Scandium is even more prevalent in the sun and several other stars than it is on Earth.

Scandium is found in ores of wolframite in Norway and thortveitite in Madagascar. It is also found in granite pegmatites and monazites. It is common in many of the ores where tin and tungsten are also found.

History

In 1871 Mendeleev recognized that there was a gap in his periodic table for an element with an atomic weight of 44 (the actual atomic weight of Sc is 44.9+). He considered this yet-to-be discovered element similar to boron and named it "ekaboron" with the symbol Eb. He made other predictions about ekaboron, including its specific gravity and solubility—which proved to be accurate once it was discovered. It was not until 1879 that Lars Fredrik Nilson (1840–1899) of Sweden discovered scandium in the mineral euxenite, naming it after his native Scandinavia—thus, the symbol Sc. Nilson was in the process of identifying rare-earths, and because scandium has some properties of a rare-earth, he considered it a new rare-earth. Although Nilson is credited with discovering scandium, it was actually Per Teodor Cleve (1840–1905) who identified this new element as the one Mendeleev had named "ekaboron."

Common Uses

Scandium was not produced in any quantities until the late 1930s. Its light weight, resistance to corrosion, and high melting point made it especially useful in the aerospace industries. In the early 1940s contractors for the U.S. Air Force appropriated almost all of the scandium metal for use in the construction of military aircraft. The pure metal form is produced by the electrolysis of a salt of scandium, $ScCl_3$. The metal has found some other uses in high-intensity lamps that produce a natural spectrum, making it useful for stadium lighting. It is also used in nickel alkaline storage batteries. Several compounds are used as catalysts to speed up chemical reactions. The radioactive scandium-46 (with a half-life of 83.8 days) emits beta radiation. This property makes it useful as a radioactive tracer in the petroleum industry to monitor **fractionation** products.

Examples of Compounds

Not very many compounds of scandium have been found or produced. It has an oxidation state of +3, thus its metallic ion is Sc^{+++} which means it combines with anions with a −1 oxidation state as follows:

Scandium chloride ($ScCl_3$): $Sc^{3+} + 3Cl^{1-} \rightarrow ScCl_3$. As mentioned, this compound is used in an electrolytic procedure to produce metallic scandium.

When scandium ions react with anions with a −2 oxidation state, different compounds result, including, for example, the following two compounds:

Scandium oxide (Sc_2O_3): $2Sc^{3+} + 3O^{2-} \rightarrow Sc_2O_3$. Scandium oxide is used to prepare scandium fluoride (ScF_3), which is also used as an electrolyte to produce scandium metal.

Scandium sulfate [$Sc_2(SO_4)_3$]: $2Sc^{3+} + 3SO_4^{2-} \rightarrow Sc_2(SO_4)_3$. This compound is used in the germination of seeds for agricultural plants.

Hazards

As with other metals, the transition metals and many of their compounds are toxic, and their powdered or gaseous forms should not be ingested or inhaled. In addition, all but one of the isotopes of scandium are radioactive and should be handled by experienced personnel.

TITANIUM

SYMBOL: Ti   **PERIOD:** 4   **GROUP:** 4 (IVB)   **ATOMIC NO:** 22
ATOMIC MASS: 47.88 amu   **VALENCE:** 2, 3, and 4   **OXIDATION STATE:** +2, +3, and +4
   NATURAL STATE: Solid
ORIGIN OF NAME: It was named after "Titans," meaning the first sons of the Earth as
   stated in Greek mythology.
ISOTOPES: There are 23 known isotopes of titanium. All but five are radioactive, ranging
   from Ti-38 to Ti-61, and have half-lives varying from a few nanoseconds to a few hours.
   The percentages of the five stable isotopes found in nature are as follows: $^{46}Ti = 8.25\%$,
   $^{47}Ti = 7.44\%$, $^{48}Ti = 73.72\%$, $^{49}Ti = 5.41\%$, and $^{50}Ti = 5.18\%$.

ELECTRON CONFIGURATION

Energy Levels/Shells/Electrons	Orbitals/Electrons
1-K = 2	s2
2-L = 8	s2, p6
3-M = 10	s2, p6, d2
4-N = 2	s2

Properties

Positioned at the top of group 4 (IVB), titanium heads up a group of metals sometimes referred to as the "titanium group." Members of this group have some similar properties. Titanium's density is 4.5 g/cm^3, which makes it heavier than aluminum but not as heavy as iron. Its melting point is high at 1,660°C, and its boiling point is even higher at 3287°C.

Titanium metal is harder than steel but much lighter and does not corrode in seawater, which makes it an excellent alloy metal for use in most environmental conditions. It is also paramagnetic, which means that it is not responsive to magnetic fields. It is not a very good conductor of heat or electricity.

Characteristics

As the first element in group 4, titanium has characteristics similar to those of the other members of this group: Zr, Hf, and Rf. Titanium is a shiny, gray, malleable, and ductile metal capable of being worked into various forms and drawn into wires.

Abundance and Source

Titanium is the ninth most abundant element found in the Earth's crust, but not in pure form. It is found in two minerals: rutile, which is titanium dioxide (TiO_2), and ilmenite ($FeTiO_3$). It is also found in some iron ores and in the slag resulting from the production of iron. The mineral rutile is the major source of titanium production in the United States. Although titanium is widely spread over the crust of the Earth, high concentrations of its minerals are scarce. In the past it was separated from it ores by an expensive process of chemical reduction that actually limited the amount of metal produced. A two-step process involves heating rutile with carbon and chlorine to produce titanium tetrachloride—TiO_2 + C + $2Cl_2$ $\Delta\rightarrow$ $TiCl_4$ + CO_2—which is followed by heating the titanium tetrachloride with magnesium in an inert atmosphere: $TiCl_4$ + 2Mg $\Delta\rightarrow$ Ti + 2 $MgCl_2$. As recently as the year 2000, a method of electrolysis was developed using titanium tetrachloride in a bath of rare-earth salts. This process can be used on a commercial scale that makes the production of titanium much less expensive. Titanium was, and still is, a difficult element to extract from its ore.

Titanium is found throughout the universe and in the stars, the sun, the moon, and the meteorites that land on Earth.

History

In 1791 Reverend William Gregor (1761–1817), an amateur mineralogist, discovered an odd black sandy substance in his neighborhood. Because it was somewhat magnetic, he calculated that it was almost 50% magnetite (a form of iron ore). Most of the remainder of the sample was a reddish-brown powder he dissolved in acid to produce a yellow substance. Thinking he had discovered a new mineral, he named it "menachanite," after the Menachan region in Cornwall where he lived. During this period, Franz Joseph Muller (1740–1825) also produced a similar substance that he could not identify. In 1793 Martin Heinrich Klaproth (1743–1817), who discovered several new elements and is considered the father of modern analytical chemistry, identified the substance that Gregor called a mineral as a new element. Klaproth named it "titanium," which means "Earth" in Latin. (As previously noted, the name also refers to the titans of Greek mythology.)

Common Uses

Given titanium's lightness, strength, and resistance to corrosion and high temperatures, its most common use is in alloys with other metals for constructing aircraft, jet engines, and missiles. Its alloys also make excellent armor plates for tanks and warships. It is the major metal used for constructing the stealth aircraft that are difficult to detect by radar.

Titanium's noncorrosive and lightweight properties make it useful in the manufacture of laboratory and medical equipment that will withstand acid and halogen salt corrosion. These same properties make it an excellent metal for surgical pins and screws in the repair of broken bones and joints.

It has many other uses as an **abrasive,** as an ingredient of cements, and as a paint **pigment** in the oxide form and in the paper and ink industries, in batteries for space vehicles, and wherever a metal is needed to resist chlorine (seawater) corrosion.

Examples of Compounds

Titanium's ions, in the three different oxidation states, react with chlorine as follows:
Titanium chloride (II): $Ti^{2+} + 2Cl^- \rightarrow TiCl_2$ (titanium dichloride)
Titanium chloride (III): $Ti^{3+} + 3Cl^- \rightarrow TiCl_3$ (titanium trichloride)
Titanium chloride (IV): $Ti^{4+} + 4Cl^- \rightarrow TiCl_3$ (titanium tetrachloride)
The same sequence using the three oxidations states of titanium occurs with oxygen to form II, III, and IV titanium oxides.

Titanium (IV) dioxide (TiO_2), also known as rutile, is one of the best-known compounds used as a paint pigment. It is ideal for paints exposed to severe temperatures and marine climates because of its inertness and self-cleaning attributes. It is also used in manufacture of glassware, ceramics, enamels, welding rods, and floor coverings.

Titanium (IV) tetrachloride ($TiCl_4$) produces a dense white smoke-like vapor when exposed to moist air. It is used as smoke screens and for skywriting, as well in theatrical productions where fog or smoke is required for the scene.

Titanium (IV) carbide (TiC) is used as an additive to carbide to make high-temperature cutting tools, cements, abrasives, and coatings.

Titanium (IV) nitride (TiN) is a hard brittle metal that is used in alloys, to manufacture semiconducting instruments, as cements, and as an abrasive.

Titanium (II) bromide (TiB_2) is used for metallurgy, high-temperature electrical wiring, electronics, computers, high temperature-resistant coatings, and super alloys, including strong lightweight aluminum alloys.

Hazards

Almost all of titanium's compounds, as well as the pure metal when in powder form, are extremely flammable and explosive. Titanium metal will ignite in air at 1200°C and will burn in an atmosphere of nitrogen. Titanium fires cannot be extinguished by using water or carbon dioxide extinguishers. Sand, dirt, or special foams must be used to extinguish burning titanium.

VANADIUM

SYMBOL: V   **PERIOD:** 4   **GROUP:** 5 (VB)   **ATOMIC NO:** 23
ATOMIC MASS: 50.9415 amu   **VALENCE:** 2, 3, 4, and 5   **OXIDATION STATE:** +2, +3, +4, and +5   **NATURAL STATE:** Solid
ORIGIN OF NAME: Named after the Scandinavian mythological goddess Vanadis because of the many colors exhibited by vanadium's compounds.
ISOTOPES: There are 27 isotopes of vanadium. Only vanadium-51 is stable and makes up 99.75% of the total vanadium on Earth. The other 0.25% of the vanadium found on Earth is from the radioisotope vanadium-50, which has such a long half-life of $1.4 \times 10^{+17}$ years that it is considered stable. The other radioactive isotopes have half-lives ranging from 150 nanoseconds to one year.

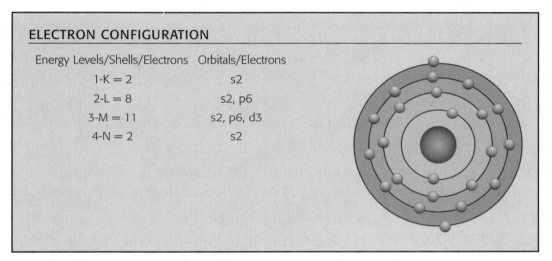

ELECTRON CONFIGURATION

Energy Levels/Shells/Electrons	Orbitals/Electrons
1-K = 2	s2
2-L = 8	s2, p6
3-M = 11	s2, p6, d3
4-N = 2	s2

Properties

Vanadium is a silvery whitish-gray metal that is somewhat heavier than aluminum, but lighter than iron. It is ductile and can be worked into various shapes. It is like other transition metals in the way that some electrons from the next-to-outermost shell can bond with other elements. Vanadium forms many complicated compounds as a result of variable valences. This attribute is responsible for the four oxidation states of its ions that enable it to combine with most nonmetals and to at times even act as a nonmetal. Vanadium's melting point is 1890°C, its boiling point is 3380°C, and its density is 6.11 g/cm³.

Characteristics

Vanadium is an excellent alloy metal with iron that produces hard, strong, corrosion-resistant steel that resists most acids and alkali. It is even more resistant to seawater corrosion than is stainless steel. Vanadium is difficult to prepare in a pure form in large amounts. Impure forms seem to work as well as a very pure form of the metal when used as an alloy. When worked as a metal, it must be heated in an inert atmosphere because it will readily oxidize.

Abundance and Source

Vanadium is not found in its pure state. Small amounts of vanadium can be found in phosphate rocks and some iron ores. Most of it is recovered from two minerals: vanadinite, which is a compound of lead and chlorine plus some vanadium oxide, and carnotite, a mineral containing uranium, potassium, and an oxide of vanadium. Because of its four oxidation states and its ability to act as both a metal and a nonmetal, vanadium is known to chemically combine with over 55 different elements.

Vanadium's principal ores are roscoelite, patronite, vanadinite, and carnotite, which are found in the states of Idaho, Montana, Arkansas, and Arizona as well as in Mexico and Peru. It is also a by-product from the production of phosphate ores.

History

Two unrelated discoveries of vanadium seem to have occurred. When he was experimenting with iron in 1830, Nils Gabriel Sefstrom (1787–1845), a Swedish chemist and mineralogist, identified a small amount of a new metal. Because vanadium compounds have beautiful colors, he named this new metal after Vanadis, the mythological goddess of youth and beauty in his native country, Scandinavia.

Earlier, in 1801 in Mexico, Andres Manuel del Rio, a chemist and mineralogist, discovered an unusual substance he called erythronium, but he was told it was similar to chromium. Only later was it found to be vanadium. Some early references credit Del Rio with vanadium's discovery, but the most recent references list Sefstrom as the discoverer.

Sometime later in 1869, vanadium metal was isolated from its ores by Henry Enfield Roscoe (1833–1915), but Sefstrom had already received credit for the discovery of the element vanadium.

Common Uses

The major use of vanadium is as an alloying metal to make a strong and corrosion-resistant form of steel that is well suited for structures such as nuclear reactors. It does not absorb neutrons or become "stretched" by heat and stress, as does normal stainless steel, thus making vanadium ideal for the construction of nuclear reactors.

Some of its compounds, particularly the oxides, are used in chemical industries as catalysts to speed up organic chemical reactions. The yellow-brown vanadium pentoxide (V_2O_5) is used as a catalyst to facilitate the production of sulfuric acid by the contact process. Vanadium pentoxide is also used as a photographic developer, to dye textiles, and in the production of artificial rubber. When combined with glass, it acts as a filter against **ultraviolet** rays from sunlight.

Examples of Compounds

The ions of vanadium are V^{2+}, V^{3+}, V^{4+}, and V^{5+}, and they enable the formation of four different oxides of vanadium: *vanadium(II) oxide* (VO), *vanadium(III) oxide* (V_2O_3), *vanadium(IV) oxide* (VO_2), and *vanadium(V) oxide* or *vanadium pentoxide* (V_2O_5).

The same sequence of four vanadium ions can combine with chlorine and fluorine to form related compounds.

Vanadium carbide (VC) is used to alloy iron to produce high-speed, high-temperature cutting tools for cutting metals and other hard substances.

Vanadium sulfate ($VOSO_4$) acts as a catalyst as well as a reducing agent. It is used to color glass and ceramics and as a mordant (fixing dyes to textiles).

Vanadium pentoxide (V_2O_5) is a reddish-yellow powder extracted from minerals using strong acids or alkalies. In addition to being used as a catalyst for many organic chemical reactions, it is used in photography and in UV-protected windowpanes and to color ceramics and dye cloth.

Hazards

Vanadium powder, dust, and most of its oxide compounds are explosive when exposed to heat and air. They are also toxic when inhaled. Vanadium chloride compounds are strong irritants to the skin and poisonous when ingested.

Many of its compounds must be stored in a dry, oxygen-free atmosphere or in containers of inert gas. Protective clothing and goggles should be worn when handling vanadium, as well as with most of the other transition elements.

CHROMIUM

SYMBOL: Cr **PERIOD:** 4 **GROUP:** 6 (VIB) **ATOMIC NO:** 24
ATOMIC MASS: 51.996 amu **VALENCE:** 2, 3, and 6 **OXIDATION STATE:** +2, +3, and +6 **NATURAL STATE:** Solid
ORIGIN OF NAME: From the Greek word *chroma* or chromos, meaning "color," because of the many colors of its minerals and compounds.
ISOTOPES: There are 26 isotopes of the element chromium; four are stable and found in nature, and the rest are artificially produced with half-lives from a few microseconds to a few days. The four stable isotopes and their percentage of contribution to the total amount of chromium on Earth are as follows: ^{50}Cr = 4.345%, ^{52}Cr = 83.789%, ^{53}Cr = 9.501%, and ^{54}Cr = 2.365%. Cr-50 is radioactive but has such a long half-life—$1.8 \times 10^{+17}$ years—that it is considered to contribute about 4% to the total amount of chromium found on Earth.

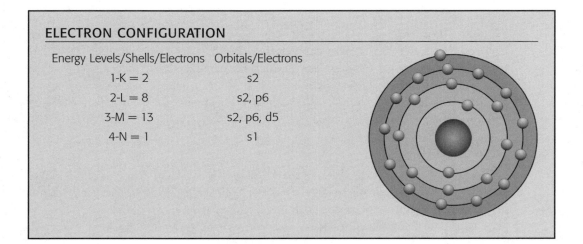

ELECTRON CONFIGURATION

Energy Levels/Shells/Electrons	Orbitals/Electrons
1-K = 2	s2
2-L = 8	s2, p6
3-M = 13	s2, p6, d5
4-N = 1	s1

Properties

Chromium is a silvery white/gray, hard, brittle noncorrosive metal that has chemical and physical properties similar to the two preceding elements in period 4 (V and Ti). As one of the transition elements, its uses its M shell rather than its outer N shell for valence electrons when combining with other elements. Its melting point is 1,857°C, its boiling point is 2,672°C, and its density is 7.19 g/cm^3.

Characteristics

Chromium is a hard, brittle metal that, with difficulty, can be **forged,** rolled, and drawn, unless it is in a very pure form, in which case the chromium is easier to work with. It is an excellent alloying metal with iron. Its bright, silvery property makes it an appropriate metal to provide a reflective, non-corrosive attractive finish for electroplating.

Various compounds of chromium exhibit vivid colors, such as red, chrome green, and chromate yellow, all used as pigments.

Abundance and Source

Chromium is the 21st most common element found in the Earth's crust, and chromium oxide (Cr_2O_3) is the 10th most abundant of the oxide compounds found on Earth. It is not found in a free metallic state.

The first source of chromium was found in the mineral crocoite. Today it is obtained from the mineral chromite ($FeCr_2O_4$), which is found in Cuba, Zimbabwe, South Africa, Turkey, Russia, and the Philippines. Chromite is an ordinary blackish substance that was ignored for many years. There are different grades and forms of chromium ores and compounds, based on the classification of use of the element. Most oxides of chromium are found mixed with other metals, such as iron, magnesium, or aluminum.

Astronauts found that the moon's basalt rocks contain several times more chromium than is found in basalt rocks of Earth.

History

Historically, chromium ore was known as Siberian red lead, which was used to make bright red paints. The source was soon identified as the mineral crocoite, and analysis indicated that it also contained lead. In 1797 the French chemist Louis-Nicolas Vauquelin (1763–1829) discovered chromium while studying some minerals that were collected in Siberia. To isolate the pure metal from its oxide, he first dissolved the lead out of the mineral with hydrochloric acid (HCl), leaving crystals of chromium oxide, which he then heated. To his surprise he ended up with crystals of pure chromium metal.

Common Uses

The best-known use of chromium is for the plating of metal and plastic parts to produce a shiny, reflective finish on automobile trim, household appliances, and other items where a bright finish is considered attractive. It also protects iron and steel from corrosion.

It is used to make alloys, especially stainless steel for cookware, and items for which strength and protection from rusting and high heat are important.

Its compounds are used for high-temperature electrical equipment, for tanning leather, as a mordant (fixes the dyes in textiles so that they will not run), and as an antichalking agent for paints.

Some research has shown that, even though most chromium compounds are toxic, a small trace of chromium is important for a healthy diet for humans. A deficiency produces diabetes-like symptoms, which can be treated with a diet of whole-grain cereal, liver, and brewer's yeast.

Chromium's most important radioisotope is chromium-51, which has a half-life of about 27 days. It is used as a radioisotope **tracer** to check the rate of blood flowing in constricted arteries.

Some chromium compounds (e.g., chromium chloride, chromic hydroxide, chromic phosphate) are used as catalysts for organic chemical reactions.

In 1960 the first ruby laser was made from a ruby crystal of aluminum oxide (Al_2O_3). These crystals contain only a small amount of chromium, which stores the energy and is responsible for the laser action. A small amount of chromium found in the mineral corundum is responsible for the bright red color of the ruby gemstone.

Examples of Compounds

There are many useful compounds of chromium. As with its preceding two partners in period four, chromium's varied oxidation states are responsible for different forms of oxide compounds. For instance, there are the following compounds:

Chromium(II) oxide: $Cr^{2+} + O^{2-} \rightarrow CrO$. Used in plating metals.
Chromic(III) oxide: $2Cr^{3+} + 3O^{2-} \rightarrow Cr_2O_3$. Green paint pigment.
Chromium (IV) oxide: $Cr^{6+} + 3O^{2-} \rightarrow CrO_3$. Solution = chromic acid.
In a similar fashion, chromium can form compounds with the halides:
Chromium(II) chloride: $Cr^{2+} + 2Cl^{1-} \rightarrow CrCl_2$. Chromium plating.
Chromium(III) fluoride: $Cr^{+3} + 3F^{1-} \rightarrow CrF_3$. High-temperature alloys.
Other compounds:
Lead chromate ($PbCrO_4$): known as a chrome yellow pigment.
Chromium carbide (Cr_3C_2): resists oxidation at high temperatures.
Chromium nitrate [$Cr(NO_3)_3$]: corrosion inhibitor, very explosive.

Hazards

Most of the compounds of chromium are hazardous when inhaled and irritating when in contact with the skin. Even though chromium may be a necessary trace element in our diets, many of its compounds are very toxic when ingested. Some are very explosive when shocked or heated (e.g., chromium nitrate) or when in contact with organic chemicals. Dust from the mining of chromium ores, which is found in **igneous rocks,** is carcinogenic and can cause lung cancer, even when small amounts are inhaled. Workers in industries that produce and use chromium are subject to bronchogenic cancer if precautions are not taken.

MANGANESE

SYMBOL: Mn PERIOD: 4 GROUP: 7 (VIIB) ATOMIC NO: 25
ATOMIC MASS: 54.9380 amu VALENCE: 2, 3, 4, 6, and 7 OXIDATION STATE: +2, +3, +4, +6, and +7 NATURAL STATE: Solid

ORIGIN OF NAME: The name manganese is derived from the mineral magnesite (or dolomite, a compound of magnesium carbonate), which was mined in the region of Magnesia of ancient Greece.

ISOTOPES: There are 30 isotopes of manganese, ranging from Mn-44 to Mn-69, with only one being stable: Mn-55 makes up 100% of the element in the Earth's crust. All the other isotopes are artificially radioactive with half-lives ranging from 70 nanoseconds to 3.7×10^6 years. Artificial radioisotopes are produced in nuclear reactors, and because most radioactive isotopes are not natural, they do not contribute to the element's natural existence on Earth.

ELECTRON CONFIGURATION

Energy Levels/Shells/Electrons	Orbitals/Electrons
1-K = 2	s2
2-L = 8	s2, p6
3-M = 13	s2, p6, d5
4-N = 2	s2

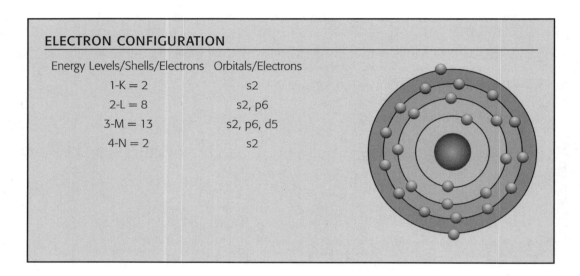

Properties

Manganese is a reactive metal that has several oxidation states (2, 3, 4, 6, and 7) that are responsible for its varied chemical compounds. The chemical and physical properties of manganese are similar to the properties of its companions in group 7—technetium ($_{43}$Tc) and rhenium ($_{75}$Re).

Pure manganese is a gray-white metal that is somewhat similar to iron, located just to the right of it in period 4. Manganese is a reactive element that, over time, will decompose in cold water and will rust (oxidize) in moist air. It has magnetic properties but is not as magnetic as iron. Its melting point is 1,233°C, its boiling point is 1962°C, and its density is 7.44 g/cm³.

Characteristics

There are four allotropic forms of manganese, which means each of its allotropes has a different crystal form and molecular structure. Therefore, each allotrope exhibits different chemical and physical properties (see the forms of carbon—diamond, carbon black, and graphite). The alpha (α) allotrope is stable at room temperature whereas the gamma (γ) form is soft, bendable, and easy to cut. The delta Δ allotrope exists only at temperatures above 1,100°C. As a pure metal, it cannot be worked into different shapes because it is too brittle. Manganese is responsible for the color in amethyst crystals and is used to make amethyst-colored glass.

Abundance and Source

Manganese minerals are widely distributed across the Earth as oxides, silicates, and carbonates. Manganese is the 11th most abundant element found in the Earth's crust, and manganese oxide minerals are the 10th most abundant compounds in the Earth's crust. Pure manganese is found in meteorites that land on Earth's surface. Its minerals psilomelane, pyrolusite, rhodichrosite, and manganite (manganese ore) are found in most countries. It is also found in low-grade iron ores and in **slag** as a by-product of iron **smelting.** Manganese ores are found in India, Brazil, the Republic of South Africa, Gabon, Australia, and Russia, as well as in the state of Montana in the United States.

Huge amounts (i.e., more than 10^{12} tons) of manganese cover vast regions of the ocean beds, and more than 10^7 tons are newly deposited each year. When recovered nodules (lumps) are removed and dried, they contain between 15% and 35% Mn, which is below the requirement for commercial mining. In addition, smaller amounts of cobalt, nickel, and copper are found in these manganese "nodules." Proposals to mine the nodules have been suggested, but no large quantities have been recovered. Mining of the ocean floors for manganese will probably not occur until the sources on Earth become more scarce and expensive to exploit, which is not likely to happen any time soon. Meanwhile, these nodule deposits serve as a reserve for several important metals.

In addition to reduction of its ores in furnaces, manganese can be produced by electrolysis. The electrolyte is manganese sulfate that is produced by treating ore with sulfuric acid, ($2MnO_2 + 2H_2SO_4 \rightarrow 2MnSO_4 + O_2 + 2H_2O$). The anode is lead alloy, and the cathode is made of a steel alloy. Pure (about 99%) manganese metal collects at the cathode, and in the process, the manganese sulfate is converted back to sulfuric acid, which can be reused to react with more MnO_2 ore.

History

The Swedish chemist Carl Wilhelm Scheele (1742–1786) discovered many gases for which he did not receive credit, including oxygen, which he discovered two years before Joseph Priestly (1733–1804). However, because Scheele did not publish his findings in time, he did not get credit.

Scheele's work led to the discovery of manganese, for which the Swedish chemist Johan Gottlieb Gahn (1745–1818) received credit in 1774. During this year Gahn used a blowpipe in a Bunsen burner flame to reduce pyrolusite ore (manganese dioxide, MnO_2) mixed with charcoal (carbon) to form a minute amount of grayish manganese metal.

Common Uses

Manganese does not have many uses, but the utility it does have is important. Manganese's major use is as an alloy with other metals. When alloyed with iron, it makes steel that is stronger, stiffer, tougher, harder, and longer-wearing. At one time railroad rails were made of iron and were replaced about once a year because of the "softness" of pure iron. Later it was discovered that with the addition of about 20% manganese to iron, it was possible to produce an extremely hard steel rail that would last over 20 years. When alloyed with aluminum, along with small amounts of antimony and copper, it forms a ferromagnetic alloy widely used in many industries. Manganese dioxide (MnO_2, pyrolusite) is used as a depolarizer in dry cells (that's the black "stuff" inside flashlight batteries). It is used as a drying agent in black paints and varnishes and as a bleaching agent for glass and oils.

Manganese is also used in the production of aluminum and other metals to produce light, impressively hard tools that can withstand high temperatures, such as the tools used to cut metal on **lathes.**

A small amount of manganese (1.5 to 5.0 milligrams) is required in the diets of humans for normal development of tendon, bones, and some **enzymes,** and it is thought to be necessary for the body to utilize vitamin B1. Manganese is found in many foods, including peas, beans, bran, nuts, coffee, and tea.

Examples of Compounds

Due to its different oxidation states, manganese can form a number of different compounds with oxygen: for instance, manganese monoxide(II) (MnO), manganese sesquinoxide(III) (Mn_2O_3), manganese dioxide(IV) (MnO_2), and manganese heptoxide(VII) (Mn_2O_7).

Potassium permanganate ($KMnO_4$) is a purplish crystal-like oxidizing compound used as an antiseptic and disinfectant to inhibit the growth of harmful skin microorganisms and bacteria. Before antibiotics were available, it was used as a treatment for trench mouth and impetigo.

Trench mouth (necrotizing ulcerative gingivitis), also called "Vincent's infection," usually affects young adults and is considered a form of periodontal disease. If untreated, it can lead to the loss of gum tissue and eventually loss of teeth. Today, there are more effective treatments for trench mouth than $KMnO_4$.

Impetigo is a skin infection that usually occurs in young children living in unsanitary conditions in warm climates. It is caused either by *Streptococcus* or *Staphylococcus* bacteria or both together. It produces a tan scab that may disappear in about 10 days, and should be treated with an antibiotic because it is contagious and can cause kidney damage.

Manganese carbonate ($MnCO_3$) is used in the production of iron ore and as a chemical reagent.

Manganese gluconate [$Mn(C_6H_{11}O_7)$] is used as a food additive, a vitamin, and a dietary supplement (also manganese glycerophosphate). It is found in green leafy vegetables, legumes, (peas and beans), and brewer's yeast.

Manganous monoxide (MnO) is used for textile printing, ceramics, paints, colored (green) glass, bleaching, and fertilizers. It is also used as dietary supplement and reagent in analytical chemistry.

Manganese chloride ($MnCl_2$) is used in the pharmaceutical industry as a dietary supplement and is added to fertilizers.

Hazards

The dust and fumes from the powder form of most manganese compounds, especially the oxides, are very toxic to plants, animals, and humans. Even inhaling small amounts is toxic. The powder form of manganese metal is flammable, and manganese fires cannot be extinguished with water. They must be smothered by sand or dry chemicals.

IRON

SYMBOL: Fe **PERIOD:** 4 **GROUP:** 8 (VIII) **ATOMIC NO:** 26
ATOMIC MASS: 55.847 amu **VALENCE:** 2 and 3 **OXIDATION STATE:** −2, +2, +3, and
  +6 **NATURAL STATE:** Solid

ORIGIN OF NAME: The name "iron" or "iren" is Anglo-Saxon, and the symbol for iron (Fe) is from *ferrum,* the Latin word for iron.

ISOTOPES: There are 30 isotopes of iron ranging from Fe-45 to Fe-72. The following are the four stable isotopes with the percentage of their contribution to the element's natural existence on Earth: Fe-54 = 5.845%, Fe-56 = 91.72%, Fe-57 = 2.2%, and Fe-58 = 0.28%. It might be noted that Fe-54 is radioactive but is considered stable because it has such a long half-life ($3.1 \times 10^{+22}$ years). The other isotopes are radioactive and are produced artificially. Their half-lives range from 150 nanoseconds to 1×10^5 years.

ELECTRON CONFIGURATION

Energy Levels/Shells/Electrons	Orbitals/Electrons
1-K = 2	s2
2-L = 8	s2, p6
3-M = 14	s2, p6, d6
4-N = 2	s2

Properties

Pure iron is a silvery-white, hard, but malleable and ductile metal that can be worked and forged into many different shapes, such as rods, wires, sheets, ingots, pipes, framing, and so on. Pure iron is reactive and forms many compounds with other elements. It is an excellent reducing agent. It oxidizes (rusts) in water and moist air and is highly reactive with most acids, releasing hydrogen from the acid. One of its main properties is that it can be magnetized and retain a magnetic field.

The iron with a valence of 2 is referred to as "ferrous" in compounds (e.g., ferrous chloride = $FeCl_2$). When the valence is 3, it is called "ferric" (e.g., ferric chloride = $FeCl_3$).

Iron's melting point is 1,535°C, its boiling point is 2,750°C, and its density is 7.873 g/cm³.

Characteristics

Iron is the only metal that can be tempered (hardened by heating, then quenching in water or oil). Iron can become too hard and develop stresses and fractures. This can be corrected by annealing, a process that heats the iron again and then holds it at that temperature until the stresses are eliminated. Iron is a good conductor of electricity and heat. It is easily magnetized, but its magnetic properties are lost at high temperatures. Iron has four allotropic states. The alpha form exists at room temperatures, while the other three allotropic forms exist at varying higher temperatures.

Iron is the most important construction metal. It can be alloyed with many other metals to make a great variety of specialty products. Its most important alloy is steel.

An interesting characteristic of iron is that it is the heaviest element that can be formed by fusion of hydrogen in the sun and similar stars. Hydrogen nuclei can be "squeezed" in the sun to form all the elements with atomic numbers below cobalt ($_{27}$Co), which includes iron. It requires the excess fusion energy of supernovas (exploding stars) to form elements with proton numbers greater than iron ($_{26}$Fe).

Abundance and Source

Iron is the fourth most abundant element in the Earth's crust (about 5%) and is the ninth most abundant element found in the sun and stars in the universe. The core of the Earth is believed to consist of two layers, or spheres, of iron. The inner core is thought to be molten iron and nickel mixture, and the outer core is a transition phase of iron with the molten magma of the Earth's mantle.

Iron's two oxide compounds (ferrous(II) oxide—FeO) and (ferric(III) oxide—Fe_2O_3) are the third and seventh most abundant compounds found in the Earth's crust.

The most common ore of iron is hematite that appears as black sand on beaches or black seams when exposed in the ground. Iron ores (ferric oxides) also vary in color from brownish-red to brick red to cherry red with a metallic shine. Small amounts of iron and iron alloys with nickel and cobalt were found in meteorites (siderite) by early humans. This limited supply was used to shape tools and crude weapons.

Even though small amounts of iron compounds and alloys are found in nature (iron is not found in its pure metallic state in nature), early humans did not know how to extract iron from ores until well after they knew how to smelt gold, tin, and copper ores. From these metals, they then developed bronze alloy—thus the Bronze Age (about 8000 BCE). It is assumed that the reason iron was not more widely used is that early humans were unable to make fires hot enough to smelt the iron for it ores. There was a long transition period form the Bronze Age to the Iron Age. It is generally assumed that the Iron Age did not arrive until about 1900BCE, during the height of the Bronze Age, and iron was not the primary metal extracted by early humans until about 1000 BCE or later.

There are several grades of iron ores, including hematite (brown ferric oxide) and limonite (red ferric oxide). Other ores are pyrites, chromite, magnetite, siderite, and low-grade taconite. Magnetite (Fe_3O_4) is the magnetic iron mineral/ore found in South Africa, Sweden, and parts of the United States. The "lodestone," a form of magnetite, is a natural magnet.

Iron ores are found in many countries. A major but diminishing source in the United States is the Mesabi Range in Minnesota, which has produced over two billion tons of ore since it was first opened in 1884. Iron ores are also found in Alabama and Pennsylvania. Iron is found throughout most of the universe, in most of the stars, and in our sun, and it probably exists on the other planets of our solar system.

History

The exact history of the discovery of iron is unclear. There is a reference in the Old Testament mythology of Genesis that refers to a descendant of Adam as an "iron instructor of every artificer in brass and iron." Archeological artifacts made from smelted iron are known from about 3000 BCE. Iron seems to have been introduced as the last stage in the "age of metals." Only small quantities were smelted until near the end of the Bronze Age in about 1500 BCE. Iron was difficult to smelt from its ores because of the high temperatures required. It requires a temperature of 1,083°C to melt copper or tin, but 1,535°C to melt iron. Therefore, it was not extensively used to manufacture tools and weapons until about 1000 BCE as the art of metallurgy developed. Metallurgy involves the smelting of iron ores, making of iron alloys, and working of iron into different shapes. Iron ores are found in many regions of the world. As metallurgy developed in Asia Minor, China, Turkey, Egypt, Cyprus, Greece, and the Iberian Peninsula, so did the spread of related uses of iron. Later, iron metallurgy techniques spread to European nations. This later Iron Age production of tools, vehicles, art, and coins, as well as weapons, created opportunities and benefits that resulted in the building of permanent towns and cities—and thus the spread of civilization.

During the American Revolutionary War, General George Washington made cannonballs and other iron products from the iron ore mined in the areas of Pine Grove Furnace and Cornwall, located in south-central Pennsylvania. The old furnace used to produce iron for cannonballs still exists in the Pine Grove Furnace State Park. Until recently, the Bethlehem Steel Company used the iron ore mined from the old Cornwall deposits near Lebanon, Pennsylvania.

No one person is given credit for the discovery or identification of iron's minerals/ores or the technologies involved in the sophisticated metallurgical processes used today. Many people have contributed to the understanding of the chemistry and how to make practical use of this important element. Modern civilization, with its many skyscrapers, large ships, trains, cars, and so forth, would not be possible without the knowledge and uses of the many iron and steel products manufactured today.

Common Uses

Smelting of iron from its ore occurs in a blast furnace where carbon (coke) and limestone are heated with the ore that results in the iron in the ore being reduced and converted to molten iron, called "pig iron." Melted pig iron still contains some carbon and silicon as well as some other impurities as it collects in the bottom of the furnace with molten slag floating atop the iron. Both are tapped and drained off. This process can be continuous since more ingredients can be added as the iron and slag are removed from the bottom of the furnace. This form of iron is not very useful for manufacturing products, given that it is brittle and not very strong.

One of the major advances in the technology of iron smelting was the development of the Bessemer process by Henry Bessemer (1813–1898). In this process, compressed air or oxygen is forced through molten pig iron to oxidize (burn out) the carbon and other impurities. Steel is then produced in a forced oxygen furnace, where carbon is dissolved in the iron at very high temperatures. Variations of hardness and other characteristics of steel can be achieved with the addition of alloys and by annealing, quench hardening, and tempering the steel.

Powder metallurgy (**sintering**) is the process whereby powdered iron or other metals are combined together at high pressure without high heat to fit molded forms. This process is used to produce homogenous (uniform throughout) metal parts.

One of the most useful characteristics of iron is its natural magnetism, which it loses at high temperatures. Magnetism can also be introduced into iron products by electrical induction. **Magnets** of all sizes and shapes are used in motors, atom smashers, CT scanners, and

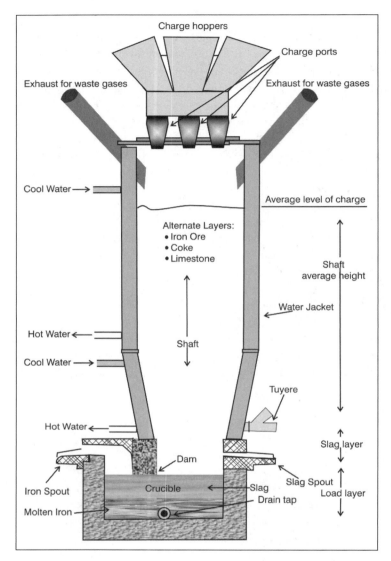

Charge hoppers

Charge ports

Exhaust for waste gases

Exhaust for waste gases

Cool Water →

Average level of charge

Alternate Layers:
• Iron Ore
• Coke
• Limestone

Shaft average height

Water Jacket

Hot Water ←

Shaft

Cool Water →

Tuyere

Hot Water ←

Slag layer

Dam

Slag Spout

Iron Spout

Crucible

Slag

Load layer

Drain tap

Molten Iron

Figure 4.2: Artist's depiction of a blast furnace used for the production of iron.

TV and computer screens, to name a few uses. Super magnets can be formed by adding other elements (see cobalt) to high-quality iron.

Iron is an important element making up hemoglobin in the blood, which carries oxygen to the cells of our bodies. It is also very important as a trace element in the diet, assisting with the oxidation of foods to produce energy. We need about 10 to 18 milligrams of iron each day, as a trace mineral. Iron is found in liver and meat products, eggs, shellfish, green leafy vegetables, peas, beans, and whole grain cereals. Iron deficiency may cause anemia (low red blood cell count), weakness, fatigue, headaches, and shortness of breath. Excess iron in the diet can cause liver damage, but this is a rare condition.

Examples of Compounds

Iron oxides [Fe_2O_3; Fe_3O_4; $Fe(OH)$; and FeO] present in several individual colors such as black, brown, metallic brown, red, and yellow. These various oxides of iron are used as catalysts, as coloring in glass and ceramic manufacturing, and as **pigments** and in laundry blueing and the manufacture of steel.

Ferrous sulfate ($FeSO_4$) is also known as iron sulfate or iron vitriol. It is used in the production of various chemicals, such as sulfur dioxide and sulfuric acid.

Pyrite (FeS_2) is more commonly known as fool's gold. It is used as an iron ore and in the production of sulfur chemicals such as sulfuric acid.

Ferrous chloride ($FeCl_2$) is used in pharmaceutical preparations, for sewage treatment, and as a mordant (which fixes dyes so that they will not run) in textiles.

Hazards

Iron dust from most iron compounds is harmful if inhaled and toxic if ingested. Iron dust and powder (even filings) are flammable and can explode if exposed to an open flame. As mentioned, excessive iron in the diet may cause liver damage.

COBALT

SYMBOL: Co   **PERIOD:** 4   **GROUP:** 9 (VIII)   **ATOMIC NO:** 27

ATOMIC MASS: 58.9332 amu   **VALENCE:** 2 and 3   **OXIDATION STATE:** +2 and +3   **NAT-URAL STATE:** Solid

ORIGIN OF NAME: Cobalt was given the name *kobolds* (or kolalds, or kololos) by German miners. It means "goblin" (see "History" for more on this story).

ISOTOPES: There are 33 isotopes of cobalt, ranging from Co-48 to Co-75, with half-lives ranging from a few nanoseconds to 5.272 years for cobalt-60. Cobalt-59 is the only stable isotope that constitutes almost all (roughly 100%) of the element's natural presence on Earth. All the other isotopes are radioactive and are created artificially in nuclear reactors or nuclear explosions.

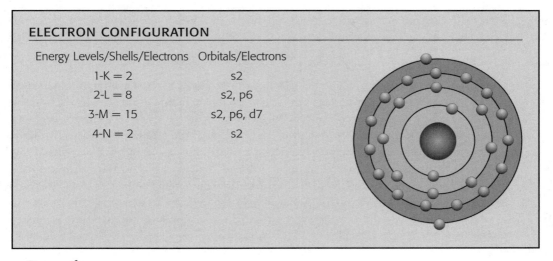

ELECTRON CONFIGURATION

Energy Levels/Shells/Electrons	Orbitals/Electrons
1-K = 2	s2
2-L = 8	s2, p6
3-M = 15	s2, p6, d7
4-N = 2	s2

Properties

Cobalt is a bluish steel-gray metal that can be polished to a bright shine. It is brittle and is not malleable unless alloyed with other metals. It is magnetic, and when alloyed with aluminum and nickel, it is called *alnico* metal, which acts as a super-magnet with many uses in industry. Chemically and physically, cobalt acts much as do its two partners, iron (Fe) and nickel (Ni), located on each side of it in period 4 on the periodic table. In particular, iron, cobalt, and nickelare unique in that they possess natural magnetic properties. Cobalt's melting point is 1,495°C, its boiling point is 2,927°C, and its density is 8.86 g/cm^3.

Characteristics

Cobalt has the highest Curie point of any metal or alloy of cobalt. The Curie point is the temperature at which an element will lose its magnetism before it reaches its melting point.

Cobalt's Curie point is 1,121°C, and its melting point is 1,495°C. About 25% of all cobalt mined in the world is used as an alloy with other metals. The most important is the alloy alnico, which consists of nickel, aluminum, and cobalt. Alnico is used to make powerful permanent magnets with many uses, such as CT, PET, and MRI medical instruments. It is also used for electroplating metals to give a fine surface that resists oxidation.

Abundance and Source

Cobalt is the 32nd most abundant element on Earth even though it makes up only 0.003% of the Earth's crust. It is not found in the free metallic state, despite being widely distributed in **igneous rocks** as minerals. Its two most common mineral ores are cobaltite ($CoAsS$) and erythrite [$Co_3(AsO_4)_2$]. These ores are placed in blast furnaces to produce cobalt arsenide (Co_2As), which is then treated with sulfuric acid to remove the arsenic. Finally, the product cobalt tetraoxide (Co_3O_4) is reduced by heat with carbon ($Co_3O_4 + C \rightarrow 3Co + 2CO_2$), resulting in cobalt metal.

Cobalt is also found in seawater, meteorites, and other ores such as linnaeite, chloanthite, and smaltite, and traces are found mixed with the ores of silver, copper, nickel, zinc, and manganese. Cobalt ores are found in Canada and parts of Africa, but most of the cobalt used in the United States is recovered as a by-product of the mining, smelting, and refining of the ores of iron, nickel, lead, copper, and zinc.

History

Ancient people in Egypt and the Middle East were aware of a mineral that could be used to make a highly prized deep blue glass. They were unaware that this material contained the element cobalt. Sixteenth-century chemists identified a mineral that they named "zaffer," and that was used to produce blue pigments and blue glass. They believed, mistakenly, that it contained the element bismuth. Additionally, German miners and the early smelters of this period also mistakenly believed that cobalt ores were really copper ores. They attributed their difficulties in extracting copper from this particular ore to mountain gnomes (*kobolds,* meaning "goblins" in German) who had placed an evil spell on the ore to prevent it from producing copper. Interestingly, the element nickel presented a similar problem for superstitious miners. Because the ores were difficult to mine as well as smelt, many accidents occurred, adding to the mythology of the Earth spirits, the kobolds.

In 1739 Georg Brandt (1694–1768) of Sweden investigated a bluish mineral he wanted to distinguish from the element bismuth, which was mined in the same regions where cobalt minerals were found. Cobalt was the first metal to be discovered that was not known to ancient alchemists. Brandt is credited with discovering the element cobalt even though it was a known mineral for many centuries. He isolated the new element, and its name relating to kobold goblins has been used ever since. Brandt is considered one of the first chemists of this era who was no longer under the influence of the old alchemists.

Common Uses

Cobalt has many practical uses.

Historically, as well as today, different compounds of cobalt have been used for their colors known as cobalt blue, cerulean, new blue, smalt, cobalt yellow, and green.

For many centuries cobalt was used to color glass, pottery, and porcelain and as an enamel. It is also used as a dye and paint pigment.

As mentioned, cobalt alloyed with iron and nickel is used to make powerful permanent magnets that are used in many industries.

A major use is as an alloy with chromium to produce high-speed machine-cutting tools that are resistant to high temperatures.

A cobalt alloy of copper and tungsten, called "stellite," also maintains its hardness at high temperatures, making it an ideal alloy for high-speed drills and cutting tools.

The radioisotope cobalt-60, with a half-life of 5.27 years (1925.3 days) through beta (β) emission, decays to form the stable element nickel-60. It is used to test welds and metal casts for flaws, to irradiate food crops to prolong freshness, as a portable source of ionizing gamma (γ) radiation, for radiation research, and for a medical source of **radiation** to treat cancers and other diseases.

Cobalt is an important trace element for proper human nutrition. It is also a natural component of vitamin B_{12}.

Examples of Compounds

There are dozens of useful cobalt compounds. A few common examples follow:

Cobalt blue [$Co(AlO_2)_2$], also known as cobalt ultramarine or azure blue, is a compound of aluminum oxide and cobalt. It is used as a pigment that mixes well with both oil and water. It also has cosmetic uses for eye shadow and in grease paint. Cobalt blue is one of the most durable blue pigments in that it resists weathering in paints and holds up to other wear and tear.

Cobaltic oxide (Co_2O_3) is also known as cobalt oxide or cobalt black. It is dark gray to black and is used in pigments, ceramic glazes, and semiconductors.

Cobalt chloride ($CoCl_2$) is used to manufacture vitamin B_{12}, even though the compound itself can cause damage to red blood cells. It is also used as a dye mordant (to fix the dye to the textile so that it will not run). It is also of use in manufacturing solid lubricants, as an additive to fertilizers, as a chemical reagent in laboratories, and as an absorbent in gas masks, electroplating, and the manufacture of vitamin B_{12}.

Cobaltous nitrate [$Co(NO_3)_2 \cdot 6H_2O$], also known as cobalt nitrate, is a red crystal that absorbs moisture. It is used in inks, pigments, animal feed, soil enhancers, and hair dyes.

Hazards

The dust and powder of cobalt metal, ores, and some compounds, such as cobaltous nitrate [$Co(NO3)2 \cdot 6H_2O$], are flammable and toxic if inhaled. Cobaltous acetate [$Co(C_2H_3O_2)_2 \cdot 4H_2O$], which is soluble in water, is not allowed to be used in food products because of its toxicity.

Cobalt is found in most natural foods. Although a necessary trace element, it is toxic to humans if ingested in large amounts. The human body does excrete in urine excessive amounts of cobalt compounds such as found in vitamin B_{12}.

Cobaltous chromate ($CoCrO_4$) is brownish-yellow to grayish-black (the color depends on its purity) is a dangerous carcinogen (causes cancer).

Some years ago, a cobalt additive was used by some beer makers to maintain a foam head on their beer. Those who imbibed excessively developed what is known as "beer drinkers' syndrome," which caused some deaths from enlarged and flabby hearts.

NICKEL

SYMBOL: Ni   **PERIOD:** 4   **GROUP:** 10 (VIII)   **ATOMIC NO:** 28
ATOMIC MASS: 58.6934 amu   **VALENCE:** 2 and 4   **OXIDATION STATE:** +2 and +3
   NATURAL STATE: Solid
ORIGIN OF NAME: The name is derived from the ore niccolite, meaning "Old Nick," referred to as the devil by German miners. The niccolite mineral ore was also called "kupfernickel," which in German stands for two things; first, it is the name of a gnome (similar to Cobalt), and second, it refers to "Old Nick's false copper."
ISOTOPES: There are 31 isotopes of nickel, ranging from Ni-48 to Ni-78. Five of these are stable, and the percentage of their contribution to the element's natural existence on Earth are as follows: Ni-58 = 68.077%, Ni-60 = 26.223%, Ni-61 = 1.140%, Ni-62 = 3.634%, and Ni 64 = 0.926%. All of the other 26 isotopes of nickel are artificially made and radioactive with half-lives ranging from a few nanoseconds to 7.6×10^4 years.

ELECTRON CONFIGURATION

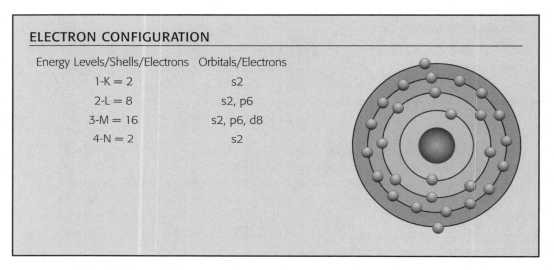

Energy Levels/Shells/Electrons	Orbitals/Electrons
1-K = 2	s2
2-L = 8	s2, p6
3-M = 16	s2, p6, d8
4-N = 2	s2

Properties

Nickel metal does not exist freely in nature. Rather, it is located as compounds in ores of varying colors, ranging from reddish-brown rocks to greenish and yellowish deposits, and in copper ores. Once refined from its ore, the metallic nickel is a silver-white and hard but malleable and ductile metal that can be worked hot or cold to fabricate many items. Nickel, located in group 10, and its close neighbor, copper, just to its right in group 11 of the periodic table, have two major differences. Nickel is a poor conductor of electricity, and copper is an excellent conductor, and although copper is not magnetic, nickel is. Nickel's melting point is 1,455°C, its boiling point is 2,913°C, and its density is 8.912 g/cm^3.

Characteristics

As mentioned, nickel is located in group 10 (VIII) and is the third element in the special triad (Fe, Co, Ni) of the first series of the transition elements. Nickel's chemical and physical properties, particularly its magnetic peculiarity, are similar to iron and cobalt.

Some acids will attack nickel, but it offers excellent protection from corrosion from air and seawater. This quality makes it excellent for electroplating other metals to form a protective coating. Nickel is also an excellent alloy metal, particularly with iron, for making stainless steel as well as a protective armor for military vehicles. It is malleable and can be drawn through dies to form wires. About one pound of nickel metal can be drawn to about 200 miles of thin wire.

Abundance and Source

Nickel is the 23rd most abundant element found in the Earth's crust. It is somewhat plentiful but scattered and makes up one-hundredth of 1% of igneous rocks. Nickel metal is found in meteorites (as are some other elements). It is believed that molten nickel, along with iron, makes up the central sphere that forms the core of the Earth.

There are several types of nickel ores. One is the major ore for nickel called pentlandite (NiS • 2FeS), which is iron/nickel sulfide. Another is a mineral called niccolite (NiAs), discovered in 1751 and first found in a mining area of Sweden. By far, the largest mining area for nickel is located in Ontario, Canada, where it is recovered from what is thought to be a very large meteorite that crashed into the Earth eons ago. This large nickel deposit is one reason for the theory of the Earth's core being molten nickel and iron, given that both the Earth and meteorites were formed during the early stages of the solar system. Some nickel ores are also found in Cuba, the Dominican Republic, and Scandinavia. Traces of nickel exist in soils, coal, plants, and animals.

History

Nickel, as with a few other minerals and compounds such as cobalt, was known to the ancients and was used to add color to their glass and ceramics. Nickel minerals produced green glass, and cobalt minerals were known for coloring glass blue. Early miners in Germany had trouble smelting copper from some of the ores because they kept getting other metals (see entry for cobalt). This caused them so much trouble that they named the new, non-copper element "kupfernickel," or "Old Nick's copper," which meant the "devil's copper."

In 1751 Baron Axel Fredrick Cronstedt (1722–1765) used some of the techniques he had learned from his teacher, Georg Brandt (1694–1768), to separate a "new" metal from copper-like ore mined in Sweden. He expected to obtain pure copper; instead, he ended up with a silver-white metal that did not have the chemical and physical properties of copper. He named this newly identified metal "nickel," shortened from the German name the early miners had given the ore: "kupfernickel."

Common Uses

The most common use of nickel is as an alloy metal with iron and steel to make stainless steel, which contains from 5% to 15% nickel. The higher the percentage of nickel in stainless steel, the greater the steel's resistance to corrosion—particularly when exposed to seawater. Nickel is also alloyed with copper to make Monel metal, which was widely used before stainless steel became more economical and practical. It was used for many purposes as varied as household appliances and general manufacturing. Nickel is also used to electroplate other metals to provide a noncorrosive protective and attractive finish.

The 5-cent coin of the United States is named after the metal nickel. It is composed of 25% nickel and 75% copper (maybe it should have been named "copper," but that name was already used in England for a coin). Thousands of years ago, a variety of metals, mainly gold and silver, were used to make coins. Because of the shortage and price increase of some coinage metals such as gold, silver, and copper, the United States mint now uses less of these metals and substitutes more iron, zinc, nickel, or steel to make the coins of today.

As mentioned, nickel is one of the three unique magnetic metals (Fe, Co, Ni) and is alloyed with iron, cobalt, and aluminum to make powerful Alnico magnets.

When nickel is alloyed with chromium, it forms what is known as Ni-Chrome, which has a rather high resistance to electrical currents, meaning that electron flow is impeded in the conductor to the point that it results in the rapid vibration of the atoms and molecules of the metal conductor, thus producing heat. This property, along with the element's high melting point and the fact that it can be drawn into different size wires, makes nickel ideal in the manufacture of heating elements in toasters and other appliances that convert electrical energy into heat. Because of their superior heat conductivity and resistance to corrosion, nickel alloys are excellent metals for manufacturing cookware.

A more recent use of nickel is in the manufacture of the rechargeable nickel-chrome electric cell. One of the **electrodes** in this type of cell (battery) is nickel (II) oxide ($Ni^{2+} + O^{2-} \rightarrow NiO$). (Note: When two or more cells are combined in an electrical circuit, they form a battery, but when just one is referred to, it is called a cell.) Although the electrical output of a Ni-Chrome cell is only 1.4 volts (as compared to 1.5 volts dry cells), Ni-Chrome has many uses in handheld instruments such as calculators, computers, electronic toys, and other portable electronic devices.

Powdered nickel metal acts as a catalyst for the hydrogenation of vegetable oils. (See the entry for hydrogen for more on hydrogenation.)

Examples of Compounds

There are many compounds of nickel, existing in a variety of colors from black to green to yellow. A few examples follow:

Nickel chloride ($NiCl_2$) is used for **electroplating** nickel onto the surfaces of other metals and as a chemical reagent in laboratories.

Nickel oxide (NiO) is produced from nickel minerals to form nickel oxide when heated to 400°C, which is then reduced at a temperature of 600°C, resulting in the formation of nickel oxide. It is used as electrodes in fuel cells.

Nickel phosphate ($3Ni^{2+} + 2P^{3-} \rightarrow Ni_3P_2$) is a greenish powder used in electroplating.

Nickel-silver is not a compound of nickel, but rather an alloy of nickel and copper and zinc that has a silvery appearance. It is used in silver and chrome plate metals.

Nickel sulfate ($NiSO_4$) exists in different states depending on its hydrated forms (where water molecules bond with ions in suspended substances). Nickel sulfate can be in the form of greenish-yellow, blue, or green crystals, depending upon the degree of hydration. It is used in nickel-plating iron and copper, as a catalyst, as a mordant in the textile industry, and as a coating for other substances.

Hazards

Nickel dust and powder are flammable. Most nickel compounds, particularly the salts, are toxic. $NiSO_4$ is a known carcinogen.

Although nickel is not easily absorbed in the digestive system, it can cause toxic reactions and is a confirmed carcinogen in high concentration in the body. Nickel workers can receive severe skin rashes and lung cancer from exposure to nickel dust and vapors.

Nickel is stored in the brain, spinal cord, lungs, and heart. It can cause coughs, shortness of breath, dizziness, nausea, vomiting, and general weakness.

COPPER

SYMBOL: Cu   **PERIOD:** 4   **GROUP:** 11 (IB)   **ATOMIC NO:** 29

ATOMIC MASS: 63.546 amu   **VALENCE:** 1 and 2   **OXIDATION STATE:** +1 and +2   **NAT-URAL STATE:** Solid

ORIGIN OF NAME: Copper's name comes from the Latin word *cuprum* or cyprium, which is related to the name "Cyprus," the island where it was found by the ancient Romans.

ISOTOPES: There are 32 known isotopes of copper, ranging from Cu-52 to Cu-80. Only two of these 32 isotopes of copper are stable, and together they make up the amount of natural copper found in the Earth's crust in the following proportions: Cu-63 = 69.17% and Cu-65 = 30.83%. All the other isotopes of copper are radioactive and are artificially produced with half-lives ranging from a few nanoseconds to about 61 hours.

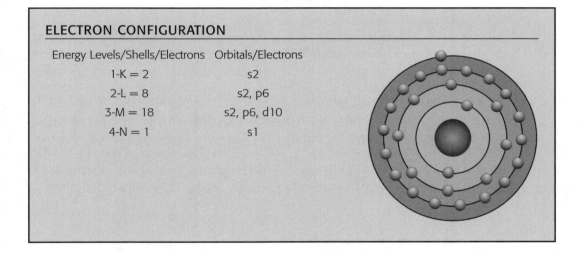

ELECTRON CONFIGURATION

Energy Levels/Shells/Electrons	Orbitals/Electrons
1-K = 2	s2
2-L = 8	s2, p6
3-M = 18	s2, p6, d10
4-N = 1	s1

Properties

Native copper has a distinctive reddish/brown color. Its first oxidation state (+1) forms compounds with copper ions named "cuprous," also referred to as "copper(I)," and these ions are easily oxidized with elements in group 16 (e.g., oxygen and sulfur) and elements in group 17 (the halogens).

Copper's second oxidation state (+2) forms cupric compounds, also referred to as copper(II), which are more stable than copper(I) compounds. For example, copper in both oxidation states can combine with fluorine: for copper(I) or cuprous fluoride, $Cu^+ + F^- \rightarrow CuF$; and for copper(II) or cupric fluoride, $Cu^{2+} + 2F \rightarrow CuF_2$.

Copper's melting point is 1,083°C, its boiling point is 2,567°C, and its density is 8.94 g/cm^3.

Characteristics

Copper, a versatile metal relatively easy to find, has made it useful for humans for many centuries. It is malleable, ductile, and easily formed into many shapes such as ingots, pipes, wire, rods, tubing, sheets, powder, shot, and coins. Although copper is resistant to weak acids, it will dissolve in strong or hot acids. It resists atmospheric corrosion better than does iron. One reason is that it forms a bluish-green film (called *patina*) over its surface when exposed to moist air or seawater. This coating of copper carbonate and copper sulfate provides a protective layer for the underlying metal that makes it ideal for use on boats, roofs, pipes, and coins. The surfaces of some copper church steeples and the Statue of Liberty have now oxidized to form a pleasing patina.

One of copper's most useful characteristics is that it is an excellent **conductor** of electricity and heat.

Abundance and Source

Copper is the 26th most abundant element on Earth, but it is rare to find pure metallic deposits. It is found in many different types of mineral ores, many of which are close to the surface and easy to extract. It is found in two types of ores: (1) sulfide ores, such as covellite, chalcopyrite, bornite, chalcocite, and enargite; and (2) oxidized ores, such as tenorite, malachite, azurite, cuprite, chrysocolla, and brochanite.

It is found in most countries of the world, but only a few high-grade deposits are cost-effective to mine. Examples of some of its ores are cuprite (CuO_2), tenorite (CuO), malachite [$CuCO_3 \cdot Cu(OH)_2$], chalcocite (Cu_2S), covellite (CuS), bornite (Cu_6FeS_4), and chalcopyrite, also known as copper pyrite.

Copper ores are found worldwide, in Russia, Chile, Canada, Zambia, and Zaire and, in the United States, in Arizona, Michigan, Montana, Nevada, New Mexico, Tennessee, and Utah. High-grade ores of 99% pure metal were found in the United States (and other countries), but many of these native ore deposits have been mined over the past hundred years and are now exhausted. Even so, many low-grade ores with concentrations of 10% to 80% pure copper still exist and await a technology that will make them more profitable for exploitation.

History

Native copper was used for decorations by prehistoric humans. A few copper beads dating back to 9000 BCE were found in Iraq. Copper ores are easy to find and are relatively easy to refine. This may be one reason copper is probably the earliest mined and refined metal by humans. At least 7,000 years ago, humans learned to smelt copper from high-grade ores. Following that, most civilizations learned how to obtain, refine, and use copper. Ancient metallurgists soon learned, most likely by accident, that pure copper was too soft to use in weapons or tools for cutting unless it was mixed with other substances. When mixed with arsenic, and later with tin ore, a copper alloy was produced that would keep a sharper edge. It was called bronze. About 2,500 years ago copper was mixed with about 5% to 45% zinc to produce an alloy known as brass. Artifacts of copper tools made by American Indians over two thousand years ago have been found in the Great Lakes region of the United States. Today, there are other alloys of copper including copper-aluminum and copper-nickel.

Common Uses

Copper, being easy to mine and refine, has become a very versatile metal over the course of civilization. Early in human history, it was discovered that soft copper could be made harder and stronger when alloyed with other metals. Copper was and still is important to technology and the development of civilizations. Over the past several thousand years, brass has found multiple uses, such as in coins, cooking utensils, and many types of instruments and hardware that are resistant to corrosion. Even today, brass is used to make musical instruments and bathroom, kitchen, and marine hardware. The U.S. one-cent penny was originally made of copper, but today the penny is made of zinc with a coating of copper. Copper is also an alloy metal used as a substitute for some of the silver in several other U.S. coins.

Some common uses are in electrical wiring and components of electronic equipment, roofing, and pipes and plumbing and in the manufacturing of alloys such as brass, bronze, Monel metal, electroplating, jewelry, cooking utensils, insecticides, marine paints, cosmetics, and wood preservatives.

Copper is second only to silver as an excellent conductor of electricity. This factor and its availability made it essential for the expansion of modern technologies. It was, and still is, a desired metal for wires to carry electricity, but the rapid expansion of modern communications would require more copper than could be made economically available. The solution has been to use optical fiberglass transmission cables as a substitute for copper wire. In addition, and even more important, is the recent explosive growth of wireless transmission as a substitute for copper wire in the communication industries.

Examples of Compounds

As previously mentioned, copper has two oxidation states. Compounds formed by +1 copper are known as "**cuprous** compounds," and those with a +2 oxidation state are "**cupric** compounds." Both oxidation states may be found in related compounds.

Copper(I) carbonate ($Cu^{1+} + CO_3^{2-} \rightarrow Cu_2O_3$) is known as cuprous carbonate since copper's ion is +1; *copper(II) carbonate* ($Cu^{2+} + CO_3^{2-} \rightarrow CuCO_3$) is known as cupric carbonate, which is also known as the green copper mineral malachite, used in pigments, as an insecticide, as a cosmetic astringent, and as a plant fungicide to prevent smut.

Copper(I) chloride (CuCl) or cuprous chloride is a white powder used as an absorbing agent for carbon dioxide gas in enclosed breathing areas such as space vehicles.

Copper(II) chloride ($CuCl_2$) or cupric chloride, is a brownish-yellow **hygroscopic** powder, or it may be formed as green **deliquescent** crystals. It is used in the dyeing and printing of textiles, as a disinfectant, as red pigment in the glass and ceramic industries, and for green-colored pyrotechnics, wood preservative, fungicide, deodorizer, water purification, feed additive, and electroplating baths.

There are many other compounds of copper used in electroplating and used as preservatives, pigments, and fungicides and insecticides.

Hazards

Copper dust and powder, as well as a few of its compounds, are flammable, or even explosive when ignited in contained areas. Many of copper's compounds are extremely toxic and poisonous either with skin contact or when inhaled or ingested and should be handled by professionals in controlled environments. Even so, both plants and animals, including humans, require traces of copper for the proper metabolism of their foods.

ZINC

SYMBOL: Zn   **PERIOD:** 4   **GROUP:** 12 (IIB)   **ATOMIC NO:** 30
ATOMIC MASS: 65.39 amu   **VALENCE:** 2   **OXIDATION STATE:** +2.   **NATURAL STATE:** Solid
ORIGIN OF NAME: Although ancients used zinc compounds, the name "zinc" is assumed to be derived from the German word *zinn,* which was related to tin.
ISOTOPES: There are 38 isotopes of zinc, ranging in atomic weights from Zn-54 to Zn-83. Just four of these are stable, and those four, plus one naturally radioactive isotope (Zn-70) that has a very long half-life ($5 \times 10^{+14}$ years), make up the element's existence on Earth. Their proportional contributions to the natural existence of zinc on Earth are as such: Zn-64 = 48.63%, Zn-66 = 27.90%, Zn-67 = 4.10%, Zn- 68 = 18.75%, and Zn-70 = 0.62%. All the other isotopes are radioactive and artificially produced.

ELECTRON CONFIGURATION

Energy Levels/Shells/Electrons	Orbitals/Electrons
1-K = 2	s2
2-L = 8	s2, p6
3-M = 18	s2, p6, d10
4-N = 2	s2

Properties

Zinc is a whitish metal with a bluish hue. As an electropositive metal, it readily gives up its two outer electrons located in the N shell as it combines with nonmetal elements. Zinc foil will ignite in moist air, and zinc shavings and powder react violently with acids. Zinc's melting point is 419.58°C, its boiling point is 907°C, and its density is 7.14 g/cm³.

Note: Zinc is not always included as one of the metals in the first series of the transition elements, but it is the first element in group 12 (IIB).

Characteristics

Zinc is malleable and can be machined, rolled, die-cast, molded into various forms similar to plastic molding, and formed into rods, tubing, wires, and sheets. It is not magnetic, but it does resist **corrosion** by forming a hard oxide coating that prevents it from reacting any further with air. When used to coat iron, it protects iron by a process called "galvanic protection," also known as "sacrificial protection." This protective characteristic occurs because the air will react with the zinc metal coating, which is a more electropositive (reactive) metal than is the coated iron or steel, which is less electropositive than zinc. In other words, the zinc is oxidized instead of the underlying metal. (See the section under "Common Uses of Zinc" for more on galvanization.)

Abundance and Source

Zinc is the 24th most abundant on Earth, which means it makes up only about 0.007% of the Earth's crust. Even so, humans have found many uses for it over the past thousands of years.

It is not found in its pure metallic form in nature but is refined from the mineral (compound) zinc sulfide ($ZnSO_4$) known as the ores sphalerite and zincblende. It is also recovered from minerals and ores known as willemite, hydrozincite, smithsonite, wurtzite, zincite, and Franklinite. Zinc ores are found in Canada, Mexico, Australia, and Belgium, as well as in the United States. Valuable grades of zinc ores are mined in Colorado and New Jersey.

History

Zinc minerals were known and being used by about 3,000 years ago when they were used as an alloy with copper to form brass. Also, zinc was not recognized as a metallic element until many years later. It was not recognized as a unique metal for many years because it was always mixed with other metals, such as lead, arsenic, antimony, and bismuth. There is some evidence that zinc metal was produced in India 1,500 years ago by heating calamine ($ZnCO_3$) with sheep's wool. Theophrastus Bombastus von Hohenheim (1493–1541), better known as Paracelsus, who used several metals to treat a variety of diseases, is often credited with the discovery of zinc. In 1659 Johann Rudolf Glauber (1604–1668) recognized that different elements produced varying colored flames when heated. He used the "glass bead" technique to identify zinc as well as many other metals that had often been confused with each other. Glauber also produced many medicines of his day, including his *sal mirabile,* or "wonderful salt," which he advertised as a cure-all because of its laxative properties. Today, it is called Glauber salt (sodium sulfate).

The person generally recognized as producing and identifying metallic zinc is Andreas Sigismund Marggraf (1709–1782), a German analytical chemist who in 1746 heated calamine with charcoal. (Marggraf also discovered beet sugar.)

The zinc that is produced today starts as the zinc sulfide (ZnS) minerals zinc blende or sphalerite or from zinc carbonate ($ZnCO_3$) known as smithsonite or calamine. In the electrolytic process, these minerals are dissolved in water to form the electrolyte in the cell where the zinc cations are attracted and collected at the cathode and deposited as a dull, brittle type of zinc.

Common Uses

Today about one-third of all the zinc metal is used for the process known as **galvanization.** This process provides a protective coating of zinc on other metals. A thin layer of zinc oxidizes in air, thus providing a galvanic corrosion protection to the iron or steel item that it coats. Several processes are used to galvanize other metals. One is the "hot dip" method wherein the outer surface of the item to be galvanized is "pickled" and then immersed into a molten zinc bath. A second method plates iron or steel items with a thin coat of zinc by electrolysis. Because of zinc's relatively high reactivity, it does not need to be actually coated onto the surface of some metals to protect them. For instance, a small disk made of zinc is sometimes attached to the rudder or other parts of an oceangoing ship. This zinc metal is "sacrificed," as it oxidizes and corrodes, while the iron rudder does not corrode and remains protected by the galvanic process.

Zinc is used with copper and other metals to produce alloys of brass, bronze, and special **die**-casting alloys (e.g., copper, aluminum, nickel, and titanium).

It is used in dry cell (batteries), in **fungicides** (to kill fungi and molds), for roofing, and in wrapping wires for protection.

Zinc is an important trace element required for all healthy plants and animals. Zinc is found in proteins, such as meats, fish, eggs, and milk. About 10 to 15 milligrams of zinc is required per day, and it may be taken as a dietary supplement. Zinc helps the blood in our bodies move the waste gas—carbon dioxide—to the lungs and helps prevent macular degeneration (loss of vision).

Examples of Compounds

There are hundreds of compounds of zinc, some of which have found practical uses over the past several thousands of years.

Zinc acetate [$Zn(C_2H_3O_2)_2$] is used as a mordant for dyeing cloth, as a wood preservative, as a laboratory agent, and as a dietary supplement.

Zinc chloride is an example of zinc's oxidation state of +2 combining with an −1 anion. ($Zn^{2+} + 2Cl^{1-} \rightarrow ZnCl_2$). Zinc chloride is used as an organic catalyst. It is deliquescent, which makes it an excellent dehydrating and drying agent. It is used in electroplating other metals, as an antiseptic, as a component of some deodorants, and as an astringent. It is also used for fireproofing materials and as a food preservative. Zinc chloride is also used in embalming and taxidermy fluids.

Zinc oxide is an example of zinc(II) combining with a −2 anion. ($Zn^{2+} + O^{2-} \rightarrow ZnO$). Zinc oxide is also added to paints as a pigment and mold inhibitor and is known as zinc white when it is used as an oil paint by artists. It is used for cosmetics (ointment to protect nose and lips from ultraviolet sunlight), as a seed treatment, and as a dietary supplement.

Zinc dust (Zn) is not a compound but a gray powder that is used as a pigment and acts as an excellent reducing agent and catalyst. It is dangerous because it can explode when exposed to moist air and may heat up and ignite spontaneously. When mixed with dry, powdered sulfur, it makes an excellent dry propellant-type rocket fuel, but is dangerous to handle.

Zinc sulfate ($ZnSO_4 \cdot 7H_2O$) is also known as zinc vitriol or white copper. In addition to being used to make rayon, zinc sulfate is used as a wood preservative, a dietary supplement, an animal feed, and as a mordant to prevent dyes from running in printed textiles. It can also be used to stanch bleeding.

Zinc sulfide (ZnS) is used as a pigment and to make white glass, rubber, and plastics. It is an ingredient in pesticides, luminous paints, and X-ray and television screens.

Hazards

As mentioned, zinc dust and powder are very explosive. When zinc shavings are placed in acid or strong alkaline solutions, hydrogen gas is produced, which may explode. Many of zinc's compounds are toxic if inhaled or ingested.

A deficiency of zinc in humans will retard growth, both physically and mentally, and contribute to anemia. It is present in many foods, particularly proteins (meat). A balanced diet provides an adequate amount of zinc. Not more than 50 milligrams per day of dietary zinc supplement should be taken, given that high levels of zinc in the body are toxic. Human bodies contain about two grams of zinc. A deficiency of zinc can cause a lack of taste and can delay growth as well as cause retardation in children.

Zinc **intoxication** can occur both from inhaling zinc fumes and particles, mainly in industrial processes, and from orally ingesting an excess of zinc in dietary supplements. Zinc intoxication can cause stomach pains, vomiting, and bleeding. Excess zinc also can cause premature birth in pregnant women.

PERIODIC TABLE OF THE ELEMENTS

TRANSITION ELEMENTS

GROUPS / PERIODS	1 IA	2 IIA	3 IIIB	4 IVB	5 VB	6 VIB	7 VIIB	8 VIII	9 VIII	10 VIII	11 IB	12 IIB	13 IIIA	14 IVA	15 VA	16 VIA	17 VIIA	18 VIIIA
1	1 **H** 1.0079																	2 **He** 4.00260
2	3 **Li** 6.941	4 **Be** 9.01218											5 **B** 10.81	6 **C** 12.011	7 **N** 14.0067	8 **O** 15.9994	9 **F** 18.9984	10 **Ne** 20.179
3	11 **Na** 22.9898	12 **Mg** 24.305											13 **Al** 26.9815	14 **Si** 28.0855	15 **P** 30.9738	16 **S** 32.066(6)	17 **Cl** 35.453	18 **Ar** 39.948
4	19 **K** 39.0983	20 **Ca** 40.08	21 **Sc** 44.9559	22 **Ti** 47.88	23 **V** 50.9415	24 **Cr** 51.996	25 **Mn** 54.9380	26 **Fe** 55.847	27 **Co** 58.9332	28 **Ni** 58.69	29 **Cu** 63.546	30 **Zn** 65.39	31 **Ga** 69.72	32 **Ge** 72.59	33 **As** 74.9216	34 **Se** 78.96	35 **Br** 79.904	36 **Kr** 83.80
5	37 **Rb** 85.4678	38 **Sr** 87.62	39 **Y** 88.9059	40 **Zr** 91.224	41 **Nb** 92.9064	42 **Mo** 95.94	43 **Tc** (98)	44 **Ru** 101.07	45 **Rh** 102.906	46 **Pd** 106.42	47 **Ag** 107.868	48 **Cd** 112.41	49 **In** 114.82	50 **Sn** 118.71	51 **Sb** 121.75	52 **Te** 127.60	53 **I** 126.905	54 **Xe** 131.29
6	55 **Cs** 132.905	56 **Ba** 137.33	★	72 **Hf** 178.49	73 **Ta** 180.948	74 **W** 183.85	75 **Re** 186.207	76 **Os** 190.2	77 **Ir** 192.22	78 **Pt** 195.08	79 **Au** 196.967	80 **Hg** 200.59	81 **Tl** 204.383	82 **Pb** 207.2	83 **Bi** 208.980	84 **Po** (209)	85 **At** (210)	86 **Rn** (222)
7	87 **Fr** (223)	88 **Ra** 226.025	▲	104 **Unq** (261)	105 **Unp** (262)	106 **Unh** (263)	107 **Uns** (264)	108 **Uno** (265)	109 **Une** (266)	110 **Uun** (267)	111 **Uuu** (272)	112 **Uub**	113 **Uut**	114 **Uuq**	115 **Uup**	116 **Uuh**	117 **Uus**	118 **Uuo**

6 ★ Lanthanide Series (RARE EARTH)

57 **La** 138.906	58 **Ce** 140.12	59 **Pr** 140.908	60 **Nd** 144.24	61 **Pm** (145)	62 **Sm** 150.36	63 **Eu** 151.96	64 **Gd** 157.25	65 **Tb** 158.925	66 **Dy** 162.50	67 **Ho** 164.930	68 **Er** 167.26	69 **Tm** 168.934	70 **Yb** 173.04	71 **Lu** 174.967

7 ▲ Actinide Series (RARE EARTH)

89 **Ac** 227.028	90 **Th** 232.038	91 **Pa** 231.036	92 **U** 238.029	93 **Np** 237.048	94 **Pu** (244)	95 **Am** (243)	96 **Cm** (247)	97 **Bk** (247)	98 **Cf** (251)	99 **Es** (252)	100 **Fm** (257)	101 **Md** (258)	102 **No** (259)	103 **Lr** (260)

Transition Elements: Second Series—Period 5, Groups 3 to 12

Introduction

Period 5 (group 3 [IIIB] to group 12 [IIB]) is located in the second row of the transition elements and represents 10 of the transition metals to nonmetals found in the periodical table of chemical elements. This period is also known to include some of the so-called rare-earth elements. Most of the rare-earths are found in the lanthanide series, which follows barium (period 6, group 3). (Check the periodic table to locate the major rare-earth elements in the lanthanide series. These are addressed in a later section of the book.)

YTTRIUM

SYMBOL: Y   **PERIOD:** 5   **GROUP:** 3 (IIIB)   **ATOMIC NO:** 39
ATOMIC MASS: 88.9059 amu   **VALENCE:** 3   **OXIDATION STATE:** +3
   NATURAL STATE: Solid
ORIGIN OF NAME: Yttrium was originally found with other elements in a mineral called gadolinite that was discovered in a mine near the Swedish the town of Ytterby.
ISOTOPES: There are 50 isotopes of Yttrium. Only one is stable (Y-89), and it constitutes 100% of the element's natural existence on Earth. The other isotopes range from Y-77 to Y-108 and are all produced artificially in nuclear reactions. The radioactive isotopes have half-lives ranging from 105 nanoseconds to 106.65 days.

ELECTRON CONFIGURATION

Energy Levels/Shells/Electrons	Orbitals/Electrons
1-K = 2	s2
2-L = 8	s2, p6
3-M = 18	s2, p6, d10
4-N = 9	s2, p6, d1
5-O = 2	s2

Properties

Yttrium is always found with the rare-earth elements, and in some ways it resembles them. Although it is sometimes classified as a rare-earth element, it is listed in the periodic table as the first element in the second row (period 5) of the transition metals. It is thus also classified as the lightest in atomic weight of all the rare-earths. (Note: Yttrium is located in the periodic table just above the element lanthanum (group 3), which begins the lanthanide rare-earth series.

Yttrium dissolves in weak acids and also dissolves in strong alkalis such as potassium hydroxide. It will also decompose in water.

Yttrium's melting point is 1,522°C, its boiling point is 5,338°C, and its density is 4.469 g/cm^3.

Characteristics

Yttrium ($_{39}$Y) is often confused with another element of the lanthanide series of rare Earths—Ytterbium ($_{70}$Yb). Also confusing is the fact that the rare-earth elements terbium and erbium were found in the same minerals in the same quarry in Sweden. Yttrium ranks second in abundance of all 16 rare-earth, and Ytterbium ranks 10th. Yttrium is a dark silvery-gray lightweight metal that, in the form of powder or shavings, will ignite spontaneously. Therefore, it is considered a moderately active rare-earth metal.

Abundance and Source

Yttrium is the 27th most abundant element found on Earth, so it is not exactly correct to think of it as "rare"—rather just difficult to find and extract from all the other similar elements found in its minerals.

The mineral gadolinite that was discovered in a quarry near Ytterby, Sweden, was analyzed as $(Ce,La,Nd,Y)_2FeBe_2Si_2O_{10}$. Today most yttrium is recovered from the ores of the mineral monazite, which is a dark, sandy mixture of elements $[(Ce,La,Th,Nd,Y)PO_4]$ and contains about 50% rare-earths, including about 3% yttrium. The yttrium is separated from the other rare-earths first by magnetic and flotation processes, which are followed by an iron-exchange displacement process. Yttrium's ions are combined with fluorine ions that are then reduced by using calcium metal that yields yttrium metal ($3Ca + 2YF_3 \rightarrow 2Y + 3CaF_2$). This reduction process produces high-purity yttrium that can be formed into ingots, crystals, sponge, powder, and wires.

History

In 1788 Bengt Reinhold Geijer (1758–1815), a Swedish mineralogist, analyzed a new mineral that resembled thick tar of coal in a quarry near Ytterby, Sweden, which was close to Stockholm. Geijer speculated that it might contain some tungsten. In either 1789 or 1794 (both dates are given) Johan Gadolin (1760–1850), a Finnish chemist and mineralogist, analyzed this black earth mineral and found that it contained 23% silicon dioxide, 4.5% beryllium oxide, 16.5% iron oxide, and about 55.5% of a new oxide he called yttria. Most references credit Gadolin rather than Geijer with the discovery of yttrium. An interesting note is that nearly a century later, the black mineral from which Gadolin obtained the new element yttrium was named in his honor, namely "gadolinite," and the element gadolinium is obtained from gadolinite.

In early 1828, Friedrich Wohler (1800–1882) obtained yttrium metal by reducing yttrium chloride with potassium (YCl_3 + K → 2KCl + Y). He is also given credit for yttrium's discovery.

Common Uses

Although yttrium metal by itself is not very useful, it has many unusual applications when combined as an alloy or as a compound with other elements. For example, when combined with iron, it is known as garnet ($Y_3Fe_5O_{12}$), which is used as a "filter" in microwave communication systems. When garnets are made with aluminum instead of iron, they form semiprecious garnet gemstones ($Y_3Al_5O_{12}$) that resemble diamonds. Aluminum garnets are referred to as "YAG" solid-state lasers because they are capable of intensifying and strengthening a single frequency of light energy that is focused through a crystal of garnet. This produces a very powerful narrow band of light waves of a single color (microwave frequency). YAG-type lasers have found uses in the medical industry and as a cutting tool for metals.

When combined with oxygen and europium, yttrium produces the red phosphor used as a coating in color television screens to produce the bright red color. Yttrium is also used as an alloy metal and as a high-temperature coating on iron and steel alloys. It is used as a substance to deoxidize (remove the oxygen) during the production of nonferrous metals such as vanadium. Yttrium has the ability to "capture" neutrons, making it useful in the nuclear power industry. It is also used in the production of several types of semiconductors.

Examples of Compounds

Because yttrium has just one oxidation state (+3), it joins with oxygen to form *yttrium oxide* ($2Y^{3+}$ + $3O^{2-}$ → Y_2O_3), which is used to produce the red colors in TV and computer screens.

Yttrium arsenide (YAs) is used in the production of high-grade semiconductors. Since it is extremely toxic, special handling and facilities are required for its use in computer industries.

Yttrium chloride (YCl_3) decomposes at the relatively low temperature of 100°C. This makes it useful as a reagent in chemical laboratories.

The compound consisting of yttrium, copper, and barium oxide, commonly called *compound 1-2-3,* was formed in 1987 by research scientists at the universities of Alabama and Houston. It had limited superconducting capabilities. It has been known for some time that conductors of electricity such as copper resist, to some extent, the flow of electrons at normal temperatures, but at temperatures near absolute zero (zero Kelvin = –273°C), this resistance to the flow of electrons in some materials is reduced or eliminated. The 1-2-3 compound proved to be superconducting at just 93°K, which is still much too cold to be used for everyday transmission of electricity at normal temperatures. Research continues to explore compounds that may achieve the goal of high-temperature superconductivity.

Hazards

As a powder or in fine particles, yttrium is flammable and may spontaneously ignite in moist air. Some of its compounds, particularly those used in the semiconductor and electrical industries, are very toxic if inhaled or ingested and should only be used under proper conditions.

ZIRCONIUM

SYMBOL: Zr **PERIOD:** 5 **GROUP:** 4 (IVB) **ATOMIC NO:** 40
ATOMIC MASS: 91.224 amu **VALENCE:** 2, 3, and 4 **OXIDATION STATE:** +4 (+2 and +3 with halogens) **NATURAL STATE:** Solid
ORIGIN OF NAME: The name "zirconium" was derived from the Arabic word *zargun,* which means "gold color." Known in biblical times, zirconium mineral had several names (e.g., jargoon, jacith, and hyacinth). Later, the mineral was called "zirconia," and the element was later named "zirconium."
ISOTOPES: Zirconium has 37 isotopes, ranging from Zr-79 to Zr-110. Four of them are stable, and one is a naturally radioactive isotope, with a very long half-life. All five contribute to the element's natural existence on Earth. The stable isotopes are the following: Zr-90 = 1.45%, Zr-91 = 11.22%, Zr-92 = 17.15%, and Zr-94 = 17.38%. The one natural radioactive isotope is considered stable: Zr-96, with a half-life of $2.2 \times 10^{+19}$ years, contributes 2.80% to zirconium's total existence on Earth. All of the other isotopes are artificially radioactive and are produced in nuclear reactors or particle accelerators. They have half-lives ranging from 150 nanoseconds to $1.53 \times 10^{+6}$ years.

ELECTRON CONFIGURATION

Energy Levels/Shells/Electrons	Orbitals/Electrons
1-K = 2	s2
2-L = 8	s2, p6
3-M = 18	s2, p6, d10
4-N = 10	s2, p6, d2
5-O = 2	s2

Properties

Zirconium can be a shiny grayish crystal-like hard metal that is strong, ductile, and malleable, or it can be produced as an undifferentiated powder. It is reactive in its pure form. Therefore, it is only found in compounds combined with other elements—mostly oxygen. Zirconium-40 has many of the same properties and characteristics as does hafnium-72, which is located just below zirconium in group 4 of the periodic table. In fact, they are more similar than any other pairs of elements in that their ions have the same charge (+4) and are of the same general size. Because zirconium is more abundant and its chemistry is better known than hafnium's, scientists extrapolate zirconium's properties for information about hafnium. This also means that one "twin" contaminates the other, and this makes them difficult to separate.

Zirconium's melting point is 1,852°C, its boiling point is 4,377°C, and its density is 6.506 g/cm^3.

Characteristics

Zirconium is insoluble in water and cold acids. Although it is a reactive element, it resists corrosion because of its rapid reaction with oxygen, which produces a protective film of zirconium oxide (ZrO_2) that protects any metal with which it is coated. Zirconium is best known as the gemstone zircon. Although there are different types of zircons, the most recognized is the hard, clear, transparent zircon crystal that has a very high index of refraction, which means it can bend light at great angles. These zircon crystals (zirconium sulfate, $ZrSiO_4$) are cut with facets to resemble diamonds.

Another characteristic that makes zirconium useful is the production of "zircaloy," which does not absorb neutrons as does stainless steel in nuclear reactors. Thus, it is ideal to make nuclear fuel tubes and reactor containers. Zircaloy is the blend (alloy) of zirconium and any of several corrosion resistant metals.

Abundance and Source

Zirconium is not a rare element. It is found over most of Earth's crust and is the 18th most abundant element, but it is not found as a free metal in nature.

It is found in the ores baddeleyite (also known as zirconia) and in the oxides of zircons, elpidite, and eudialyte.

History

Several minerals containing zirconium were known in ancient times, one of which, jacinth, is mentioned several times in the Bible. It was not until 1789 that Martin Heinrich Klaproth (1743–1817), a German analytical chemist who also discovered uranium, identified zirconium after many others before him had failed. Klaproth analyzed the mineral jargoon ($ZnSiO_4$), as did other scientists, and found that it contained 25% silica, 5% iron oxide, and 70% zirconia. The other scientists confused zirconia with alumina (aluminum oxide, Al_2O_3). Klaproth used more refined techniques and correctly identified the element zirconium.

Zirconium was isolated from other compounds in 1824 by Baron Jöns Jacob Berzelius (1779–1848), a Swedish chemist, but it was not produced in pure form until 1914 because of the difficulty in separating it from hafnium.

Common Uses

About 90% of all the zirconium produced in the United States is used in the nuclear electrical power industry. Since it does not readily absorb neutrons, it is a desired metal in the manufacture of nuclear reactors and their fuel tubes, but it must be free of its "twin" hafnium for these purposes. Zirconium is also used as an alloy with steel to make surgical instruments.

Zirconium dioxide (ZrO_2) as an **abrasive** is used to make grinding wheels and special sandpaper. It is also used in ceramic glazes, in enamels, and for lining furnaces and high-temperature molds. It resists corrosion at high temperatures, making it ideal for crucibles and other types of laboratory ware. ZrO_2 is used as a "getter" to remove the last trace of air when producing vacuum tubes.

As mentioned, zircon ($ZrSiO_4$) has many forms, but the most used is the transparent crystal that is cut to resemble a diamond. There is even one form of zirconium used in medicine: zirconium carbonate ($3ZrO_2 \bullet CO_2 \bullet H_2O$), which, as a lotion, can be used to treat poison ivy infections.

When zirconium is alloyed with niobium, it becomes superconductive to electricity at temperatures near absolute zero Kelvin (–273°C).

Examples of Compounds

Zirconium's common oxidation state is +4, but when combined with chlorine and other halogens, it can exist in +2 and +3 oxidation states, as follows:

Zirconium dichloride: $Zr^{2+} + 2Cl^{1-} \rightarrow ZrCl_2$

Zirconium trichloride: $Zr^{3}+ + 3Cl^{1-} \rightarrow ZrCl_3$

Zorconium tetrachloride: $Zr^{4+} + 4Cl^{1-} \rightarrow ZrCl_4$

Zirconium oxide (ZrO_2) is the most common compound of zirconium found in nature. It has many uses, including the production of heat-resistant fabrics and high-temperature electrodes and tools, as well as in the treatment of skin diseases. The mineral baddeleyite (known as zirconia or ZrO_2) is the natural form of zirconium oxide and is used to produce metallic zirconium by the use of the Kroll process. The Kroll process is used to produce titanium metal as well as zirconium. The metals, in the form of metallic tetrachlorides, are reduced with magnesium metal and then heated to "red-hot" under normal pressure in the presence of a blanket of inert gas such as helium or argon.

Zirconium carbide (ZrC) is used for light bulb filaments, for **cladding** metals to protect them from corrosion, in making adhesives, and as a high-temperature lining for refractory furnaces.

Zirconium sulfate [$Zr(SO_4)_2$] is an ingredient in lubricants that do not disintegrate at high-temperatures. It is also used for tanning leather to make it white and as a **chemical reagent** and catalyst in chemical laboratories.

Zirconium-95 is the most important of the artificial radioactive isotopes of zirconium. It is placed in pipelines to trace the flow of oil and other fluids as they flow through the pipes. It is also used as a catalyst in petroleum-**cracking** plants that produce petroleum products from crude oil.

Zirconium carbonate ($3ZrO_2 \bullet CO_2 \bullet H_2O$), when used as an additive to lotion, is an effective treatment for skin exposed to poison ivy.

Zirconium silicate ($ZrSiO_4$) is one form of the mineral whose crystals when polished are known as cubic zircons, which resemble diamond gemstones.

Hazards

There is disagreement relative to the dangers of the elemental form of zirconium. Some say that the metal and gemstone forms are harmless, but there is some evidence that the vapors and powder forms of the metal may be carcinogenic. Also, several zirconium compounds can produce allergic reactions in humans and have proven to be toxic to the skin or lungs if inhaled.

The fine powder and dust of zirconium are explosive, especially in the presence of nonmetals that oxidize these forms of zirconium.

NIOBIUM

SYMBOL: Nb **PERIOD:** 5 **GROUP:** 5 (VB) **ATOMIC NO:** 41
ATOMIC MASS: 92.906 amu **VALENCE:** 3 and 5 **OXIDATION STATE:** +3 and +5 (also +2 and +3 as oxides) **NATURAL STATE:** Solid

ORIGIN OF NAME: Niobium is named after the Greek mythological figure Niobe who was the daughter of Tantalus. Tantalus was a Greek god whose name is the source of the word "tantalize," which implies torture: he cut up his son to make soup for other gods.

ISOTOPES: There are 49 isotopes of niobium, ranging from Nb-81 to Nb-113. All are radioactive and made artificially except niobium-93, which is stable and makes up all of the element's natural existence in the Earth's crust.

ELECTRON CONFIGURATION

Energy Levels/Shells/Electrons	Orbitals/Electrons
1-K = 2	s2
2-L = 8	s2, p6
3-M = 18	s2, p6, d10
4-N = 12	s2, p6, d4
5-O = 1	s1

Properties

Niobium is a soft grayish-silvery metal that resembles fresh-cut steel. It is usually found in minerals with other related metals. It neither **tarnishes** nor oxidizes in air at room temperature because of a thin coating of niobium oxide. It does readily oxidize at high temperatures (above 200°C), particularly with oxygen and halogens (group 17). When alloyed with tin and aluminum, niobium has the property of superconductivity at 9.25 Kelvin degrees.

Its melting point is 2,468°C, its boiling point is 4,742°C, and its density is 8.57 g/cm^3.

Characteristics

Some of niobium's characteristics and properties resemble several other neighboring elements on the periodic table, making them, as well as niobium, difficult to identify. This is particularly true for tantalum, which is located just below niobium on the periodic table.

Niobium is not attacked by cold acids but is very reactive with several hot acids such as hydrochloric, sulfuric, nitric, and phosphoric acids. It is **ductile** (can be drawn into wires through a die) and malleable, which means it can be worked into different forms.

Abundance and Source

Niobium is the 33rd most abundant element in the Earth's crust and is considered rare. It does not exist as a free elemental metal in nature. Rather, it is found primarily in several mineral ores known as columbite (Fe, Mn, Mg, and Nb with Ta) and pyrochlore [(Ca, Na)$_2$Nb$_2$O$_6$(O, OH, F)]. These ores are found in Canada and Brazil. Niobium and tantalum [(Fe, Mn)(Ta, Nb)$_2$O$_6$] are also products from tin mines in Malaysia and Nigeria. Niobium is a chemical "cousin" of tantalum and was originally purified by its separation through the process known as fractional crystallization (separation is accomplished as a result of the different rates at which some elements crystallize) or by being dissolved in special **solvents.** Today most of the niobium metal is obtained from columbite and pyrochlore through a complicated refining process that ends with the production of niobium metal by electrolysis of molten niobium potassium fluoride (K_2NbF_7).

History

Niobium has a rather confusing history, starting in 1734 when the first governor of Connecticut, John Winthrop the Younger (1681–1747), discovered a new mineral in the iron mines of the New England. He named this new mineral "columbite." Although he did not know what elements the mineral contained, he believed it contained a new and as yet unidentified element. Hence, he sent a sample to the British Museum in London for analysis. It seems that the delivery was mislaid and forgotten for many years until Charles Hatchett (1765–1847) found the old sample and determined that, indeed, a new element was present. Hatchett was unable to isolate this new element that he named columbium, which was derived from the name of Winthrop's mineral.

The story became more complicated when in 1809 the English scientist William Hyde Wollaston (1766–1828) analyzed the sample mineral and declared that columbium was really the same element as tantalum ($_{73}$Ta). This error is understandable given that the level of analytical equipment available to scientists in those days was fairly primitive. Also, tantalum and niobium are very similar metals that are usually found together and thus are difficult to separate for analysis.

However, the story does not end there. It was not until 1844 when Heinrich Rose (1795–1864) "rediscovered" the element by producing two similar acids from the mineral: niobic acid and pelopic acid. Rose did not realize he had discovered the old "columbium," so he gave this "new" element the name niobium. Twenty years later, Jean Charles Galissard de Marignac (1817–1894) proved that niobium and tantalum were two distinct elements. Later, the Swedish scientist Christian Wilhelm Blomstrand (1826–1899) isolated and identified the metal niobium from its similar "twin," tantalum.

Common Uses

Refined niobium metal is most useful as an alloy with other metals. It is used to produce special stainless steel alloys, to make high-temperature magnets, as special metals for rockets and missiles, and for high- and low-temperature–resistant ceramics. Stainless steel that has been combined with niobium is less likely to break down under very high temperatures. This physical attribute is ideal for construction of both land- and sea-based nuclear reactors.

Niobium has special **cryogenic** properties. It can withstand very cold temperatures, which improves its ability to conduct electricity. This characteristic makes it an excellent metal for low-temperature electrical **superconductors.**

Niobium alloyed with germanium becomes a superconductor of electricity that does not lose its superconductivity at 23.2° Kelvin as large amounts of electrical current are passed through it, as do some other superconductive alloys. In the pure metallic state, niobium wires are also superconductors when the temperatures are reduced to near absolute zero (–273°C). Niobium alloys are also used to make superconductive magnets as well as jewelry.

Examples of Compounds

Niobium in its +5 oxidation state forms both oxygen and halogen compounds (Niobium in oxidation states of +2, +3 and +4 also forms compounds—for example, niobium(II) dioxide and niobium(IV) tetraoxide):

Niobium (V) pentoxide: $2Nb + 5O_2 \rightarrow 2Nb_2O_5$.
Niobium (V) pentachloride: $Nb + 5Cl \rightarrow NbCl_5$.
Nibium (V) pentafluoride: $Nb + 5F \rightarrow NbF_5$.

Niobium carbide (NbC) is used to make hard-tipped tools and special steels and to coat graphite in nuclear reactors.

Niobium silicide (NbSi$_2$) is used as a lining for high-temperature refractory furnaces.

Niobium-uranium alloy has a high **tensile** strength, making it ideal in the manufacture of fuel rods for **nuclear reactors** that resist separation.

Niobium alloys are components of experimental supermagnets that are being tested to "drive" super-fast forms of ground transportation.

Hazards

Niobium is not considered reactive at normal room temperatures. However, it is toxic in its physical forms as dust, powder, shavings, and vapors, and it is carcinogenic if inhaled or ingested.

MOLYBDENUM

SYMBOL: Mo **PERIOD:** 5 **GROUP:** 6 (VIB) **ATOMIC NO:** 42
ATOMIC MASS: 95.94 amu **VALENCE:** 6 **OXIDATION STATE:** +6 (also lower states of +2, +3, +4, and +5) **NATURAL STATE:** Solid
ORIGIN OF NAME: Molybdenum is derived from the Greek word *molybdos,* meaning lead. At one time, the mineral molybdaena (later called molybdenite) was believed to be a variety of lead ore.
ISOTOPES: There are 36 isotopes of molybdenum, ranging in atomic weights from Mo-83 to Mo-115. Of the seven isotopes considered stable, one (Mo-100) is radioactive and is considered stable because it has such a long half-life ($0.95 \times 10^{+19}$ years). The proportions of the seven stable isotopes contributing to molybdenum's natural existence on Earth are as follows: Mo-92 = 14.84%, Mo-94 = 9.25%, Mo-95 = 15.92%, Mo-96 = 16.68%, Mo-97 = 9.55%, Mo-98 = 24.13%, and Mo-100 = 9.63%.

ELECTRON CONFIGURATION

Energy Levels/Shells/Electrons	Orbitals/Electrons
1-K = 2	s2
2-L = 8	s2, p6
3-M = 18	s2, p6, d10
4-N = 13	s2, p6, d5
5-O = 1	s1

Properties

Molybdenum is in the middle of the triad elements of group 6. These three metals (from periods 4, 5, and 6) are chromium, molybdenum, and tungsten, which, in their pure states, are relatively hard, but not as hard as iron. They are silvery-white as pure metals, and they have similar oxidation states. Their electronegativity is also similar—Cr = 1.6, Mo = 1.8, and W = 1.7—which is related to their reactivity with nonmetals.

Molybdenum is malleable and ductile, but because of its relatively high melting point, it is usually formed into shapes by using powder metallurgy and sintering techniques.

Molybdenum's melting point is 2,617°C, boiling point = 4,612°C, and its density is 10.22 g/cm³.

Characteristics

Given that molybdenum is located between chromium and tungsten in group 6, it chemically resembles a cross between these two partner elements. The three related elements do not occur as free elements in nature, but rather are found in minerals and ores. Their metal (elemental) radius size increases from chromium = 44 to molybdenum = 59 to tungsten = 60, which is related to their electronegativity and results in their using electrons in shells inside the outer shell during metallic bonding. This is a major characteristic of the transition of elements from metals to nonmetals.

Molybdenum oxidizes at high temperatures but not at room temperatures. It is **insoluble** in acids and **hydroxides** at room temperatures. At room temperatures, all three metals (chromium, molybdenum, and tungsten) resist atmospheric corrosion, which is one reason chromium is used to plate other metals. They also resist attacks from acids and strong alkalis, with the exception of chromium, which, unless in very pure form, will dissolve in hydrochloric acid (HCl).

Abundance and Source

Molybdenum is the 54th most abundant element on Earth. It is relatively rare and is found in just 126 ppm in the Earth's crust. Its major ore is molybdenite (MoS_2), which is mined in Colorado in the United States and is found too in Canada, Chile, China, England, Norway, Sweden, Mexico, and Australia. Moldybdenum is also found in two less important ores: wulfenite ($PbMoO_4$) and powellite ([$Ca(MoW)O_4$]. These ores are usually found in the same sites along with tin and tungsten ores.

Molybdenite ore is very similar to **graphite,** and they have been mistaken for each other in the past.

History

Peter Jacob Hjelm (1746–1813) is given credit for discovering molybdenum in 1781 despite the fact that his paper was not published until 1890. He followed the advice of Carl Wilhelm Scheele (1742–1786), who isolated and identified molybdenum, but incorrectly thought it was an element related to lead.

Although some reference works do give Scheele credit, most do not credit him for the discovery of either molybdenum or the other elements he "discovered," such as oxygen and manganese.

Scheele did not receive credit for discovering oxygen two years before Joseph Priestley (1733–1804) announced his discovery and was given the credit. Scheele's publisher was negligent in getting his work published in time. (There is a lesson in this story for all young scientists—keep completed and accurate records of all your lab work and observations, and when you are sure of your experimental results, make sure to *publish*.)

The name "molybdenum" is derived from the Greek word for lead, *molybdos,* which stands for any black minerals that historically could be used for writing. This also explains why the Greek word *plumbago* or "black lead" was used for graphite.

Common Uses

The high melting point of molybdenum is the major determinant of how it is used. Its chief use is as an alloy in the manufacture of engines of automobiles. "Moly-steel" contains up to 8% molybdenum and can withstand high pressures and the relatively rapid changes of engine temperatures (e.g. cold engine to hot and back again without the metal warping and with the ability to withstand excessive expansion and contraction).

Its high melting point also makes it useful for metal electrodes in glassmaking furnaces. Molybdenum's high resistance to electricity makes it useful in high-temperature filament wires and in the construction of parts for missiles, spacecrafts, and nuclear power generators.

Molybdenum is also used as a catalyst in petroleum refining, as a pigment for paints and printer's ink, and as a high-temperature lubricant (molybdenum disulphide-MoS_2) for use by spacecraft and high-performance automobiles.

In hospitals, radioisotope Mo-99, which decays into technetium-99, is given internally to cancer patients as a "radioactive cocktail." Radioactive Tc-99 is absorbed by tissues of cancer patients, and then x-ray-like radiation is used to produce pictures of the body's internal organs.

Examples of Compounds

Molybdenum boride (Mo_2B) is used to braze (weld) special metals and for noncorrosive electrical connectors and switches. It is also used to manufacture high-speed cutting tools and noncorrosive, abrasion-resistant parts for machinery.

Molybdenum pentachloride ($MoCl_5$) is used as a **brazing** and **soldering flux** and to make fire-retardant resins.

Molybdenum trioxide (MnO_3) is a compound used to make enamels adhere to metals.

Molybdenum's ions can exhibit lower oxidation states as follows:

Molybdenium(II) chloride ($Mo^{2+} + 2Cl^{1-} \rightarrow MoCl_2$), and

Molybdenium(III) chloride ($Mo^{3+} + 3Cl^{1-} \rightarrow MoCl_3$.

An odd occurrence takes place when molybdenum oxide (MoO_3) is heated in a vacuum along with some powdered molybdenum metal. Reduction occurs, resulting in extreme forms of compounds of molybdenum and oxygen. These odd molecules have formulas such as, $Mo_{17}O_{47}$ and Mo_8O_{23}.

Many different compounds of molybdenum exhibit a variety of vivid colors—from violet to blue to brown.

Hazards

The powder and dust forms of molybdenum are flammable. The fumes from some of the compounds should not be inhaled or ingested.

Only some of the compounds of molybdenum are toxic, particularly in the powder or mist form. Small traces of molybdenum are essential for plant and animal nutrition.

TECHNETIUM

SYMBOL: Tc   **PERIOD:** 5   **GROUP:** 7 (VIIB)   **ATOMIC NO:** 43
ATOMIC MASS: 98.9062 amu   **VALENCE:** 4, 6, and 7   **OXIDATION STATE:** + 4, +6, and +7 (also +2, +3, and +5)   **NATURAL STATE:** Solid
ORIGIN OF NAME: Technetium's name was derived from the Greek word *technetos,* meaning "artificial."
ISOTOPES: There are 47 isotopes. None are stable and all are radioactive. Most are produced artificially in cyclotrons (particle accelerators) and nuclear reactors. The atomic mass of its isotopes ranges from Tc-85 to Tc-118. Most of technetium's radioactive isotopes have very short half-lives. The two natural radioisotopes with the longest half-lives—Tc-98 = $4.2 \times 10^{+6}$ years and Tc-99 = $2.111 \times 10^{+5}$ years—are used to establish technetium's atomic weight.

ELECTRON CONFIGURATION

Energy Levels/Shells/Electrons	Orbital/Electrons
1-K = 2	s2
2-L = 8	s2, p6
3-M = 18	s2, p6, d10
4-N = 13	s2, p6, d5
5-O = 2	s2

Properties

As the central member of the triad of metals in group 7, technetium (period 5) has similar physical and chemical properties as its partners manganese (period 4) above it and rhenium (period 6) below it. The sizes of their atomic radii do not vary greatly: Mn = 127, Tc = 136, and Re = 137. Neither does their level of electronegativity vary significantly: Mn = 1.5, Tc = 1.9, and Re = 1.9.

Technetium metal is grayish-silver and looks much like platinum. As with most transition elements, technetium in pure form is a noncorrosive metal. It requires only 55 ppm of technetium added to iron to transform the iron into a noncorroding alloy. Because of technetium's radioactivity, its use as an alloy metal for iron is limited so as to not expose humans to unnecessary radiation.

Technetium's melting point is 2,172°C, its boiling point is 4,877°C, and its density is 11.50 g/cm³.

Characteristics

Technetium was the first element, not found on Earth, to be artificially produced by bombarding molybdenum with **deuterons.**

The major characteristic of technetium is that it is the only element within the 29 transition metal-to-nonmetal elements that is artificially produced as a uranium-fission product in nuclear power plants. It is also the lightest (in atomic weight) of all elements with no stable isotopes. Since all of technetium's isotopes emit harmful radiation, they are stored for some time before being processed by solvent extraction and ion-exchange techniques. The two long-lived radioactive isotopes, Tc-98 and Tc-99, are relatively safe to handle in a well-equipped laboratory.

Since all of technetium's isotopes are produced artificially, the element's atomic weight (atomic mass units) is determined by which isotopes are selected for the calculation.

Abundance and Source

Technetium is the 76th most abundant element, but it is so rare that it is not found as a stable element on Earth. All of it is artificially produced. Even though natural technetium is so scarce that it is considered not to exist on Earth, it has been identified in the light spectrum from stars. Using a spectroscope that produces unique lines for each element, scientists are able to view several types of stars. The resulting spectrographs indicate that technetium exists in the stars and thus the universe, but not on Earth as a stable element.

It was the first new element to be produced artificially from another element experimentally in a laboratory. Today, all technetium is produced mostly in the nuclear reactors of electrical generation power plants. Molybdenum-98 is bombarded with neutrons, which then becomes molybdenum-99 when it captures a neutron. Since Mo-99 has a short half-life of about 66 hours, it decays into Tc-99 by beta decay.

History

Mendeleev, who developed the periodic table, recognized a gap in group 7 of the transition metals between manganese (Mn-55) and ruthenium (Re-186). Using his system for anticipating unknown elements in his table, he used the word **eka,** which means "first" in Sanskrit, to name missing elements. Thus, element eka-manganese was suggested to fill the atomic number 43 gap based on Mendeleev's speculation that this missing element would have chemical and physical properties similar to manganese, located just above it in group 7. Some other examples of eka elements predicted by their placement on the periodic tables (some were accurate, others just close) were eka-aluminum, eka-gallium, eka-boron, eka-scandium, and eka-silicone. (Note: Mendeleev used atomic weights instead of atomic numbers for the elements in his original periodic table.)

This system gave several scientists clues as to what to look for, but because there are no stable (nonradioactive) atoms of element 43 (eka-manganese) on Earth, they had to find new techniques for identifying element 43. Many scientists claimed to have discovered an element with atomic number 43 and even gave it names such as *davyum, illmenium, lucium,* and *nipponium.* None proved to be the correct element. About this time, it was known that Enrico Fermi (1901–1954) had changed one element to another by bombarding the element's nuclei with deuterons, the atomic nuclei of heavy hydrogen (^{2}H), which have 1 proton and 1 neutron. This was an example of artificial transmutation of one element to another element, a technique long sought by ancient alchemists who unsuccessfully tried to turn lead into gold.

Technetium was discovered in platinum ore shipped from Columbia through X-ray spectroscopy by Walter Noddack and Ida Tacke in Berlin. In 1937 Emilio Gino Segre (1905–1989) and Carlo Perrier (dates unknown) knew about Fermi's work and decided that if technetium did not exist on Earth, they could make it using Fermi's technique. They bombarded molybdenum ($_{42}$Mo) with deuterons in a cyclotron, which added a proton to each of molybdenum's nuclei, and thus created technetium ($_{43}$Tc). It worked. Even though their sample was extremely small (10^{-1}gram), it was enough to verify that a new element had been synthesized in the laboratory. They were given the privilege of naming the first artificially produced element, just as other scientists had had for naming naturally occurring elements they had discovered. In 1939 they named it "technetium," from the Greek word for "artificial."

Common Uses

Technetium is one of the few artificially produced elements that has practical industrial applications. One is that a very small amount (55-ppm) added to iron creates a corrosion-resistant alloy metal. This property is shared with many of the other transition metallic elements, but not with other artificially produced elements that have higher atomic numbers and are radioactive.

A radioisotope of technetium is widely used in nuclear medicine. The patient is injected with saline solution containing Tc-99^m (the superscript "m" means that the isotope is unstable and that its nuclei holds more energy than the regular Tc-99 nuclei into which it decays). This means that the Tc-99^m will start to emit energy and will finally decay and change to the regular nuclei of Tc-99 when injected into the patient. This energy is in the form of very penetrating gamma rays (a strong type of X-rays). The radioactive solution of Tc-99^m may be combined with other elements that are absorbed by certain organs of the human body being diagnosed or treated. For instance, adding tin to the solution targets the red blood cells, whereas phosphorus in the solution concentrates the radioactive solution in heat muscles. The gamma rays are strong enough to expose an X-ray film that depicts the internal image of the organ under examination. This procedure is safe because Tc-99^m has a half-life of only 6.015 hours, and the Tc-99 has a half-life of over 200,000 years. However, the radioactivity will be harmless in less than a day because the body rapidly eliminates the residual radioactive solution.

Technetium is also used as an alloy metal to produce super-strong magnets that are super-cooled to near absolute zero to improve their efficiency. Powerful magnets are used in imaging equipment and possibly in future magnetic driven trains. Its radioactivity makes it useful as a tracer in the production of metals and tracing flowing fluids in pipelines.

Examples of Compounds

Technetium-99^m is produced in commercial quantities in nuclear reactors by bombarding molybdenum with large numbers of neutrons. A simplified version of the radioactive decay reaction follows:

$$^{99}MoO_4^{2-} \rightarrow \beta \rightarrow {}^{99m}TcO_4^{1-} \rightarrow \gamma \rightarrow {}^{99}TcO_4^{1-}$$

It might be noted that the oxidation states for technetium can easily change from one to the other. The oxidation states of Tc(III), Tc(IV), and Tc(V) for various compounds can be adjusted to target different organs when used for medical diagnostics and treatments.

Only a few compounds of technetium have been made. Some examples follow:

Technetium(IV) dioxide: $Tc^{4+} + 2O^{2-} \rightarrow TcO_2$.

Technetium(VI) chloride: $Tc^{6+} + 6Cl^{1-} \rightarrow TcCl_6$.

Tectnetium(VII) septoxide: $2Tc^{7+} + 7O^{2-} \rightarrow Tc_2O_7$.

It might be noted that there are several forms of technetium oxides. Their formula depends on the oxidation state of Tc ions; an example is NH_4TcO_4.

Hazards

The hazards of technetium are the same as for all radioactive elements. Excessive exposure to radiation can cause many kinds of tissue damage—from sunburn to radiation poisoning to death.

RUTHENIUM

SYMBOL: Ru **PERIOD:** 5 **GROUP:** 8 (VIII) **ATOMIC NO:** 44

ATOMIC MASS: 101.07 amu **VALENCE:** 3 **OXIDATION STATE:** +3 (also +4, +5, +6, and +8) **NATURAL STATE:** Solid

ORIGIN OF NAME: "Ruthenium" is derived from the Latin word *Ruthenia* meaning "Russia," where it is found in the Ural Mountains.

ISOTOPES: There are 37 isotopes for ruthenium, ranging in atomic mass numbers from 87 to 120. Seven of these are stable isotopes. The atomic masses and percentage of contribution to the natural occurrence of the element on Earth are as follows: Ru-96 = 5.54%, Ru-98 = 1.87%, Ru-99 = 12.76%, Ru-100 = 12.60%, Ru-101 = 17.06%, Ru-102 = 31.55%, and Ru-104 = 18.62%.

ELECTRON CONFIGURATION

Energy Levels/Shells/Electrons	Orbitals/Electrons
1-K = 2	s2
2-L = 8	s2, p6
3-M = 18	s2, p6, d10
4-N = 15	s2, p6, d7
5-O = 1	s1

Properties

Ruthenium is a rare, hard, silvery-white metallic element located in group 8, just above osmium and below iron, with which it shares some chemical and physical properties. Both ruthenium and osmium are heavier and harder than pure iron, making them more brittle and difficult to refine. Both ruthenium and osmium are less **tractable** and malleable than iron. Although there are some similar characteristics between ruthenium and iron, ruthenium's properties are more like those of osmium. Even so, ruthenium is less stable than osmium. They are both rare and difficult to separate from minerals and ores that contain other elements. These factors make it more difficult to determine ruthenium's accurate atomic weight.

The oxidation state of +8 for ruthenium and its "mate" osmium is the highest oxidation state of all elements in the transition series. Ruthenium's melting point is 2,310°C, its boiling point is 3,900°C, and its density is 12.45 g/cm^3.

Characteristics

Ruthenium also belongs to the **platinum group**, which includes six elements with similar chemical characteristics. They are located in the middle of the second and third series of the transition elements (groups 8, 9, and 10). The platinum group consists of ruthenium, rhodium, palladium, osmium, iridium, and platinum.

Ruthenium is a hard brittle metal that resists corrosion from all acids but is vulnerable to strong alkalis (bases). Small amounts, when alloyed with other metals, will prevent corrosion of that metal.

Abundance and Source

Ruthenium is a rare element that makes up about 0.01 ppm in the Earth's crust. Even so, it is considered the 74th most abundant element found on Earth. It is usually found in amounts up to 2% in platinum ores and is recovered when the ore is refined. It is difficult to separate from the leftover residue of refined platinum ore.

Ruthenium is found in South America and the Ural Mountains of Russia. There are some minor platinum and ruthenium ores found in the western United States and Canada. All of the radioactive isotopes of ruthenium are produced in nuclear reactors.

History

There is a long and mixed history for the claims of discovery for ruthenium. In 1748 Antonio de Ulloa (1716–1795), a Spanish scientist and explorer, reported finding a special metal in South America. It was silvery-gray and denser than gold, but it did not have the attractive luster of gold or silver. He did know that he had located a new element along with the platinum metal.

In 1807 Jedrzej Sniadecki (1768–1807), a Polish scientist, was the first to isolate the new element and assign it the atomic number 44. He named his new element "Vesta" after a large asteroid and wrote several papers describing his discovery. The Paris Commission, the scientific society of the day, ignored Sniadecki's work because he was not famous and the commission could not replicate his work. Rather than argue with the commission, Sniadecki dropped his claim. Regardless, some sources give Sniadecki credit for the discovery of ruthenium.

In 1827 or 1828 Gottfried Wilhelm Osann (1796–1866), a well-known German scientist, found in the Ural mountains of Russia what he claimed were several new elements in platinum ores, including ruthenium. However, after his announcement, nothing more happened and he did not withdraw his claim, as did Sniadecki. Some sources now give Osann credit for discovering ruthenium.

It was not until 1844 that Karl Karlovich Klaus (1796–1864), a well-known Russian (Estonian) scientist of the day, separated enough ruthenium from platinum to be able to correctly identify its properties. He is the person most sources credit as the "discoverer" of $_{44}$Ru.

If Sniadecki had not withdrawn his claim, he may have, eventually, been credited with the discovery of ruthenium. However, as history stands, Osann is credited with first finding ruthenium and Klaus for adequately identifying its properties to determine that it fit into the "hole" in the periodic table for element 44.

Common Uses

Since ruthenium is rare and difficult to isolate in pure form, there are few uses for it. Its main uses are as an alloy to produce noncorrosive steel and as an additive to jewelry metals such as platinum, palladium, and gold, making them more durable.

It is also used as an alloy to make electrical contacts harder and wear longer, for medical instruments, and more recently, as an experimental metal for direct conversion of solar cell material to electrical energy.

Ruthenium is used as a catalyst to affect the speed of chemical reactions, but is not altered by the chemical process. It is also used as a drug to treat eye diseases.

Examples of Compounds

The most common oxidation state for ruthenium is +3 as the metal ion Ru^{3+}:
Ruthenium(III) trichloride: $Ru^{3+} + 3Cl \rightarrow RuCl_3$.
This compound is used for technical analysis in chemistry laboratories. It is highly toxic.
Ruthenium(III) hydroxide: $Ru^{3+} + 3(HO)^{1-} \rightarrow Ru(OH)_3$.
Examples of oxides of ruthenium with higher oxidation states follow:
Ruthenium(IV) oxide: $Ru^{4+} + 2O^{2-} \rightarrow RuO_2$.
Ruthenium(VIII) oxide: $Ru^{8+} + 4O^{2-} \rightarrow RuO_4$.

Hazards

The main hazard is the explosiveness of ruthenium fine power or dust. The metal will rapidly oxidize (explode) when exposed to oxidizer-type chemicals such as potassium chloride at room temperature. Most of its few compounds are toxic and their fumes should be avoided.

RHODIUM

SYMBOL: Rh   **PERIOD:** 5   **GROUP:** 9 (VIII)   **ATOMIC NO:** 45
ATOMIC MASS: 102.906 amu   **VALENCE:** 3   **OXIDATION STATE:** +3 (also +4 in compounds)   **NATURAL STATE:** Solid
ORIGIN OF NAME: Named after the Greek word *rhodon,* which means "rose," because of the reddish color of its salt compounds.
ISOTOPES: There are 52 isotopes of rhodium, ranging from Rh-89 to Rh-122. All are produced artificially with relatively short half-lives except one stable isotope, Rh-103, which constitutes 100% of the element's existence in the Earth's crust.

ELECTRON CONFIGURATION

Energy Levels/Shells/Electrons	Orbitals/Electrons
1-K = 2	s2
2-L = 8	s2, p6
3-M = 18	s2, p6, d10
4-N = 16	s2, p6, d8
5-O = 1	s1

Properties

Rhodium is a hard shiny-white metal that resists corrosion from oxygen, moisture, and acids at room temperatures. As a member of group 8 (VIII), $_{45}$Rh shares many chemical and physical properties with cobalt ($_{27}$Co) just above it and iridium ($_{77}$Ir) below it in the vertical group. Therefore, it is considered one of the elements that are transitory between metals and nonmetals. It is rare and only found in combination with platinum ores.

Rhodium's melting point is 1,966°C, its boiling point is 3,727°C, and its density is 12.41 g/cm^3.

Characteristics

Rhodium is one of the six platinum transition elements that include Ru, Rh, Pd, Os, Ir, and Pt. Of these metals, rhodium has the highest electrical and thermal conductivity. Although a relatively scarce metal, rhodium makes an excellent electroplated surface that is hard, wears well, and is permanently bright—ideal for plating the reflectors in automobile headlights.

Abundance and Source

Rhodium is rare, but not as rare as ruthenium. It makes up only 1 part in 20 million of the elements found in the Earth's crust. Even so, it is considered the 79th most abundant element and is found mixed with platinum ore, and to a lesser extent, it is found with copper and nickel ores. It is found in Siberia, South Africa, and Ontario, Canada.

Rhodium is recovered from platinum and other ores by refining and purification processes that start by dissolving the other platinum group metals and related impurities with strong acids that do not affect the rhodium itself. Any remaining platinum group elements are removed by oxidation and bathing the mixture in chlorine and ammonia.

Rhodium is usually produced as a powder and can be formed by either casting or powder metallurgy.

History

William Hyde Wollaston (1766–1828), who had also discovered palladium ($_{46}$Pd) in the early 1800s, announced in 1803 his discovery of another metal that he had isolated from platinum ores found in South America. His procedures were rather complicated but resulted in the discovery of several platinum-related metals. He used these newly discovered metals, rhodium and iridium, to alloy with platinum to make improved laboratory vessels that were harder and more durable than pure platinum as well as noncorrosive. Wollaston derived great wealth from this refining process, the details of which were not disclosed until his death in 1828.

Common Uses

Rhodium is commercially used as an alloy metal with other metals to form durable high-temperature electrical equipment, **thermocouples,** electrical contacts and switches, and laboratory **crucibles.**

Because of its high reflectivity, it is used to electroplate jewelry, silverware, optical instruments, mirrors, and reflectors in lighting devices.

When rhodium is combined with platinum and palladium, the elements together form the internal metals of automobile **catalytic converters,** which convert hot unburned hydrocarbon exhaust gases to less harmful CO_2 and H_2O. Similar alloys are used to manufacture high-temperature products such as electric coils for metal refining furnaces and high-temperature spark plugs.

Examples of Compounds

The most common oxidation state of rhodium is +3, but it can also exhibit ions of +2 and +4 in certain compounds.

Rhodium(III) trichloride: $Rh^{3+} + 3Cl^{1-} \rightarrow RhCl_3.$

Rhodium(III) oxide: $Rh^{3+} + 3O^{2-} \rightarrow Rh_2O_3.$

Rhodium(IV) oxide: $Rh^{4+} + 2O^{2-} \rightarrow RhO_2.$

Several other complicated compounds exist as well:

Sodium rhodium chloride ($Na_3RhCl_6 \bullet 12H_2O$) is the hydrate form of rhodium that was first produced by Wollaston. A similar compound is formed as *potassium rhodium sulfate* ($K_3Rh(SO_4)_3$, which is a rose-colored crystal.

Of some interest are the different colored crystals resulting from compounding rhodium with the halogens: RhF_3 = red, $RhCl_3$ = red, $RhBr_3$ = brown, and ReI_3 = black.

Hazards

The powder and dust of rhodium metal are flammable in air. Some of the compounds may cause skin irritations. It is best to use approved laboratory procedures when handling any of the six elements in the platinum family of metals.

PALLADIUM

SYYMBOL: Pd   **PERIOD:** 5   **GROUP:** 10 (VIII)   **ATOMIC NO:** 46
ATOMIC MASS: 106.42 amu   **VALENCE:** 2, 3, and 4   **OXIDATION STATE:** +2, +3, and +4   **NATURAL STATE:** Solid
ORIGIN OF NAME: Palladium is named after the asteroid Pallas, which was discovered at about the same time as the element. Pallas is the name of two mythological Greek figures, one male and the other female.

ISOTOPES: There are 42 isotopes of palladium, ranging from Pd-91 to Pd-124. All but six are radioactive and artificially produced in nuclear reactors with half-lives ranging from 159 nanoseconds to $6.5 \times 10^{+6}$ years. The six stable isotopes of palladium and their proportional contribution to their existence in the Earth's crust are as follows: Pd-102 = 1.02%, Pd-104 = 11.14%, Pd-105 = 22.23%, Pd-106 = 27.33%, Pd-108 = 26.46%, and Pd-110 = 11.72%.

ELECTRON CONFIGURATION

Energy Levels/Shells/Electrons	Orbitals/Electrons
1-K = 2	s2
2-L = 8	s2, p6
3-M = 18	s2, p6, d10
4-N = 18	s2, p6, d10
5-O = 0	s0

Properties

Palladium is the middle element in group 10 of the transition elements (periods 4, 5, and 6). Many of its properties are similar to nickel located above it and platinum just below it in this group.

Palladium is a soft, silvery-white metal whose chemical and physical properties closely resemble platinum. It is mostly found with deposits of other metals. It is malleable and ductile, which means it can be worked into thin sheets and drawn through a die to form very thin wires. It does not corrode. Its melting point is 1,554°C, its boiling point is 3,140°C, and its density is 12.02 g/cm³.

Characteristics

One of palladium's unique characteristics is its ability to absorb 900 times its own volume of hydrogen gas. When the surface of the pure metal is exposed to hydrogen gas (H_2), the gas molecules break into atomic hydrogen. These hydrogen atoms then seep into the holes in the crystal structure of the metal. The result is a metallic hydride ($PdH_{0.5}$) that changes palladium from an electrical conductor to a semiconductor. The compound palladium dichloride ($PdCl_2$) also has the ability to absorb large quantities of carbon monoxide (CO). These characteristics are useful for many commercial applications. Palladium is the most reactive of all the platinum family of elements (Ru, Rh, Pd, Os, Is, and Pt.)

Abundance and Source

Palladium is considered a rare metal, making up only about 1 part per 100 million parts of the Earth's crust. It is considered the 77th most abundant element on Earth, although it is seldom found in pure states. Rather, it is mixed with other metals or in compounds of palladium.

It was originally found in gold ores from Brazil, where the miners thought the gold was contaminated by what they referred to as "white gold." Later, it was considered an alloy combination of palladium and gold.

Deposits of ores containing palladium, as well as other metals, are found in Siberia and the Ural Mountains of Russia, Canada, and South Africa, as well as in South America.

History

Several scientists in the early 1800s were aware that there were other elements mixed with platinum and other ores (i.e., nickel, copper, silver, and gold). They had difficulty separating the metals because these elements (metals) were so similar in their physical and chemical characteristics.

In 1803 William Hyde Wollaston (1776–1828), an English chemist who also discovered rhodium, isolated palladium at the time he analyzed the platinum and gold ores sent to him from Brazil. Dissolving the platinum in **aqua regia** acid, Wollaston then treated the residue with mercuric cyanide to produce the compound of palladious cyanide that was reduced by burning it to extract metallic palladium.

Common Uses

Palladium's ability to absorb large amounts of hydrogen makes it an excellent catalyst for chemical reactions as well as catalytic converters for internal combustion engines. Palladium is also an excellent catalyst for **cracking** petroleum fractions and for hydrogenation of liquid vegetable oils into solid forms, such as corn oil into margarine. It is also used to purify hydrogen gas by passing raw H_2 gas under pressure through thin sheets of palladium, where the pure hydrogen passes through the metal's crystal structure, leaving behind impurities. Palladium is used to manufacture CO-monitoring devices because of its ability to absorb carbon monoxide.

Palladium is used in the manufacture of surgical instruments, electrical contacts, springs for watches and clocks, high-quality spark plugs, and special wires and as "white gold" in jewelry. Because it is noncorrosive, it is used as a coating for other metals and to make dental fillings and crowns.

Examples of Compounds

The most common oxidation states of palladium are +2 and +4. Some examples follow:

Palladium(II) oxide: $Pd^{2+} + O^{2-} \rightarrow PdO$.

Palladium(IV) dioxide: $Pd^{4+} + 2O^{2-} \rightarrow PdO_2$.

When combining with the halogens, the oxidation states of +2 or +4 for palladium are used.

Palladium(IV) fluoride: $Pd^{4+} + F^{1-} \rightarrow PdF_4$, a brick-red color.

Palladium(II) chloride: $Pd^{2+} + Cl^{1-} \rightarrow PdCl_2$. This compound is a dark brown color and is used to coat other metals without the need for electrolysis. It is also used in photography, to make indelible inks, and as a catalyst in analytical chemistry (used to speed up or slow down chemical reactions).

Palladium(II) bromide: $Pd^{2+} + Br^{1-} \rightarrow PdBr_2$, brownish-black color.

Palladium(II) iodide: $Pd^{2+} + I^{1-} \rightarrow PdI_2$, black in color.

Palladium sodium chloride: $NaPdCl_2$. This compound is used to test for the presence of such gases as carbon monoxide, illuminating and cooking gas, and ethylene and for the presence of iodine.

Hazards

Palladium is not combustible except as fine powder or dust. Several of palladium's compounds are oxidizing agents, and some react violently with organic substances.

SILVER

SYMBOL: Ag   **PERIOD:** 5   **GROUP:** 11 (1B)   **ATOMIC NO:** 47

ATOMIC MASS: 107.868 amu   **VALENCE:** 1   **OXIDATION STATE:** +1   **NATURAL STATE:** Solid

ORIGIN OF NAME: Silver's modern chemical symbol (Ag) is derived from its Latin word *argentum,* which means silver. The word "silver" is from the Anglo-Saxon world "siolfor." Ancients who first refined and worked with silver used the symbol of a crescent moon to represent the metal.

ISOTOPES: There are 59 isotopes of silver, ranging from Ag-93 to Ag-130 with half-lives from a few milliseconds to a few days to 418 years. All but two of these 59 isotopes are radioactive and are produced synthetically. The two stable isotopes found in nature are Ag-107 and Ag-109. These two make up 100% of the element's existence in the Earth's crust.

ELECTRON CONFIGURATION

Energy Levels/Shells/Electrons	Orbitals/Electrons
1-K = 2	s2
2-L = 8	s2, p6
3-M = 18	s2, p6, d10
4-N = 18	s2, p6, d10
5-O = 1	s1

Properties

Silver is located in group 11 (IB) of period 5, between copper (Cu) above it in period 4 and gold (Au) below it in period 6. Thus, silver's chemical and physical properties are somewhat similar to these two group 11 partners.

Silver is a soft, while, lustrous metal that can be worked by pounding, drawing through a die, rolling, and so forth. It is only slightly harder than gold. It is insoluble in water, but it will dissolve in hot concentrated acids. Freshly exposed silver has a mirror-like shine that slowly darkens as a thin coat of tarnish forms on its surface (from the small amount of natural hydrogen sulfide in the air to form silver sulfide, AgS). Of all the metals, silver is the best conductor of heat and electricity. This property determines much of its commercial usefulness. Its melting point is 961.93°C, its boiling point is 2,212°C, and its density is 10.50 g/cm³.

Characteristics

Silver is somewhat rare and is considered a commercially precious metal with many uses. Pure silver is too soft and usually too expensive for many commercial uses, and thus it is alloyed with other metals, usually copper, making it not only stronger but also less expensive. The purity of silver is expressed in the term "fitness," which describes the amount of silver in the item. Fitness is just a multiple of 10 times the silver content in an item. For instance, sterling silver should be 93% (or at least 92.5%) pure silver and 7% copper or some other metal. The fitness rating for pure silver is 1000. Therefore, the rating for sterling silver is 930, and most sliver jewelry is rated at about 800. This is another way of saying that most silver jewelry is about 20% copper or other less valuable metal.

Many people are fooled when they buy Mexican or German silver jewelry, thinking they are purchasing a semiprecious metal. These forms of "silver" jewelry go under many names, including Mexican silver, German silver, Afghan silver, Austrian silver, Brazilian silver, Nevada silver, Sonara silver, Tyrol silver, Venetian silver, or just the name "silver" with quotes around it. None of these jewelry items, under these names or under any other names, contain any silver. These metals are alloys of copper, nickel, and zinc.

Abundance and Source

Silver is the 66th most abundant element on the Earth, which means it is found at about 0.05 ppm in the Earth's crust. Mining silver requires the movement of many tons of ore to recover small amounts of the metal. Nevertheless, silver is 10 times more abundant than gold. And though silver is sometimes found as a free metal in nature, mostly it is mixed with the ores of other metals. When found pure, it is referred to as "native silver." Silver's major ores are argentite (silver sulfide, Ag_2S) and horn silver (silver chloride, $AgCl$). However, most silver is recovered as a by-product of the refining of copper, lead, gold, and zinc ores. Although silver is mined in many countries, including the United States, Mexico, and Canada, most silver is recovered from the electrolytic processing of copper ores. Silver can also be recovered through the chemical treatment of a variety of ores.

History

Silver is probably one of the first metallic elements used by humans. It was known to primitive humans before 5000 BCE, at about the time copper and gold were also found in native "free" forms. Silver jewelry was found in tombs that are over 6,000 years old.

It was not until about 4000 BCE that humans learned how to obtain these base metals from ores by using heat. Both the refining of and the uses for silver as jewelry are described in ancient Egyptian writings as well as the Old Testament of the Bible. During these ancient times, silver was more valuable than gold primarily because it was more difficult to find in its natural state as well as to extract from its ores.

Common Uses

During the Middle Ages and even during the period of the early settlement of North America, silversmiths were considered important craftsmen in the community. They created many items such as dinner flatware, vases, pouring vessels, and eating utensils, as well as jewelry required by well-to-do households.

Silver has a multitude of uses and practical applications both in its elemental metallic form and as a part of its many compounds. Its excellent electrical conductivity makes it ideal for use in electronic products, such a computer components and high-quality electronic equipment. It would be an ideal metal for forming the wiring in homes and transmission lines, if it were more abundant and less expensive.

Metallic silver has been used for centuries as a coinage metal in many countries. The amount of silver now used to make coins in the United States has been reduced drastically by alloying other metals such as copper, zinc, and nickel with silver.

In the mid-1800s, Thomas Wilberger Evans (1823–1897) introduced the practice of using an amalgam of silver, which is a solution-type alloy (mixture) of mercury and silver. (Sometimes tin is also added.) In essence, the mercury is used to "cement" or bind the silver so that it can be used as a filling for decayed teeth. Due to the toxic nature of mercury, fillings are now made with mercury-free, nontoxic compounds.

Silver is used as a catalyst to speed up chemical reactions, in water purification, and in special high-performance batteries (cells). Its high reflectivity makes it ideal as a reflective coating for mirrors.

Several of its compounds were not only useful but even essential for the predigital photographic industry. Several of the silver salts, such as silver nitrate, silver bromide, and silver chloride, are sensitive to light and, thus, when mixed with a gel-type coating on photographic film or paper, can be used to form light images. Most of the silver used in the United States is used in photography.

Photochromic (transition) eyeglasses that darken as they are exposed to sunlight have a small amount of silver chloride imbedded in the glass that forms a thin layer of metallic silver that darkens the lens when struck by sunlight. This photosensitive chemical activity is then reversed when the eyeglasses are removed from the light. This chemical reversal results from a small amount of copper ions placed in the glass. This reaction is repeated each time the lenses are exposed to sunlight.

Crystals of silver iodide (AgI), in addition to being useful in photographic processing, are used to "seed" clouds. The atmospheric conditions (humidity and such) must be right for this to work because the tiny crystals act as "nuclei" on which moisture can condense with the expectation that the small droplets will become heavy enough to drop to earth—as rain.

Examples of Compounds

Some examples of silver with its main oxidation state of +1 follow:

Silver(I) oxide: $2Ag^{1+} + O^{2-} \rightarrow Ag_2O$. Catalysts and lab reagent.

Silver(I) chloride: $Ag^{1+} + Cl^{1-} \rightarrow AgCl$. Used in photographic films, to coat and silver glass, as an antiseptic, and to absorb **infrared** light in lenses.

Other commercially valuable compounds of silver are as follows:

Silver nitrate ($AgNO_3$): Photographic emulsions, antiseptic, silver plating, and inks.

Silver nitride (Ag_3N): Sensitive to shock, underwater explosive upon contact with water.

Silver phosphate (Ag_3PO_4): Photo emulsions, pharmaceuticals.

Silver sulfate (Ag_2SO_4): Lab reagent, highly toxic.

Silver peroxide (Ag_2O_2): Used to manufacture silver-zinc cells (batteries).

Argyrol: Trade name for compound of silver and a protein that is used as an antiseptic to treat specific types of bacterial infections.

Hazards

Silver is not toxic as a free, elemental metal, but many of its compounds, particularly its nitrogen compounds, are toxic.

Several silver salts—in particular, $AgNO_3$—are deadly when ingested, even in small amounts. When ingested, the silver compounds are slowly absorbed by the body, and the skin turns bluish or black, a condition referred to as "argyria." In the past, the eyes of newborn babies were swabbed with dilute silver nitrate to prevent blindness from STDs (sexually transmitted diseases, in particular gonorrhea). This procedure is no longer performed.

Several silver compounds are extremely explosive—for example, silver picrate, silver nitride, silver peroxide, silver perchlorate, and silver permanganate. They are used in various types of explosives and as concussion caps for rifle and pistol ammunition.

CADMIUM

SYMBOL: Cd   **PERIOD:** 5   **GROUP:** 12 (IIB)   **ATOMIC NO:** 48

ATOMIC MASS: 112.41 amu   **VALENCE:** 2   **OXIDATION STATE:** +2   **NATURAL STATE:** Solid

ORIGIN OF NAME: The word cadmium is from the Latin word *cadmia* or the Greek word kadmeia, meaning the zinc oxide ore "calamine" that contains the element cadmium.

ISOTOPES: There are 52 isotopes of cadmium. Forty-four are radioactive and artificially produced, ranging from Cd-96 to Cd-131. Of these 52 isotopes, there are five stable isotopes plus three naturally occurring radioactive isotopes with extremely long half-lives that are considered as contributing to the element's natural occurrence in the Earth's crust. The three naturally radioactive isotopes (Cd-106, Cd-113, and Cd-116) are the longest known beta emitters. They are two million years older than when the solar system was formed about 4.5 billion years ago. The five stable isotopes and their proportional contributions to the element's existence on Earth are as follows: Cd-108 = 0.89%, Cd-110 = 12.49%, Cd-111= 12.80%, Cd-112 = 24.13%, and Cd-114 = 28.73%.

ELECTRON CONFIGURATION

Energy Levels/Shells/Electrons	Orbitals/Electrons
1-K = 2	s2
2-L = 8	s2, p6
3-M = 18	s2, p6, d10
4-N = 18	s2, p6, d10
5-O = 2	s2

Properties

Cadmium is a soft, blue-white metal that is malleable and ductile although it becomes brittle at about 80°C. It is also found as a grayish-white powder. It is considered rare and is seldom found by itself as an ore. Its melting point at 320.9°C is considered low. Its boiling point is 765°C, and its density is 8.65 g/cm^3. Certain alloys of cadmium have extremely low melting points at about 70°C.

Characteristics

Although cadmium is not considered a transition element in some periodic tables, it is the central element of the triad with zinc and mercury. Zinc is just above it and mercury is below it in group 12 of the periodic table. Cadmium's chemical and physical properties are similar to its group 12 mates. Their electronegativity is very similar: Zn = 1.6, Cd = 1.7, and Hg = 1.9.

Cadmium is resistant to alkalis, but is soluble in acids, mainly nitric acid. Although it is used to electroplate steel to prevent corrosion, it will tarnish in moist air.

Abundance and Source

Cadmium is considered a rare element even though it is widely distributed over the Earth's crust. Its estimated abundance in the Earth's crust is 1.10^{-1} milligrams per kilogram. It is considered the 65th most abundant element, but it does not occur as a free metal in nature. It is usually found in relationship with other metallic ores. Its abundance is only about 1/1000th that of zinc. It is found in an ore called greenockite, which is cadmium sulfite (CdS). This ore does not have a high enough concentration of cadmium to be mined profitably. Cadmium is found along with zinc, lead, and copper ores. Today, most cadmium is obtained as a by-product from the processing and refining of zinc ores. In addition, dust and fumes from roasting zinc ores are collected by an electrostatic precipitator and mixed with carbon (coke) and sodium or zinc chloride. This residue is then treated to recover the cadmium. Other refining processes can obtain up to 40% recovery of cadmium from zinc ores.

Greenockite ore, as well as zinc and other ores, which produce cadmium as a by-product, are found in many countries, including Australia, Mexico, Peru, Zaire, Canada, Korea, and Belgium-Luxembourg and in the central and western United States.

History

Several people in the 18th and 19th centuries attempted to produce a pure form of zinc oxide for medical purposes. They were unaware that their samples contained cadmium, which at that time was an unknown element. In 1817 Friedrich Strohmeyer (1776–1835), a German chemist, analyzed a zinc compound (calamine) he believed contained zinc oxide (ZnO). However, what he really found was zinc carbonate (ZnCO$_3$), which, though at first unknown to him, contained some cadmium. Strohmeyer then treated his sample with acids until all the zinc was dissolved and thus removed. He then heated the residue with **carbon black,** resulting in a small ingot of soft, bluish-white metal that proved to be a new element—cadmium. Strohmeyer is given credit for the discovery of cadmium.

Common Uses

Cadmium alloyed with silver forms a type of solder with a low melting point. It is used to join electrical junctions and other specialized metallic components. Precautions are required since it is a toxic substance. (Note: This is not the same as common solder used to join metals, which is relatively safe.) Other cadmium alloys are used to manufacture long-wearing bearings and as thin coatings for steel to prevent corrosion.

Cadmium is a **neutron absorber,** making it useful as control rods in nuclear reactors. The rods are raised to activate the reactor and then lowered into the reactor to absorb neutrons that halt the fission reaction.

Cadmium, along with nickel, forms a nickel-cadmium alloy used to manufacture "nicad batteries" that are shaped the same as regular small dry-cell batteries. However, a major difference is that the nicads can be recharged numerous times whereas the common dry cells cannot. A minor difference between the two types of cells is that nicads produce 1.4 volts, and regular carbon-zinc-manganese dioxide dry-cell batteries produce 1.5 volts.

Wood's metal is another alloy that contains about 12.5% cadmium plus bismuth. It has a very low melting point of about 70°C, which makes it ideal for the "fuse" in overhead sprinklers in hotels and office buildings. Any fire will melt the trigger-like fuse, opening the valve to jettison water spray over the hot area that melted the Wood's metal alloy.

Several cadmium compounds—for example, CdSe, $CdCl_2$ and CdS—are used to make brilliant red-, yellow-, and orange-colored artists' oil paints.

Examples of Compounds

Cadmium has a single oxidation state of +2. Several examples follow:

Cadmium(II) bromide: $Cd^{2+} + 2Br^{1-} \rightarrow CdBr_2$. This compound is used in photography, engraving, and lithography. The other halogen elements also combine with cadmium in a similar ionic reaction as with bromine.

Cadmium chloride ($CdCl_2$), a soluble crystal, is formed when cadmium metal is treated with hydrochloric acid (Cd + 2HCl → $CdCl_2$ + H_2). $CdCl_2$ is used in dyeing and printing textiles, in electroplating baths, in photography, and as the ingredient for cadmium yellow in artists' oil paint.

Cadmium(II) oxide: $Cd^{2+} + O^{2-} \rightarrow CdO$. This is used for cadmium plating baths, electrodes for batteries (cells), ceramic glazes, and insecticides. CdO is a deadly poison and carcinogen.

Cadmium sulfate (CdS), also called "orange cadmium," is used to produce phosphors and fluorescent screens. It is also used as a pigment in inks and paints, to color ceramics glazes, in the manufacture of transistors in electronics, photovoltaic cells, and solar cells, and in fireworks.

Cadmium tungstate ($CdWO_4$) is used in fluorescent paint, X-ray screens, and scintillation counters and as a catalyst. It is very toxic when inhaled.

Many colored pigments are based on cadmium compounds. For instance, *cadmium sulfide* (CdS) and *cadmium selenide* (CdSe) are used as pigments when a durable, nonfading color is required. Red is produced by CdSe, and bright yellow is produced by CdS.

Cadmium compounds are also used for ceramics, TV and computer screens, **transistors, photovoltaic** cells, and solar cells.

Hazards

Cadmium powder, dust, and fumes are all flammable and toxic if inhaled or ingested. Cadmium and many of its compounds are carcinogenic.

Severe illness and death can occur from exposure to many cadmium compounds. It is absorbed in the **gastrointestinal** tract. However, it can be eliminated in the urine and feces in young, healthy people.

Cadmium, in trace amounts, is common in our foods, and as we age, our bodies cannot eliminate it effectively, so cadmium poisoning may result. The symptoms of mild poisoning are burning of the eyes, irritation of the mouth and throat, and headaches. As the **intoxication** increases, there may be severe coughing, nausea, vomiting, and diarrhea. There is a 15% chance of death from cadmium poisoning. The main risk from cadmium poisoning comes from industrial exposure—not from a healthy diet.

HAFNIUM

SYMBOL: Hf   **PERIOD:** 6   **GROUP:** 4 (IVB)   **ATOMIC NO:** 72
ATOMIC MASS: 178.49 amu   **VALENCE:** 2, 3, and 4   **OXIDATION STATE:** +4   **NATURAL
   STATE:** Solid
ORIGIN OF NAME: Named after Hafnia, the Latin name for the city of Copenhagen, Denmark.
ISOTOPES: There are 44 known isotopes for hafnium. Five are stable and one of the unstable isotopes has such a long half-life (Hf-174 with a $2.0 \times 10^{+15}$ years) that it is included
   as contributing 0.16% to the amount of hafnium found in the Earth's crust. The percentage contributions of the 5 stable isotopes to the element's natural existence on Earth are
   as follows: Hf-176 = 5.26%, Hf-177 = 18.60%, Hf-178 = 27.28%, Hf-179 = 13.62%,
   and Hf-180 = 35.08%.

ELECTRON CONFIGURATION

Energy Levels/Shells/Electrons	Orbitals/Electrons
1-K = 2	s2
2-L = 8	s2, p6
3-M = 18	s2, p6, d10
4-N = 32	s2, p6, d10, f14
5-O = 10	s2, p6, d2
6-P = 2	s2

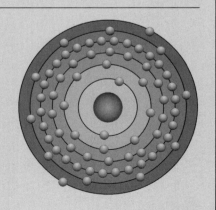

PERIODIC TABLE OF THE ELEMENTS

TRANSITION ELEMENTS

GROUPS PERIODS	1 IA	2 IIA	3 IIIB	4 IVB	5 VB	6 VIB	7 VIIB	8	9 VIII	10	11 IB	12 IIB	13 IIIA	14 IVA	15 VA	16 VIA	17 VIIA	18 VIIIA
1	1 H 1.0079																	2 He 4.00260
2	3 Li 6.941	4 Be 9.01218											5 B 10.81	6 C 12.011	7 N 14.0067	8 O 15.9994	9 F 18.9984	10 Ne 20.179
3	11 Na 22.9898	12 Mg 24.305											13 Al 26.9815	14 Si 28.0855	15 P 30.9738	16 S 32.066(6)	17 Cl 35.453	18 Ar 39.948
4	19 K 39.0983	20 Ca 40.08	21 Sc 44.9559	22 Ti 47.88	23 V 50.9415	24 Cr 51.996	25 Mn 54.9380	26 Fe 55.847	27 Co 58.9332	28 Ni 58.69	29 Cu 63.546	30 Zn 65.39	31 Ga 69.72	32 Ge 72.59	33 As 74.9216	34 Se 78.96	35 Br 79.904	36 Kr 83.80
5	37 Rb 85.4678	38 Sr 87.62	39 Y 88.9059	40 Zr 91.224	41 Nb 92.9064	42 Mo 95.94	43 Tc (98)	44 Ru 101.07	45 Rh 102.906	46 Pd 106.42	47 Ag 107.868	48 Cd 112.41	49 In 114.82	50 Sn 118.71	51 Sb 121.75	52 Te 127.60	53 I 126.905	54 Xe 131.29
6 ★	55 Cs 132.905	56 Ba 137.33	★	72 Hf 178.49	73 Ta 180.948	74 W 183.85	75 Re 186.207	76 Os 190.2	77 Ir 192.22	78 Pt 195.08	79 Au 196.967	80 Hg 200.59	81 Tl 204.383	82 Pb 207.2	83 Bi 208.980	84 Po (209)	85 At (210)	86 Rn (222)
7 ▲	87 Fr (223)	88 Ra 226.025	▲	104 Unq (261)	105 Unp (262)	106 Unh (263)	107 Uns (264)	108 Uno (265)	109 Une (266)	110 Uun (267)	111 Uuu (272)	112 Uub	113 Uut	114 Uuq	115 Uup	116 Uuh	117 Uus	118 Uuo

6 ★ Lanthanide Series (RARE EARTH)	57 La 138.906	58 Ce 140.12	59 Pr 140.908	60 Nd 144.24	61 Pm (145)	62 Sm 150.36	63 Eu 151.96	64 Gd 157.25	65 Tb 158.925	66 Dy 162.50	67 Ho 164.930	68 Er 167.26	69 Tm 168.934	70 Yb 173.04	71 Lu 174.967
7 ▲ Actinide Series (RARE EARTH)	89 Ac 227.028	90 Th 232.038	91 Pa 231.036	92 U 238.029	93 Np 237.048	94 Pu (244)	95 Am (243)	96 Cm (247)	97 Bk (247)	98 Cf (251)	99 Es (252)	100 Fm (257)	101 Md (258)	102 No (259)	103 Lr (260)

©1996 R.E. KREBS

Properties

Hafnium is a **ductile** metal that looks and feels much like stainless steel, but it is significantly heavier than steel. When freshly cut, metallic hafnium has a bright silvery shine. When the fresh surface is exposed to air, it rapidly forms a protective oxidized coating on its surface. Therefore, once oxidized, hafnium resists corrosion, as do most transition metals, when exposed to the air. Chemically and physically, hafnium is very similar to zirconium, which is located just above it in group 4 on the periodic table. In fact, they are so similar that it is almost impossible to secure a pure sample of either one without a small percentage of the other. Each will contain a small amount of the other metal after final refining.

Hafnium's melting point is 2,227°C, its boiling point varies from about 2,500°C to 5,000°C depending on its purity, and its density is 13.29 g/cm^3. The compound hafnium nitride (HfN) has the highest melting point (over 3,300°C) of any two-element compound.

Characteristics

As the first element in the third series of the transition elements, hafnium's atomic number ($_{72}$Hf) follows the lanthanide series of rare-earths. The lanthanide series is separated out of the normal position of sequenced atomic numbers and is placed below the third series on the periodic table ($_{57}$La to $_{71}$Li). This rearrangement of the table allowed the positioning of elements of the third series within groups more related to similar chemical and physical characteristics—for example, the triads of Ti, Zr, and Hf; V, Nb, and Ta; and Cu, Ag, and Au.

Abundance and Source

Hafnium is the 47th most abundant element on Earth. Thus, it is more abundant than either gold or silver. Because hafnium and zirconium are always found together in nature, both metals are refined and produced by the Kroll process. Pure samples of either hafnium or zirconium are almost impossible to separate by the Kroll or other refining processes. Baddeleyite (ZrO$_2$), a zirconium ore, and zircon (ZrSiO$_4$) are treated with chlorine along with a carbon catalyst that produces a mixture of zirconium and hafnium tetrachlorides. These are reduced by using sodium or magnesium, resulting in the production of both metals. The molten metals are separated by the process known as **fractionation,** which depends on their different melting points and densities. As the mixture of the two metals cools during the fractionation process, the denser solidified hafnium sinks to the bottom of the vessel while the less dense zirconium (with a higher melting point than hafnium) floats on top.

History

Even though hafnium is not a scarce or rare element, it was not discovered until 1923 because of its close association with zirconium. Several scientists suspected that another element was mixed with zirconium but could not determine how to separate the two because zirconium ore contains about 50 times more zirconium than hafnium. Mendeleev predicted that there was an element with the atomic number of 72, but he predicted it would be found in titanium ore, not zirconium ore.

In 1923 Georg Karl von Hevesy (1885–1966) and Dirk Coster (1889–1950), on the advice of Danish physicist Niels Henrik Bohr (1885–1962), used X-ray spectroscopy to study the pattern of electrons in the outer shell of zirconium. Their analysis led to the discovery and identity of element 72, and thus, they are given credit for hafnium's discovery. They also named it after Hafnia, Latin name for the city of Copenhagen in Denmark, to honor Niels Bohr for his work on the quantum structure of matter and the science of spectroscopy. This discovery required some revision in the periodic table. Data on zirconium had to be reanalyzed and corrected, and the blank space of the new element with 72 protons in its nucleus could now be filled in.

Common Uses

Hafnium has a great affinity for absorbing slow neutrons. This attribute, along with its strength and resistance to corrosion, makes it superior to cadmium, which is also used for making **control rods** for nuclear reactors. This use is of particular importance for the type of nuclear reactors used aboard submarines. By moving the control rods in and out of a nuclear reactor, the fission chain reaction can be controlled as the neutrons are absorbed in the metal of the rods. The drawback to hafnium control rods is their expense: it costs approximately one million dollars for several dozen rods for use in a single nuclear reactor.

In vacuum tubes and other applications that must have gases removed, hafnium is used as a "**getter**" to absorb any trace oxygen or nitrogen in the tube, thus extending the life of the vacuum tube. Hafnium's qualities also make it ideal for filaments in light bulbs and, when mixed with rare-earth metals, as a "sparking" **misch metal.** Hafnium is also used to a lesser extent as an alloying agent for several other metals, including iron, titanium, and niobium.

Examples of Compounds

Hafnium carbide (HfC): This alloy has one of the highest melting points of any binary compound (3.890°C). It is extremely hard and resists corrosion while absorbing slow neutrons. Therefore, it is an ideal metal in the manufacture of control rods for nuclear reactors.

Hafnium oxide (HfO_2): Resists heat and corrosion, making it an ideal lining for refractory furnaces.

Following are two compounds formed by the main oxidation state of hafnium (+4), both of which are used in the refining and production of hafnium metal:

Hafnium chloride: $Hf^{4+} + 4Cl \rightarrow HfCl_4$, also known as *hafnium tetrachloride.*

Hafnium fluoride: $Hf^{4+} + 4F \rightarrow HfF_4$, also known as *hafnium tetrafluoride.*

Hazards

Although the metal hafnium is not harmful, its powder and dust are both toxic if inhaled and explosive even when wet.

TANTALUM

SYMBOL: Ta   **PERIOD:** 6   **GROUP:** 5 (VB)   **ATOMIC NO:** 73
ATOMIC MASS: 180.948 amu   **VALENCE:** 2, 3, and 5   **OXIDATION STATE:** +5   **NATU-**
   RAL STATE: Solid
ORIGIN OF NAME: Tantalum was named after Tantalus, who was the father of Niobe, the
   queen of Thebes, a city in Greek mythology. (Note: The element tantalum was originally
   confused with the element nobelium.)
ISOTOPES: There are 49 isotopes of tantalum. Only the isotope Ta-181 is stable and
   accounts for 99.988% of the total mass of the element on Earth. Just 0.012% of the

element's mass is contributed by Ta-180, which has a half-life of $1.2 \times 10^{+15}$ years and is thus considered naturally stable. The remaining 47 isotopes are all artificially produced in nuclear reactions or particle accelerators and have half-lives ranging from a few microseconds to few days to about two years.

ELECTRON CONFIGURATION

Energy Levels/Shells/Electrons	Orbitals/Electrons
1-K = 2	s2
2-L = 8	s2, p6
3-M = 18	s2, p6, d10
4-N = 32	s2, p6, d10, f14
5-O = 11	s2, p6, d3
6-P = 2	s2

Properties

Tantalum has properties similar to niobium and vanadium above it in group 5. It is a very hard and heavy metal with a bluish color when in its rough state, but if polished, it has a silvery shine. It is ductile, meaning it can be drawn into fine wires, and also malleable, meaning it can be hammered and worked into shapes. Thin strips and wires of tantalum will ignite in air if exposed to a flame.

Tantalum's melting point is 2,996°C, which is almost as high as tungsten and rhenium. It boiling point is 5,425°C, and its density is 19.3 g/cm³.

Characteristics

Tantalum is almost as chemically inert at room temperatures (it has the ability to resist chemical attacks, including hydrofluoric acid) as are platinum and gold. It is often substituted for the more expensive metal platinum, and its inertness makes it suitable for constructing dental and surgical instruments and artificial joints in the human body.

Abundance and Source

Tantalum is the 51st most abundant element found on Earth. Although it is found in a free state, it is usually mixed with other minerals and is obtained by heating tantalum potassium fluoride or by the electrolysis of melted salts of tantalum. Tantalum is mainly obtained from the following ores and minerals: columbite [(Fe, Mn, Mg)(Nb, Ta)$_2$O$_6$]; tantalite [(Fe, Mn)(Ta, Nb)$_2$O$_6$]; and euxenite [(Y, Ca, Er, La, Ce, U, Th)(Nb, Ta, Ti)$_2$O$_6$]. Tantalum's ores are mined in South America, Thailand, Malaysia, Africa, Spain, and Canada. The United States has a few small native deposits but imports most of the tantalum it uses.

Since tantalum and niobium are so similar chemically, a solvent process must be employed to separate them from the common ores. They are dissolved in a solvent, resulting in 98%

pure niobium oxide being extracted during this part of the process. This is followed by 99.5% pure tantalum oxide being extracted in a second solvent process

History

Anders Gustav Ekeberg (1767–1813) discovered tantalum in 1802, while analyzing ores sent to him by his friend Jons Jakob Berzelius (1779–1848) from the famous mineral deposits of Ytterby, Sweden. At first, it was thought that this new element was an **allotrope** (close relative) of niobium because they were so similar in physical and chemical characteristics. Ekeberg named it tantalum after the Greek King Tantalus, who was condemned to everlasting torment. The word means "to tantalize." Tantalum was not separated, analyzed, and identified as a separate element with an atomic number 72 until 1866, by Jean Charles Galissard de Marignac (1817–1894) of Switzerland. He proved that tantalum and niobium are two different and distinct elements. The first pure samples were not produced until the year 1907.

Common Uses

A mixture of tantalum carbide (TaC) and graphite is a very hard material and is used to form the cutting edge of machine tools. Tantalum pentoxide (Ta_2O_5) is **dielectric,** making it useful to make capacitors in the electronics industry. When mixed with high-quality glass, it imparts a high index of refraction, making it ideal for camera and other types of lenses.

Because of its hardness and noncorrosiveness, tantalum is used to make dental and surgical tools and implants and artificial joints, pins, and screws. The metal does not interact with human tissues and fluids. Since tantalum can be drawn into thin wires, it is used in the electronics industry, to make smoke detectors, as a getter in vacuum tubes to absorb residual gases, and as **filaments** in incandescent lamps. It has many other uses in the electronics industry.

The use of tantalum to make miniaturized electrolytic capacitors that store electric charges in devices such as cell phones and computers is becoming increasingly popular. Powdered tantalum is used in the process of **sintering** to form malleable bars and plates as well as special electrodes for the electronics industry.

As a result of their hardness, noncorrosiveness, and ductility, tantalum alloys are used to fabricate parts for nuclear reactors, missiles, and airplanes, and in industries where metal with these qualities is required.

Examples of Compounds

Tantalum pentoxide is representative of tantalum's stable oxidation state of +5: $2Ta^{5+} + 5O^{2-} \rightarrow Ta_2O_5$. Tantalum oxide is used to make optical glass for lenses and in electronic circuits.

Tantalum carbide (TaC) is one of the hardest substances known. This compound represents its oxidation state of +4 for tantalum.

Tantalum disulfide (TaS_2) is used to make solid lubricants and special noncorrosive greases.

Tantalum fluoride (TaF_5) is a catalyst used to speed up organic chemical reactions.

Tantalum pentoxide (Ta_2O_5) is used to make special optical glass, for lasers, and in electronic circuits.

Hazards

The dust and powder of tantalum are explosive. Several tantalum compounds are toxic if inhaled or ingested, but the metal itself is nonpoisonous.

TUNGSTEN

SYMBOL: W   **PERIOD:** 6   **GROUP:** 6 (VIB)   **ATOMIC NO:** 74
ATOMIC MASS: 183.85 amu   **VALENCE:** 2, 4, 5, and 6   **OXIDATION STATE:** +4, and +6
   NATURAL STATE: Solid
ORIGIN OF NAME: Tungsten was originally named "Wolfram" by German scientists, after
   the mineral in which it was found, Wolframite—thus, its symbol "W." Later, Swedish scientists named it tung sten, which means "heavy stone," but it retained its original symbol
   of "W."
ISOTOPES: There are 36 isotopes of tungsten. Five are naturally stable and therefore contribute proportionally to tungsten's existence on Earth, as follows: W-180 = 0.12%, W-
   182 = 26.50%, W-183 = 14.31%, W-184 = 30.64%, and W-186 = 28.43%. The other
   31 isotopes are man-made in nuclear reactors and particle accelerators and have half-
   lives ranging from fractions of a second to many days.

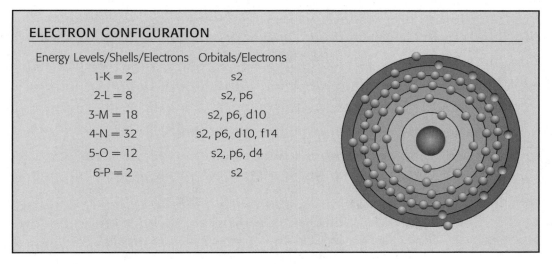

ELECTRON CONFIGURATION

Energy Levels/Shells/Electrons	Orbitals/Electrons
1-K = 2	s2
2-L = 8	s2, p6
3-M = 18	s2, p6, d10
4-N = 32	s2, p6, d10, f14
5-O = 12	s2, p6, d4
6-P = 2	s2

Properties

Extremely pure samples of tungsten are rather soft and can be cut easily with a simple saw.
Pure tungsten can be drawn into fine wires (ductile). On the other hand, if there are even a few
impurities in the sample, the metal becomes very hard and brittle. It is a very dense metal with
a whitish-to-silvery-grayish color when freshly cut. It has the highest melting point of all metals at 3,422°C, making it a useful metal where high temperatures are required. Incidentally,
the transition metals on both sides of it in period 6 ($_{73}$Ta and $_{75}$Re) have the second- and third-
highest melting points. Tungsten's boiling point is also high at 5,927°C.

Characteristics

Tungsten is considered part of the chromium triad of group six (VIB), which consists of
$_{24}$Cr, $_{42}$Mo, and $_{74}$W. These elements share many of the same physical and chemical attributes.
Tungsten's high melting point makes it unique insofar as it can be heated to the point that
it glows with a very bright white light without melting. This makes it ideal as a filament
for **incandescent** electric light bulbs. Most metals melt long before they reach the point of
incandescence.

Chemically, tungsten is rather inert, but it will form compounds with several other elements at high temperatures (e.g., the halogens, carbon, boron, silicon, nitrogen, and oxygen). Tungsten will corrode in seawater.

Abundance and Source

Tungsten is the 58th most abundant element found on Earth. It is never found in 100% pure form in nature. Its major ore is called wolframite or tungsten tetroxide, $(Fe,Mn)WO_4$, which is a mixture of iron and manganese and tungsten oxide. During processing, the ore is pulverized and treated with strong alkalis resulting in tungsten trioxide (WO_3), which is then heated (reduced) with carbon to remove the oxygen. This results in a variety of bright color changes and ends up as a rather pure form of tungsten metal: $2WO_3 + 3C \rightarrow 2WO + 3CO_2$. Or, if hydrogen is used as the reducing agent, a more pure form of metal is produced: $WO_3 + 3H_2 \rightarrow W + 3H_2O$.

Tungsten ores (oxides) are found in Russia, China, South America, Thailand, and Canada. In the United States, the ores are found in Texas, New Mexico, Colorado, California, Arizona, and Nebraska. Today, it is estimated that about 75% of all tungsten is found in China.

The tungsten ore called "scheelite" is named after Carl Wilhelm Scheele (1742–1786), who studied and experimented with tungsten minerals, but as with many of his other "near" discoveries, such as oxygen, fluorine, hydrogen sulfide, hydrogen cyanide, and manganese, he was not given credit.

History

In the mid-1700s a number of scientists experimented with and attempted to isolate element 74 by treating ores of other metals with reagents. One problem was that tungsten was often confused with tin and arsenic. It was not until 1783 that Don Fausto de Elhuyar (1755–1833) and his brother Don Juan Jose de Elhuyar isolated a substance from tin ore that they called "wolframite." They named it after the mineral in which it was found. At about the same time the Swedish named it *tung sten,* which means "heavy stone" in Swedish. This explains the potentially confusing use of W for the symbol for tungsten.

Common Uses

Since its melting temperature is over 3,400°C, tungsten is one of the few metals that can glow white hot when heated without melting. This factor makes it the second most frequently used industrial metal (the first is iron). Tungsten is used in the filaments of common light bulbs, as well as in TV tubes, cathode ray tubes, and computer monitors. Its ability to be "pulled" into thin wire makes it useful in the electronics industry. It is also used in solar energy products and X-ray equipment. Its ability to withstand high temperatures makes it ideal for rocket engines and electric-heater filaments of all kinds. Tungsten carbide is used as a substitute for diamonds for drills and grinding equipment. This attribute is important in the manufacture of exceptionally hard, high-speed cutting tools.

Examples of Compounds

Most of tungsten's stable compounds have the main oxidation state of +6 (e.g., $W^{6+} + 6Cl^{1-} \rightarrow WCl_6$), and the lower oxidation state of +4 occurs in the hard tungsten carbide (e.g., $W^{4+} + C^{4-} \rightarrow WC$).

Tungsten carbide (WC) is extremely hard and resistant to high temperatures. When cemented to tools, it is as hard as corundum (aluminum oxide) and makes excellent grinding surfaces and cutting edges for machine tools.

Tungsten disulfide (WS_2) is used as a solid lubricant that can withstand high temperatures. It is also used as a spray lubricant.

Tungsten oxide (WO_3) is used to make tungsten alloys. Tungsten oxide is also used as fireproofing for various surfaces and is used as a yellow pigment in ceramics.

Tungsten steel is an alloy that acts somewhat like molybdenum to form important steel alloys; tungsten steel is tough and hard, wears well, resists rusting, and will take a sharp cutting edge.

Tungsten compounds of calcium and magnesium have phosphorescent properties that make them useful in manufacturing fluorescent lighting fixtures.

Hazards:

Tungsten dust, powder, and fine particles will explode, sometimes spontaneously, in air. The dust of many of tungsten's compounds is toxic if inhaled or ingested.

RHENIUM

SYMBOL: Re   PERIOD: 6   GROUP: 7 (VIIB)   ATOMIC NO: 75
ATOMIC MASS: 186.207 amu   VALENCE: 4, 6, and 7   OXIDATION STATE: +4. +6, and
   +7   NATURAL STATE: Solid
ORIGIN OF NAME: Derived from the Latin word *Rhenus,* which stands for the Rhine River
   in Western Europe.
ISOTOPES: There are 45 isotopes of rhenium. Only one of these is stable: Re-185, which
   contributes 37.40% to the total amount of rhenium found on Earth. Re-187, which is
   radioactive with a very long half-life of $4.35 \times 10^{+10}$ years, contributes 62.60% to rhenium's existence on Earth. The remaining 43 isotopes are radioactive with relatively short
   half-lives and are artificially manufactured.

ELECTRON CONFIGURATION

Energy Levels/Shells/Electrons	Orbitals/Electrons
1-K = 2	s2
2-L = 8	s2, p6
3-M = 18	s2, p6, d10
4-N = 32	s2, p6, d10, f14
5-O = 13	s2, p6, d5
6-P = 2	s2

Properties

Rhenium ranges in color from silvery-white to gray to a black powder. It is a rather dense element. As a refined metal, rhenium is ductile, but because it is rather rare, its properties have not found many uses. Rhenium does have the widest range of valences. In addition to its common valences of 4, 6, and 7, it also has the uncommon valences of 2, –1, and –7.

Rhenium has a high melting point of 3,180°C, a boiling point of 5,627°C, and a density of 21.04 g/cm^3.

Characteristics

Rhenium is one of the transition elements, which range from metals to metal-like elements. Its chemical and physical properties are similar to those of technetium, which is above it in the periodic table. It is not very reactive. When small amounts are added to molybdenum, it forms a unique type of semiconducting metal. It is also noncorrosive in seawater.

Abundance and Source

Rhenium is the 78th most common element found on Earth, which makes it somewhat rare. During the early twentieth century, it required the processing of about a 1,000 pounds of earth to secure just one pound of rhenium, resulting in a price of about $10,000 per gram. Thus, there were few uses for rhenium. Later in the century, improved mining and refining techniques reduced the price. Today, the United States produces about 1,000 pounds of rhenium per year, and the world's total estimated supply is only about 400 tons.

The main sources of rhenium are the molybdenite and columbite ores. Some rhenium is recovered as a by-product of the smelting of copper sulfide (CuS) ores. Molybdenum sulfide (MoS_2) is the main ore and is usually associated with igneous rocks and, at times, metallic-like deposits. Molybdenite is found in Chile, as well as in the states of New Mexico, Utah, and Colorado in the United States.

History

Rhenium is extremely rare. It is not one of the historic, accidentally discovered elements, even though there was a "predicted" blank space—atomic number 75 on the periodic table. It was known as one of the manganese group (VIIB). Under the Mendeleevian terminology, element number 43 was known as eka-manganese and element number 75 was called dvi-manganese. The element 75 was specifically sought out by Ida Tacke Noddack (1896–1979), Walter Noddack (1893–1960), and Otto Carl Berg (1875–1939), who calculated, and predicted, some of rhenium's chemical and physical properties as they searched for number 75. In 1925, by using various analytical techniques, they concentrated some gadolinium ore 100,000 times, which was a Herculean task. This resulted in a small sample that was adequate to study and identify spectroscopically the element 75 that they named after the Rhine River.

An odd property of many elements that helped scientists theoretically determine the characteristics of undiscovered elements is that elements having an even number of protons in their nuclei (atomic number) are more commonly found on Earth than are the elements with an odd number of protons. It is unclear why elements with odd numbers of protons in their nuclei are less commonly found than are those with even numbers of protons.

Common Uses

Small quantities of rhenium are alloyed with iron to form steel that is both hard and resistant to wear and high-temperatures. Because of its high melting point, rhenium is used in many applications where long-wearing, high-temperature electrical components are required, such as electrical contacts and switches and high-temperature thermocouples. This physical quality makes rhenium alloys ideal for use in rocket and missile engines. It is also used to form the filaments in photographic flash lamps.

Rhenium's isotope (^{187}Re) has a very long half-life and decays by both beta and alpha radiation at a very steady rate. This factor makes it useful as a standard to measure the age of the universe.

Examples of Compounds

The +3 and +5 oxidation states of rhenium are demonstrated by reactions with chlorine, as in, for example, the following reactions:

Rhenium (III) chloride: $Re^{3+} + 3Cl^{1-} \rightarrow ReCl_3$.

Rhenium (V) chloride: $Re^{5+} + 5Cl^{1-} \rightarrow ReCl_5$.

The +4 and +6 oxidation states, on the other hand, are associated with fluorine. Some examples follow:

Rhenium (IV) fluoride: $Re^{4+} + 4F^{1-} \rightarrow ReF_4$.

Rhenium (VI) fluoride: $Re^{6+} + 6F^{1-} \rightarrow ReF_6$.

Rhenium heptasulfide (Re_2S_7) is one of several possible rhenium sulfur compounds. It is used as a catalyst to speed up chemical reactions.

The oxides of rhenium involve the +7 oxidation state of rhenium:

Rhenium heptoxide (Re_2O_7), for example, is explosive. There are at least six different forms of rhenium oxides.

Hazards

Rhenium is flammable in powder form. Rhenium dust and powder and many of its compounds are toxic when inhaled or ingested.

OSMIUM

SYMBOL: Os   **PERIOD:** 6   **GROUP:** 8 (VIII)   **ATOMIC NO:** 76
ATOMIC MASS: 190.2 amu   **VALENCE:** 2, 3, 4, 6, and 8   **OXIDATION STATE:** +3, +4, and +8   **NATURAL STATE:** Solid
ORIGIN OF NAME: Its name is derived from the Greek word *osme*, meaning "odor" or "smell," because of the element's objectionable smell when it is first isolated from platinum ores using aqua regia.
ISOTOPES: Osmium has 41 isotopes, five of which are stable. Two are naturally radioactive isotopes with very long half-lives. Following are the stable isotopes and their contribution to the element's natural existence in the Earth's crust: Os-187 = 1.6%, Os-188 = 13.29%, Os-189 = 16.21%, Os-190 = 26.36%, and Os-192 = 40.93%. The remain-

ing percentage of the element on Earth is in the form of the two naturally radioactive isotopes: Os-184 = 0.02% and Os-186 = 1.59%. All the other isotopes of osmium are radioactive and artificially produced in nuclear reactors and particle accelerators.

ELECTRON CONFIGURATION

Energy Levels/Shells/Electrons	Orbitals/Electrons
1-K = 2	s2
2-L = 8	s2, p6
3-M = 18	s2, p6, d10
4-N = 32	s2, p6, d10, f14
5-O = 14	s2, p6, d6
6-P = 2	s2

Properties

One of the important properties of osmium is the formation of gases when the metal is exposed to air. These fumes are extremely toxic, which limits osmium's usefulness. Osmium is a hard, tough, brittle, bluish-white metal that is difficult to use except in a powder form that oxidizes into osmium tetroxide (OsO_4), which not only has objectionable odor but also is toxic.

Osmium has a relatively high melting point of 3,054°C and a boiling point of 5,500°C, with a density of 22.61 g/cm^3.

Characteristics

Osmium is found in group 8 (VIII) of the periodic table and has some of the same chemical, physical, and historical characteristics as several other elements. This group of similar elements is classed as the platinum group, which includes Ru, Rh, and Pd of the second transition series (period 5) and Os, Ir, and Pt of the third series of transition metals (period 6).

Abundance and Source

Osmium is the 80th most abundant element on Earth. As a metal, it is not found free in nature and is considered a companion metal with iridium. It is also found mixed with platinum- and nickel-bearing ores. It is recovered by treating the concentrated residue of these ores with aqua regia (a mixture of 75% HCl and 25% HNO). The high cost of refining osmium is made economically feasible by also recovering marketable amounts of platinum and nickel.

Osmium occurs along with iridium in nature as the mineral iridosmine. It is found in Canada, Russia, and parts of Africa.

History

Osmium was discovered in 1803, at the same time as iridium, by Smithson Tennant (1761–1815). Several researchers, including Tennant, were curious about a black metallic substance that was produced when they refined platinum ore. At one time it was thought to be graphite or possibly another allotropic form of carbon. Further research indicated that this substance was not a form of carbon but a mixture of two new elements. Tennant named one of the new elements "iridium" because of its brilliant colors and named the other "osmium" because of its repugnant smell.

Common Uses

Because of its hard brittle nature, the metal osmium has few uses. However, the powdered form can be sintered under high pressure and temperatures to form some useful products, despite its toxicity and malodor. Its main use is as an alloy to manufacture devices that resist wear and stand up to constant use. As an alloy, osmium loses both its foul odor and toxicity. Some of these products are ballpoint and fountain pen tips, needles for record players, and pivot points for compass needles. Osmium alloys are also used for contact points on special switches and other devices that require reduced frictional wear.

Another use is as a stain for animal tissues that are to be examined with a microscope to improve the contrast of the specimen.

Examples of Compounds

The stable oxidation states of +3 and +4 of osmium are responsible for several different compounds, including the following:

Osmium (III) tetrachloride: $Os^{3+} + 3Cl \rightarrow OsCl_3$.

Osmium (IV) dioxide: $Os^{4+} + 2O^{2-} \rightarrow OsO_2$.

There is a special case where the +8 oxidation state of osmium is possible:

Osmium (VIII) tetraoxide ($Os^{8+} + 4O^{2-} \rightarrow OsO_4$) is a yellow crystal and probably the most important compound used as an oxidizing agent, as a biological stain in microscopy, and to detect fingerprints.

Hazards

Most of the oxides of osmium are not noxious, but are toxic if inhaled or ingested. The compound OsO_4 is extremely poisonous. It is a powerful oxidizing agent that is soluble in water and will produce serious burns in skin as it oxidizes the various layers of tissues.

IRIDIUM

SYMBOL: Ir   **PERIOD:** 6   **GROUP:** 9 (VIII)   **ATOMIC NO:** 77

ATOMIC MASS: 192.217 amu   **VALENCE:** 1, 2, 3, 4, and 6   **OXIDATION STATE:** +3 and +4   **NATURAL STATE:** Solid

ORIGIN OF NAME: The name iridium comes from the Latin word *iris,* meaning "rainbow," because of the element's highly colored salts.

ISOTOPES: There are 55 isotopes of iridium, two of which are stable and account for the element's total existence on Earth. Those two are Ir-191, which makes up 37.3% of the amount in the Earth's crust, and Ir-193, which constitutes 62.7% of iridium's existence on Earth. All the other 53 isotopes of iridium are radioactive with half-lives ranging from a few microseconds to a few hours or days and up to a few hundred years. These unstable isotopes are all artificially produced.

ELECTRON CONFIGURATION

Energy Levels/Shells/Electrons	Orbitals/Electrons
1-K = 2	s2
2-L = 8	s2, p6
3-M = 18	s2, p6, d10
4-N = 32	s2, p6, d10, f14
5-O = 15	s2, p6, d7
6-P = 2	s2

Properties

Iridium is a hard, brittle, white, metallic substance that is almost impossible to machine. It is neither ductile nor malleable. Iridium will only oxidize at high temperatures and is the most corrosive-resistant metal known. This is why it was used to make the standard meter bar that is an alloy of 90% platinum and 10% iridium.

At about the time of the French Revolution, it was decided to determine the length of the meter bar by first calculating the distance from the North Pole to the equator running through Paris. This distance was then divided into equal lengths of 1/10,000,000. A single unit of this distance was then called a "meter" ("measure" in Greek). This platinum-iridium meter bar, currently preserved in France, was for many years the standard unit of length in the metric system that is based on the decimal system. However, this metal bar is no longer used as the standard meter. Instead, the meter is now defined by scientists in terms of the length of the path traveled by light in a vacuum at the time of 1/299,792,458 of a second.

Iridium is highly resistant to attack by other chemicals and is one of the most dense elements found on Earth. Its melting point is 2,410°C, its boiling point is 4,130°C, and its density is 22.560 g/cm^3.

Characteristics

Iridium is one of the so-called platinum group of 6 transition elements (Ru, Rh, and Pd of period 5 and Os, Ir, and Pt of period 6). It is resistant to strong acids, including aqua regia. It is the only metal that can be used in equipment that must withstand temperatures up to 2,300°C or 4,170°F. Iridium can be poured into casts after it becomes molten. As it cools, it becomes crystalline and, while in this state, can be pulled into wires and formed into sheets. Unlike steel, which becomes more malleable (less brittle) after **annealing** (a process of heating followed by slowly cooling), iridium is just the opposite—it becomes more brittle and impossible to work into shapes after cooling.

Abundance and Source

Iridium is the 83rd most abundant element and is found mixed with platinum, osmium, and nickel ores. The minerals containing iridium are found in Russia, South Africa, Canada, and Alaska.

Iridium metal is separated from its other metal ores when the combined minerals are dissolved with a strong acid know as aqua regia, which is a mixture of 25% nitric acid and 75% hydrochloric acid. Aqua regia is the only acid that will dissolve platinum and gold. Once the platinum and other metals are dissolved, the iridium, which is insoluble in this strong acid, becomes the residue. The refined iridium ends up in the form of either powder or crystals.

An interesting story as to how most of the iridium appeared on Earth was explained recently by scientists who discovered a thin layer of iridium in the sediments that were laid down in the Earth's crust at the end of the Cretaceous period. This was a period about 65 million years ago when meteors and asteroids crashed into the Earth. These extraterrestrial bodies contained a high percentage of iridium. Dust from the impact spread around the Earth and blocked the sun for months, resulting in the extinction of many plants and animals, including the dinosaurs. This extensive dust cloud also deposited a thin coat of the element iridium that was contained in the fiery **bolides.**

History

Iridium and its partner osmium were discovered in 1803 by the English chemist Smithson Tennant (1761–1815). In essence, he employed the same technique to separate these elements from platinum ores that is used today to purify iridium. He dissolved the minerals with aqua regia, which left a black residue that looked much like graphite. After analyzing this shiny black residue, he identified two new elements—Ir and Os. Tennant was responsible for naming iridium after the Latin word "iris" because of the element's rainbow of colors.

Common Uses

Iridium's most common use is as an alloy metal that, when added to platinum, makes it harder and more durable. It is also mixed with other metals to make electrical contacts, thermocouples (two dissimilar metals joined to form a special type of thermometer), and instruments that will withstand high temperatures without breaking down. It is also used to make special laboratory vessels because iridium will not react with most chemical substances. An alloy of iridium and platinum is used as the standard kilogram weight because it is noncorrosive and will not oxidize and, thus, change its weight over long periods of time.

The radioisotope, iridium-192, is used to treat cancer and to take X-ray pictures of metal castings to detect flaws. Iridium is also used as a catalyst for several chemical reactions.

Examples of Compounds

The oxidation states of +3 and +4 form the most stable compounds, examples of which follow:

Iridium (III) chloride: $Ir^{3+} + 3Cl^{1-} \rightarrow IrCl_3$.
Iridium (IV) chloride: $Ir^{4+} + 4Cl^{1-} \rightarrow IrCl_4$.

More complex iridium compounds are possible as well, including the following:

Iridium potassium chloride (K_2IrCl_6) is used as a black pigment to make black porcelain kitchen and bathroom fixtures.

Iridomyrmecin ($C_{10}H_{16}O_2$) is one of the few colorless compounds of iridium. It is used to manufacture insecticides.

Iridosmine is not a compound, but an alloy of iridium, osmium, and a small amount of platinum that is used to make fine-pointed surgical instruments and needles and to form the fine tips of fountain pens. It is used worldwide to make weights because it resists oxidation better than any other alloyed metals.

Hazards

The elemental metal form of iridium is almost completely inert and does not oxidize at room temperatures. But, as with several of the other metals in the platinum group, several of iridium's compounds are toxic. The dust and powder should not be inhaled or ingested.

PLATINUM

SYMBOL: Pt **PERIOD:** 6 **GROUP:** 10 (VIII) **ATOMIC NO:** 78
ATOMIC MASS: 195.078 amu **VALENCE:** 2 and 4 **OXIDATION STATE:** +2 and +4; also +3 and +6 **NATURAL STATE:** Solid
ORIGIN OF NAME: The name "platinum" is derived from the Spanish word *platina,* which means "silver."
ISOTOPES: There are a total of 43 isotopes for platinum. Five of these are stable, and another has such a long half-life that it is considered practically stable (Pt-190 with a half-life of $6.5 \times 10^{+11}$ years). Pt-190 contributes just 0.014% to the proportion of platinum found on Earth. The stable isotopes and their contributions to platinum's existence on Earth are as follows: Pt-192 = 0.782%, Pt-194 = 32.967%, Pt-195 = 33.832%, Pt-196 = 25.242%, and Pt-198 = 7.163%. All the other isotopes are radioactive and are produced artificially. They have half-lives ranging from a few microseconds to minutes to hours, and one has a half-life of 50 years (Pt-193).

ELECTRON CONFIGURATION

Energy Levels/Shells/Electrons	Orbitals/Electrons
1-K = 2	s2
2-L = 8	s2, p6
3-M = 18	s2, p6, d10
4-N = 32	s2, p6, d10, f14
5-O = 17	s2, p6, d9
6-P = 1	s1

Properties

Platinum is classed by tradition and commercial usefulness as a precious metal that is soft, dense, dull, and silvery-white in color, and it is both malleable and ductile and can be formed into many shapes. Platinum is considered part of the "precious" metals group that includes gold, silver, iridium, and palladium. It is noncorrosive at room temperature and is not soluble in any acid except aqua regia. It does not oxidize in air, which is the reason that it is found in its elemental metallic form in nature. Its melting point is 1,772°C, its boiling point is 3,827°C, and its density is 195.09g/cm^3.

Characteristics

Platinum is the main metal in the platinum group, which consists of metals in both period 5 and period 6. They are ruthenium (Ru), rhodium (Ro), and palladium (Pd) in period 5 and osmium (Os), iridium (Ir), and platinum (Pt) in period 6. All six of these metals share some of the same physical and chemical properties. Also, the other metals in the group are usually found in platinum ore deposits.

Platinum can absorb great quantities of hydrogen gas, which makes it useful as a catalyst in industry to speed up chemical reactions.

Abundance and Source

Platinum is the 75th most abundant element and, unlike many elements, is found in its pure elemental form in nature, as are deposits of silver and gold. Platinum is widely distributed over the Earth and is mined mainly in the Ural Mountains in Russia and in South Africa, Alaska, the western United States, Columbia in South America, and Ontario in Canada. When found in the mineral sperrylite ($PtAs_2$), it is dissolved with aqua regia to form a precipitate called "sponge" that is then converted into platinum metal. It is also recovered as a by-product of nickel mining, mainly in Ontario, Canada.

History

There is archeological evidence that native platinum was known and used by the ancient Egyptians as early as the seventh century BCE. A platinum metal burial-type box or casket was found in a tomb dating from that period. There is also evidence that pre-Columbian natives of South America used platinum several centuries later. In 1735 Spanish astronomer Antonio de Ulloa (1716–1795) described a free metal he found in South America that was heavier than gold, had a higher melting point, and was noncorrosive. He named it *platina,* which is the Spanish word for "silver." He is given credit for the discovery of platinum. Samples of platinum ore from South America were sent in the 1700s for analysis in Europe, where in 1741 it was "rediscovered" by the Englishman Charles Wood (dates unknown).

An interesting bit of history regarding platinum occurred in 1978, when a defector from the Communist state of Bulgaria was assassinated when he was shot by a small platinum slug that contained a deadly poison. Because platinum is inert, there was no infection or inflammation, and the physicians did not detect the wound until it was too late to save his life.

Common Uses

Because of its chemical and physical properties, platinum has many uses. It is used widely in jewelry making and is often mixed with gold to improve gold's strength and durability.

In the early 1800s it was known that when hydrogen is passed over powdered platinum, the hydrogen ignites without being heated and without consuming the platinum. This property led to a major use of platinum as a catalyst to speed up chemical reactions. An example is present in the internal combustion engine that produces harmful gases. Gasoline fuel burns inside the engine, as compared to the steam engine, where the fuel is burned outside the engine. The catalytic converter in automobiles uses a platinum-coated ceramic grid in the exhaust system to convert unburned fuel to carbon dioxide and water. The platinum in the converter will last as long as the car since a catalyst is not consumed by the chemical reaction.

As a catalyst, platinum is used for hydrogenation of liquid vegetable oils to produce solid forms of the oil, such as margarine. It is also used in the cracking process that breaks down large crude oil molecules into smaller, more useful molecules, such as gasoline. The catalytic properties of platinum make it useful in the production of sulfuric acid (H_2SO_4) and in fuel cells that unite hydrogen and oxygen to produce electricity.

Platinum and some of its alloys have about the same expansion rate as glass, meaning that platinum wires can be inserted and sealed in glass structures such as bulbs and globes used for X-ray and other medical electronic equipment. Due to platinum's inertness and ductile property, it can be formed into wires that are inserted into the human body, such as those used in electronic cardiac pacemakers. Platinum's malleability permits it to be hammered into extremely thin sheets (only about 100 molecules thick) that are then used to plate the nose cones of missiles, fuel nozzles of jet engines, and the cutting edge of some instruments.

Platinum electrodes are attached to ocean-going ships and steel pipes and other submerged devices to help prevent corrosion from seawater. An alloy of platinum and iridium is used to form the standard weight for the kilogram, which is the basic unit of mass for the metric system. It is kept in a special vault in Paris, and all metric weights, worldwide, are judged for accuracy against this standard kilogram. Powdered platinum is used in air filters for spacecraft and highflying aircraft to convert high altitude ozone (O_3) into oxygen (O_2). Ozone is a poisonous gas found in the upper atmosphere.

More recently it was found that some compounds of platinum slow the growth of certain types of cancer. Research continues to ascertain how platinum can contribute to the reduction of ovarian and testicular tumors.

Examples of Compounds

Platinum has found many uses as an alloy with other metals, but it does form compounds with the halogens in its +2 and +4 oxidation states. It also forms many compounds with oxygen in both these oxidation states as well as in +3 and +6 oxidation states. A few examples follow:

Platinum (II) chloride: $Pt^{2+} + 2Cl^{1-} \rightarrow PtCl_2$.

Platinum (IV) fluoride: $Pt^{2+} + 4F^{1-} \rightarrow PtF_4$.

Platinum (III) oxide: $Pt^{3+} + 3O^{2-} \rightarrow Pt_2O_3$.

Platinum (IV) oxide ($Pt^{4+} + 2O^{2-} \rightarrow Pt O_2$) is also known as platinum dioxide. It is a dark-brown to black powder known as Adams catalyst that is used as a hydrogenation catalyst.

Chloroplantinic acid (H_2PtCl_6) is one of the most commercially important compounds of platinum. Its many uses include etching on zinc, making indelible ink, plating, and coloring in fine porcelains and use in photography, in mirrors, and as a catalyst.

Hazards

Fine platinum powder may explode if near an open flame. Because platinum is rather inert in its elemental metallic form, it is not poisonous to humans, but some of its compounds, particularly its soluble salts, are toxic if inhaled or ingested.

GOLD

SYMBOL: Au PERIOD: 6 GROUP: 11 (1B) ATOMIC NO: 79
ATOMIC MASS: 196.967 amu VALENCE: 1 and 3 OXIDATION STATE: +1 and +3 NATURAL STATE: Solid
ORIGIN OF NAME: The name "gold" is Anglo-Saxon as well as from the Sanskrit word *javal.* The symbol Au is from the Latin word aurum, which means "shining dawn."
ISOTOPES: There are a total of 54 isotopes of gold, only one of which is stable: Au-197, which accounts for the element's total natural existence on Earth. The remaining 53 isotopes are radioactive, are artificially produced in nuclear reactors or particle accelerators, and have half-lives ranging from a few microseconds to a few seconds to a few hours to a few days.

ELECTRON CONFIGURATION

Energy Levels/Shells/Electrons	Orbitals/Electrons
1-K = 2	s2
2-L = 8	s2, p6
3-M = 18	s2, p6, d10
4-N = 32	s2, p6, d10, f14
5-O = 18	s2, p6, d10
6-P = 1	s1

Properties

Gold is a soft, malleable, ductile, dense metal with a distinctive yellow color. It is almost a heavy as lead, and both can be cut with a knife. One ounce of gold can be beaten and pounded into a thin sheet that is only a few molecules thick and that will cover over 300 square feet of surface. Although gold is chemically nonreactive, it will react with chlorine and cyanide solutions and can be dissolved in aqua regia. Its melting point is 1,064.4°C, its boiling point is 2,808°C, and its density is 19.3 g/cm^3 (as compared to lead's density of 11.35 g/cm^3).

Characteristics

Gold is not only pleasing to look at but also pleasing to touch, which made it a desirable metal for human decoration in prehistoric days. It is still the preferred metal for jewelry making today.

Gold is classed as a heavy, noble metal located just below copper and silver in group 11 of the periodic table. Gold is a good conductor of electricity as well as an excellent heat reflector of infrared radiation, which makes it an efficient thin coating on glass in skyscrapers to reflect the heat of sunlight.

The purity of gold is measured in "carats" (one carat is equal to one part in twenty-four). The purest gold is rated at 24 carats, but it is much too soft to be used for jewelry. Good jewelry is made from 18-carat gold that is 18 parts gold and six parts alloy metal. Thus, an 18-carat gold ring is about 75% pure gold and contains about 25% of another metal, such as nickel or copper, to make it harder and more durable. Other alloy metals mixed with gold are silver, platinum, and palladium—all used to increase gold's strength and reduce its cost. Some less expensive jewelry contains 14 or 10 carats of gold (14/24 or 10/24) as well as some other alloy metals.

Abundance and Source

Gold is the 72nd most abundant element and is widely spread around the world, but it is not evenly distributed through the surface of the Earth. It is usually found in a few concentrated regions, sometimes in pure flake and nugget metallic forms. Most of it exists in conjunction with silver ore, quartz (SiO_2), and the ores of tellurium, zinc, and copper. About one milligram of gold exists in every ton of seawater (this is about 10 parts of gold per trillion parts of seawater, which amounts to a total of about 79 million tons of gold in solution). No economical method of extracting gold from seawater has been developed to recover this "treasury of the sea."

Free metallic gold is found in veins of rocks and in ores of other metals. Alluvial gold (placer deposits) is found in the sand and in the gravel at the bottom of streams where it has been deposited as a result of the movement of water over eons.

Most gold is recovered from quartz veins called "loads" and from ores that are crushed. The crushed ores are treated with a cyanide solution that dissolves the gold. This cyanide–gold solution is filtered and then treated with zinc to extract the gold. Disposal of the cyanide residue after the gold has been removed has become an environmental problem over the years. Gold can also be extracted from ores by using mercury to form an amalgam-like alloy, which is heated to drive off the mercury and recover the gold.

Gold is recovered economically in South Africa, Russia, Canada, Australia, Mexico, China, and India and in the states of California, Utah, Alaska, Nevada, and South Dakota in the United States. Small, scattered deposits of gold are found in several other states, including Florida, Arkansas, Washington, Oregon, Texas, Georgia, and the Carolinas. All the gold that has been refined in the world would form a cube with a volume of over 8,000 cubic meters.

History

Gold is considered one of the first metals used by humans. Along with other free metals, mankind discovered gold thousands of years ago. Most likely, early humans found pebble-like nuggets of metals, including gold, which they admired for their colors. There is evidence that gold was known in ancient Egypt about 6,000 years ago, and gold is mentioned several times in the Old Testament of the Bible.

The word "metal" is from the Greek word meaning "to hunt for" or "to search for," which is how early humans found these rare, heavy metals. The search for gold was one of the contributing factors in the exploration and settlement of the New World by Europeans in the 15th to the 18th centuries.

Common Uses

Gold's chemical and physical properties make it a very versatile element. Its noncorrosive nature provides protection as plating for other metals. Its malleability and ductile qualities mean it can be formed into many shapes, including very thin sheets (gold leaf) and very thin gold **diode** wires. Gold has the ability to carry electricity with little resistance, making it an excellent component for all kinds of electronic equipment. Gold leaf finds many uses in surgery, space vehicles, and works of art. Gold electronic switches do not create a dangerous spark when engaged, and they last for a long time. The element's reflective surface provides protection from infrared heat radiation as a coating on the visors of aerospace personnel, as well as on large window expanses in buildings. Its color and durability make it a major metal for the jewelry industry. Gold has been used to replace teeth for many ages, and the teeth usually last longer than the person wearing them. Gold is also the worldwide monetary standard, although the United States abandoned the gold standard in the 1930s. Even so, gold is still traded as a commodity. Small amounts of other metals are added to gold coinage for hardening purposes so that the coins will not wear out with use. Gold bars (bullion) are stored in the treasuries of most countries. Some countries maintain huge stockpiles of gold for both monetary and industrial uses.

Two forms of gold provide medical treatments. The radioactive isotope Au-198, with a short half-life of 2.7 days, is used to treat cancer and is produced by subjecting pure gold to neutrons within a nuclear reactor. A gold salt, a solution called sodium thiosulfate ($AuNa_3O_6Cl_4$), is injected as an internal treatment for rheumatoid arthritis. However, since gold and some of its compounds are toxic when ingested, this treatment may cause complications such as skin rashes and kidney failure. It is a less popular treatment, particularly with the development of newer and more effective medications.

An interesting bit of gold's history occurred when Ernest Rutherford (1871–1937) and his junior assistants used a very thin gold leaf that was only 1/50,000 of an inch thick (about 2,000 atoms thick) to perform a classic experiment. Rutherford bombarded the gold foil with alpha particles (helium nuclei), and most of the particles passed through the foil and were detected by photographic plates that were placed behind the gold foil. Some alpha particles were deflected sideways by the nuclei of the atoms in the gold foil, and some were sent in different directions and were also recorded on photographic plates. A few particles squarely hit a few gold atoms and bounced back toward the source of the alpha particles. Because the vast majority of the particles passed through the foil as though nothing could stop them, this experiment demonstrated that the gold atom was mostly empty space, with a small, dense, positively charged nucleus, surrounded by orbiting electrons. (See the section of the book titled "Atomic Structure" for more on the Rutherford experiment.)

Examples of Compounds

Because gold is a rather inert metal, it does not form many compounds. Even its oxidation states of +1 and +3 do not react with oxygen to form metallic oxides, as do most other metals. Examples of two possible compounds of gold follow:

Auric chloride: $Au + 2Cl_2 \rightarrow AuCl_4$. It is important to note that the Cl_2 used here must be hot chlorine gas.

Chlorauric acid: $Au + 4HCl + HNO_2 \rightarrow HAuCl_4 + NO + 2H_2O$. The HCl plus the HNO_2 (hydrochloric and nitric acids) are combined to produce agua regia acid, which is the only acid that can dissolve gold.

Hazards

Pure gold, if ingested, can cause skin rash or even a sloughing off of skin. It can also cause kidney damage and problems with the formation of white blood cells.

MERCURY

SYMBOL: Hg   **PERIOD:** 6   **GROUP:** 12 (IIB)   **ATOMIC NO:** 80
ATOMIC MASS: 200.59 amu   **VALENCE:** 1 and 2   **OXIDATION STATE:** +1 and +2
NATURAL STATE: Liquid
ORIGIN OF NAME: Named for the mythological Roman god of travel, Mercurius, the messenger to other gods. Its symbol Hg is from the Latin word *hydrargyrus,* meaning "liquid silver."
ISOTOPES: There are a total of 45 isotopes of the element mercury. Seven of these are stable and contribute to the total element's natural existence on Earth as follows. Hg-196 = 0.15%, Hg-198 = 9.97%, Hg-199 = 16.87%, Hg-200 = 23.10%, Hg-201 = 13.18%, Hg-202 = 29.86%, and Hg-204 = 6.87%. All the other isotopes of mercury are radioactive with half-lives of a few milliseconds, to a few seconds, to a few hours, and up to about 500 years.

ELECTRON CONFIGURATION

Energy Levels/Shells/Electrons	Orbitals/Electrons
1-K = 2	s2
2-L = 8	s2, p6
3-M = 18	s2, p6, d10
4-N = 32	s2, p6, d10, f14
5-O = 18	s2, p6, d10
6-P = 2	s2

Properties

Mercury is the only metal that is in a liquid state at room temperatures and remains liquid at temperatures well below the freezing temperature of water. Mercury is a noncombustible, heavy, silvery-colored metal that evenly expands and contracts with temperature and does not "wet" or stick to glass, which makes it ideal as a liquid for thermometers. Mercury is slightly volatile and will give off toxic fumes, especially if heated. Its has a unique melting point of $-38.83°C$, a boiling point of $3,56.73°C$, and a density of $13.5336 \ g/cm^3$.

Characteristics

Mercury is located in group 12 (IIB) below Zn and Cd. Even though mercury is at the end of the third series of transition elements, it is not always considered one of the transition elements.

Mercury forms alloys, called amalgams, with other metals such as gold, silver, zinc, and cadmium. It is not soluble in water, but will dissolve in nitric acid. It has a high electric conductivity, making it useful in the electronics industry. However, unlike most other metals, it is a poor conductor of heat. Because of its high surface tension, it does not "wet" the surfaces that it touches. This characteristic also accounts for its breakup into tiny droplets when poured over a surface. If spilled, it should not be collected with bare hands, but with a thin piece of cardboard to scoop it up.

Abundance and Source

Mercury is the 68th most abundant element. Although it can occur in its natural state, it is more commonly found as a sulfide of mercury. Its chief ore is cinnabar (HgS), which sometimes is called "vermilion" due to its red color. Historically, cinnabar was used as a red pigment. Today it is mined in Italy, Spain, and California. The best-known mercury mine is located at Almaden, Spain. It has been in continuous operation since 400 BCE.

Mercury is also found in black metacinnabar and mercury chloride. Small liquid droplets of mercury may be visible in high-grade deposits. Mercury ores are also found in Algeria, Mexico, Bosnia, and Canada as well as in Spain and California.

History

The history of mercury extends over the past four thousand years; therefore, it is impossible to attribute its discovery to any one individual. The Chinese used mercury before 2000 BCE, and vials of mercury have been found in ancient Egyptian tombs. Mercury and its ore cinnabar were well known to alchemists of the Dark and Middle Ages and were ingredients for preparing the "philosophers' stone." Alchemist-physicians used mercury to make "elixirs" to try to cure all illnesses, but this often killed the patients because mercury is a poison. It was also used in their "experiments" to convert base metals into gold. Metallic gold has an affinity for mercury, and it seems that mercury simply dissolves into the gold, turning it into a slivery color, leading the alchemists to incorrectly believe that mercury could in some way convert other metals into gold.

Joseph Priestley (1733–1804) heated mercury with air and formed a red powder (mercuric oxide) that, when heated in a test tube, produced small globs of mercury metal on the inside of the glass tube, as well as a gas that caused other substances to burn more rapidly than they did in air. Priestley did not know it at the time, but he had separated oxygen from the compound HgO.

Today, just about all mercury is produced via the reduction of cinnabar (HgS) by using a reducing agent, such as oxygen, iron, or calcium oxide (CaO). The resulting mercury vapor is passed through water where it liquefies (changes into its normal metallic state and sinks to the bottom of the water bath) while all the impurities float to the surface.

Common Uses

One of the most common uses of mercury is to make amalgams, which are solid "solutions" of various metals that can be combined without melting them together. Metals such as gold, silver, platinum, uranium, copper, lead, potassium, and sodium will form amalgams with mercury. The most common is silver-mercury amalgam, used as fillings for tooth cavities. Other types of fillings are now used since it has been proven that mercury, on contact with the skin of the mouth, can over time **leach** and act as a mild poison. Amalgams are also used to extract mercury from its ores.

Another common use is as a liquid contact in electrical "silent switches." Also, mercury-vapor produces the bluish-white light of streetlights. However, mercury-vapor lights now are being replaced by sodium-vapor lights that produce a yellowish-white light. Mercury is used in thermometers and barometers, to coat mirrors, and in the electronics industry and several other industries.

Mercury batteries (cells) consist of a zinc anode and a mercuric oxide cathode. These cells produce a steady 1.3 volts throughout the cell's lifetime.

The industrial uses of mercury are becoming more restrictive because of the element's toxicity.

Examples of Compounds

Almost all the stable mercury compounds are formed from the +1 and +2 oxidation states of the element as follows:

Mercury (I) fluoride: $2Hg^{1+} + 2F^{1-} \rightarrow Hg_2F_2$.

Mercury (II) chloride: $Hg^{2+} + 2Cl^{1-} \rightarrow HgCl_2$. This compound is used by the pharmaceutical industry and is used also as a fungicide, as a poison, in fireworks, and to control maggots.

There are many other examples of mercury compounds used in industry.

Mercury oxide (HgO) exists in two forms, red and yellow mercuric oxides, and is related to mercurous oxide (Hg_2O), which is black. All have industrial uses, ranging from antiseptics to pigments.

Mercuric sulfide (HgS) is a fine, very brilliant scarlet powder that is deadly if ingested. Also known as the mercury ore cinnabar and metacinnabar, it is used as a pigment in the manufacture of paints.

Mercury fulminate [$Hg(CNO)_2$] is very explosive and is used to manufacture blasting caps and detonators.

Hazards

Mercury metal is a very toxic and accumulative poison (it is not easily eliminated by the body). When inhaled as vapors, ingested as the metal or as part of a compound, or even absorbed when in contact with the skin, it can build up to deadly amounts. The fumes of most compounds are poisonous and must be avoided. Anyone who ingests mercury should contact a poison center immediately.

If mercury metal is spilled, it needs to be carefully gathered so as not to spread the little globs, but rather combine them for easy collection.

Many patients are having the mercury amalgams that were used as dental fillings replaced because of their potential toxicity. There are conflicting data concerning the extent of risk resulting from the dental use of mercury amalgams, but there is some evidence that bacteria in saliva of the mouth may leach out traces of mercury from the amalgam. Regardless, different metals and plastics are preferred for use today to fill dental cavities.

Another danger is the waste mercury that has been deposited by industries and agricultural chemicals in the lakes and oceans of the world. Several decades ago most of the nations of the world approved an international ban on dumping mercury into our waterways and oceans. The problem is that smaller ocean plants and animals consume mercury. Larger fish consume the mercury, and then we consume the larger animals. In turn, we receive an abundance of accumulated mercury that has built up in our seafood chain. Most nations, including the United States, have banned the use of mercury in agricultural chemicals, including most pesticides and insecticides. In addition to being poisonous, some compounds of mercury (mercury fulminate) are extremely explosive.

Mercury can be a cumulative poison, which means that minor amounts absorbed over long periods of time build up until damage to internal organs occurs. Years ago, a mercury compound was used in the manufacturing process of felt hats. Workers who came in contact with the mercury developed a variety of medical problems, including the loss of hair and teeth and loss of memory along with general deterioration of the nervous and other systems. This became known as the "mad as a hatter" syndrome because of the afflicted individuals' odd behavior.

PERIODIC TABLE OF THE ELEMENTS

TRANSITION ELEMENTS

GROUPS / PERIODS	1 IA	2 IIA	3 IIIB	4 IVB	5 VB	6 VIB	7 VIIB	8 VIII	9 VIII	10 VIII	11 IB	12 IIB	13 IIIA	14 IVA	15 VA	16 VIA	17 VIIA	18 VIIIA
1	1 H 1.0079																	2 He 4.00260
2	3 Li 6.941	4 Be 9.01218											5 B 10.81	6 C 12.011	7 N 14.0067	8 O 15.9994	9 F 18.9984	10 Ne 20.179
3	11 Na 22.9898	12 Mg 24.305											13 Al 26.9815	14 Si 28.0855	15 P 30.9738	16 S 32.066(6)	17 Cl 35.453	18 Ar 39.948
4	19 K 39.0983	20 Ca 40.08	21 Sc 44.9559	22 Ti 47.88	23 V 50.9415	24 Cr 51.996	25 Mn 54.9380	26 Fe 55.847	27 Co 58.9332	28 Ni 58.69	29 Cu 63.546	30 Zn 65.39	31 Ga 69.72	32 Ge 72.59	33 As 74.9216	34 Se 78.96	35 Br 79.904	36 Kr 83.80
5	37 Rb 85.4678	38 Sr 87.62	39 Y 88.9059	40 Zr 91.224	41 Nb 92.9064	42 Mo 95.94	43 Tc (98)	44 Ru 101.07	45 Rh 102.906	46 Pd 106.42	47 Ag 107.868	48 Cd 112.41	49 In 114.82	50 Sn 118.71	51 Sb 121.75	52 Te 127.60	53 I 126.905	54 Xe 131.29
6	55 Cs 132.905	56 Ba 137.33	★	72 Hf 178.49	73 Ta 180.948	74 W 183.85	75 Re 186.207	76 Os 190.2	77 Ir 192.22	78 Pt 195.08	79 Au 196.967	80 Hg 200.59	81 Tl 204.383	82 Pb 207.2	83 Bi 208.980	84 Po (209)	85 At (210)	86 Rn (222)
7	87 Fr (223)	88 Ra 226.025	▲	104 Unq (261)	105 Unp (262)	106 Unh (263)	107 Uns (264)	108 Uno (265)	109 Une (266)	110 Uun (267)	111 Uuu (272)	112 Uub	113 Uut	114 Uuq	115 Uup	116 Uuh	117 Uus	118 Uuo

6 ★ Lanthanide Series (RARE EARTH)

57 La 138.906	58 Ce 140.12	59 Pr 140.908	60 Nd 144.24	61 Pm (145)	62 Sm 150.36	63 Eu 151.96	64 Gd 157.25	65 Tb 158.925	66 Dy 162.50	67 Ho 164.930	68 Er 167.26	69 Tm 168.934	70 Yb 173.04	71 Lu 174.967

7 ▲ Actinide Series (RARE EARTH)

89 Ac 227.028	90 Th 232.038	91 Pa 231.036	92 U 238.029	93 Np 237.048	94 Pu (244)	95 Am (243)	96 Cm (247)	97 Bk (247)	98 Cf (251)	99 Es (252)	100 Fm (257)	101 Md (258)	102 No (259)	103 Lr (260)

Metallics—Metalloids—Semiconductors—Nonmetals

Introduction

The periodic table classifies elements into families that have similar physical properties and chemical characteristics. These families of related elements are called groups. The next several sections of this book present elements from the boron group (group 13; IIIA) through the noble gas group (group 18; VIII). The similarities and differences between the elements in each of these groups are not uniform and require some study in order to be understood in relation to the periodic table. Please note that names given to these groups in this book are not necessarily the same as used by some other references. Rather, they are descriptive as to their properties and characteristics.

Not all elements in these groups have the same properties and characteristics. For instance, in group15, nitrogen is a gas, whereas the element just below it in group 15 is phosphorous, a nonmetallic solid (semimetal). Just below phosphorous is arsenic (semimetal), followed by antimony and then bismuth, which are more metal-like. These last two, antimony and bismuth, are metals that might be considered an extension of periods 5 and 6 of the transition elements.

Even though the elements listed in groups 13 (IIIA) to 18 (VIIIA) may not have the same properties and characteristics, they do have a distinct number of electrons in their outer valence shells related to their specific group. For instance, group 13 elements have three electrons in their outer valence shell, and group 14 elements have four electrons. Group 15 elements have five electrons in their outer valence shell, and group 16 elements have six electrons. The halogens in group 17 have seven electrons, and the inert elements in group 18 have a completed outer valence shell with eight electrons. At the end of each periods of group 18 (whose elements each have eight electrons in their respective outer valence shells), the table starts over with elements containing one electron in their respective valence shells (the alkali earth metals in group 1 [IA]).

Some elements in these groups are more like metals and give up electrons; some are more like metalloids or semiconductors, which means they may act somewhat like metals and somewhat as nonmetals; and some are more like nonmetals because they gain electrons in chemical reactions.

As elements progress from group 13 to group 17, they show a shift from metallic characteristics to properties of the nonmetals, but the distinctions are not cut-and-dried. Some elements listed in groups 13, 14, 15, and 16 may have both metal-like qualities—metalloids or semiconductors—as well as a few nonmetal properties.

There are several general ways to categorize elements in groups 13 to 16. These are metals different in several ways from the transition elements. They range from **metallics** (other metals) to **metalloids** (semiconductors) to **nonmetals.** The elements in these groups are arranged according to their properties, characteristics, and the position of their electrons in their atom's outer shells. These, and other factors, determine how they are depicted in the periodic table.

The metallics are often called "other metals" and begin an arrangement on the periodic table in zigzag steps. (You may view this dark zigzag line that divides the metallics from metalloids on a copy of the Periodic table.) For the "other metals" or metallics, this zigzag line runs run from aluminum to gallium to indium to tin to thallium to lead and then ends with bismuth. Elements left of the zigzag are also called "poor metals."

Some of the metalloids are considered semiconductors. The term "metalloids" is used in this reference book because these elements do have characteristics of both metals and nonmetals, and the term "semiconductor" refers only to particular elements somewhere between metals and nonmetals. Semiconductors also have properties of both metals and nonmetals. Therefore, they have the ability to act as conductors of electricity and thermal energy (heat), as well as the ability to act as **insulators** or nonconductors of electricity and **heat,** depending upon the kind and amount of impurities their crystals contain. Again, following the zigzag steps on the periodic table, the metalloids having properties of both metals and nonmetals are as follows: boron, silicon, germanium, arsenic, antimony, tellurium, and polonium.

This book uses the vertical structure for listing the elements in groups 13 through 18 between periods 2 and 5. Therefore, the elements included as metallics, metalloids, nonmetals, and so on are arranged in a different order (vertical) according to their atomic numbers rather than following the zigzag line on the periodic table.

As mentioned, metalloids will, under certain circumstances, conduct electricity. Therefore, they are often called semiconductors. Elements listed as semiconductors or metalloids are crystalline in structure. As very small amounts of impurities are added to their crystal structure, their capability of conducting electricity or acting as insulators increases or decreases. These impurities affect the capacity of electrons to carry electric currents. The flow of electricity is restricted according to the degree and type of impurities. This is why the "semi" is included in their name.

In contrast, nonmetals in groups 16 (VIA), 17 (VIIA), and 18 (VIIIA) are characterized by being very inefficient at conducting both electricity and heat. In fact, most can be thought of as "insulators" because they are such poor conductors of electricity and heat. To confuse matters even more, some references list "semiconductors" as a special group of metals.

The Boron Group (Metallics to Semimetals): Periods 2 to 6, Group 13 (IIIA)

Introduction

The boron group (group 13; IIIA) consists of the elements boron (B), aluminum (Al), gallium (Ga), indium (In), and thallium (Ti). All have three electrons in their outer valence shell. A few exhibit metal-like characteristics by losing one or more of their outer electrons. For example, aluminum can lose one or three of its valence electrons and become a positive ion just as do other metals, but other elements in this group have characteristics more like metalloids or semiconductors.

BORON

SYMBOL: B   **PERIOD:** 2   **GROUP:** 13 (IIIA)   **ATOMIC NO:** 5
ATOMIC MASS: 10.811 amu   **VALENCE:** 3   **OXIDATION STATE:** +3
   NATURAL STATE: Solid
ORIGIN OF NAME: It is named after the Arabic word *bawraq,* which means "white borax."
ISOTOPES: There are a total of 13 isotopes of boron, two of which are stable. The stable isotope B-10 provides 19.85% of the element's abundance as found in the Earth's crust, and the isotope B-11 provides 80.2% of boron's abundance on Earth.

ELECTRON CONFIGURATION

Energy Levels/Shells/Electrons	Orbitals/Electrons
1-K = 2	s2
2-L = 3	s2, p1

Properties

Boron has only three electrons in its outer shell, which makes it more metal than nonmetal. Nonmetals have four or more electrons in their valence shell. Even so, boron is somewhat related to metalloids and also to nonmetals in period 2.

It is never found in its free, pure form in nature. Although less reactive than the metals with fewer electrons in their outer orbits, boron is usually compounded with oxygen and sodium, along with water, and in this compound, it is referred to as borax. It is also found as a hard, brittle, dark-brown substance with a metallic luster, as an **amorphous** powder, or as shiny-black crystals.

Its melting point is 2,079°C, its boiling point is 2,550°C, and its density is 2.37 g/cm^3.

Characteristics

Boron is a semimetal, sometimes classed as a metallic or metalloid or even as a nonmetal. It resembles carbon more closely than aluminum, the latter of which is located just below boron in group 13. Although it is extremely hard in its purified form—almost as hard as diamonds—it is more brittle than diamonds, thus limiting its usefulness. It is an excellent conductor of electricity at high temperatures, but acts as an insulator at lower temperatures. It is less reactive than the elements below it in group 13

Abundance and Source

Boron is the 38th most abundant element on Earth. It makes up about 0.001% of the Earth's crust, or 10 parts per million, which is about the same abundance as lead. It is not found as a free element in nature but rather in the mineral borax, which is a compound of hydrated sodium, hydrogen, and water. Borax is found in salty lakes, dry lake-beds, or alkali soils. Other naturally occurring compounds are either red crystalline or less dense, dark-brown or black powder.

Boron is also found in kernite, colemanite, and ulexite ores, and is mined in many countries, including the western United States.

History

While experimenting with **electrodes** for batteries, Sir Humphry Davy (1778–1829) discovered that some of these metal-containing compounds would break down when a current was passed through water containing a solution (**electrolyte**) of the compound. The metal was deposited on one electrode, and hydrogen gas was released from the water. He used this method to isolate barium, strontium, calcium, magnesium, and boron. Davy is given credit for discovering boron in 1808.

In 1808 two French chemists, Joseph Louis Gay-Lussac (1778–1850) and Louis-Jacques Thenard (1777–1857), experimented along the same lines as Davy and should also be given some credit for the discovery of these elements. The Frenchmen named the new element "bore," and Davy called it "boracium."

Alfred Stock (1876–1946) studied the **hydrides** of some of these metal-like elements. A hydride occurs when hydrogen gains (or shares) an electron rather than losing its single electron when it combines with metals or metallic-like elements. Stock spent years experimenting with boron hydrides (B_6H_6 and BH_3), which were used as hydrogen-based rocket fuels powerful enough to lift rockets into space.

Common Uses

Even though boron has a very simple atom with just five protons in its nucleus and only three valence electrons, it has proven to be somewhat bewildering and continues to intrigue chemists as a more-or-less exotic element. Even so, boron has found many uses and has become an important industrial chemical.

Borax is used as a cleaning agent and water softener that removes ions of elements such as magnesium and calcium that cause hard water. When these "hard" water elements are mixed with soap, they prevent soap from sudsing and form a scum or residue that is deposited on hard surfaces. Borax can eliminate this residue ring by replacing the Mg^{++} and Ca^{++} ions with the more soluable Na^+ and K^+ ions. Borax is the third most important boron compound.

A common but important compound of boron is boric acid (H_3BO_3), which is made by heating borax with an acid (either HCl or H_2SO_4). Boric acid is weak and can be used as eyewash. More importantly, it is used to manufacture heat-resistant borosilicate glass that is known by the trade name Pyrex. Pyrex is commonly used for baking utensils, so that a drastic change in temperature will not damage the glass. Pyrex is an example of one of many products developed by NASA's space program that led to everyday practical use. Boric acid is the second most important compound of the element boron. Sodium borate pentahydrate ($Na_2B_4O_7 \bullet 5H_2O$) is the most important boron compound; it is used to make fiberglass insulation.

A hydride of boron that is combined with hydrogen is an effective "booster" for rocket fuel in spacecrafts.

Boron is used as an alloy metal, and when combined with other metals, it imparts exceptional strength to those metals at high temperatures.

It is an excellent **neutron absorber** used to "capture" neutrons in nuclear reactors to prevent a runaway fission reaction. As the boron rods are lowered into the reactor, they control the rate of fission by absorbing excess neutrons. Boron is also used as an oxygen absorber in the production of copper and other metals,

Boron finds uses in the cosmetics industry (talc powder), in soaps and adhesives, and as an environmentally safe insecticide.

A small amount of boron is added as a "dope" to silicon transistor chips to facilitate or impede the flow of current over the chip. Boron has just three valence electrons; silicon atoms have four. This dearth of one electron in boron's outer shell allows it to act as a positive "hole" in the silicon chip that can be "filled" or left vacant, thus acting as a type of switch in transistors. Many of today's electronic devices depend on these types of doped-silicon semiconductors and transistors.

Boron is also used to manufacture borosilicate glass and to form enamels that provide a protective coating for steel. It is also used as medication for relief of the symptoms of arthritis.

Due to boron's unique structure and chemical properties, there are still more unusual compounds to be explored.

Examples of Compounds

Boron's +3 oxidation state permits it to join ions with oxidation states of −1, −2, and −3 as follows:

Boron fluorides: $B^{3+} + 3F^{1-} \rightarrow BCl_3$.

Boron oxide: $2B^{3+} + 3O^{2-} \rightarrow B_2O_3$.

Boron nitrate ($B^{3+} + N^{3-} \rightarrow BN$) is a white powder used to line high-temperatures furnaces, to make heating crucibles, for electrical and chemical equipment, for heat shields on spacecraft nosecones, and to make high-strength fabrics.

Boron-10, a stable isotope of boron, is used to absorb slow neutrons in nuclear reactors. It produces high-energy alpha particles (helium nuclei) during this process.

Boron carbide (B_4C) is a hard, black crystal that is used as an abrasive powder and as an additive to strengthen composite parts in aircraft.

Boric acid (boracic acid; H_3BO_3) is used for the manufacture of glass, welding, mattress batting, cotton textiles, and a weak eyewash solution.

Refined borax ($Na_2B_4O_7$) is an additive in laundry products such as soaps and water-softening compounds. Also used for cosmetics, body powders, and the manufacture of paper and leather. Borax is an environmentally safe natural herbicide and insecticide.

Hazards

Powdered or fine dust of elemental boron is explosive in air and toxic if inhaled. Several of the compounds of boron are very toxic if ingested or if they come in contact with the skin. This is particularly true of the boron compounds used for strong insecticides and **herbicides.**

ALUMINUM

SYMBOL: Al   PERIOD: 3   GROUP: 13 (IIIA)   ATOMIC NO: 13

ATOMIC MASS: 26.981538 amu   VALENCE: 3   OXIDATION STATE: +3

NATURAL STATE: Solid

ORIGIN OF NAME: From the Latin word *alumen,* or aluminis, meaning "alum," which is a bitter tasting form of aluminum sulfate or aluminum potassium sulfate.

ISOTOPES: There are 23 isotopes of aluminum, and only one of these is stable. The single stable isotope, Al-27, accounts for 100% of the element's abundance in the Earth's crust. All the other isotopes are radioactive with half-lives ranging from a few nanoseconds to $7.17 \times 10^{+15}$ years.

ELECTRON CONFIGURATION

Energy Levels/Shells/Electrons	Orbitals/Electrons
1-K = 2	s2
2-L = 8	s2, p6
3-M = 3	s2, p1

Properties

Pure metallic aluminum is not found in nature. It is found as a part of compounds, especially compounded with oxygen as in aluminum oxide (Al_2O_3). In its purified form, aluminum is a bluish-white metal that has excellent qualities of malleability and ductility. Pure aluminum is much too soft for construction or other purposes. However, adding as little as 1% each of silicon and iron will make aluminum harder and give it strength.

Its melting point is 660.323°C, its boiling point is 2,519°C, and its density is 2.699 g/cm³.

Characteristics

Alloys of aluminum are light and strong and can easily be formed into many shapes—that is, it can be extruded, rolled, pounded, cast, and welded. It is a good conductor of electricity and heat. Aluminum wires are only about 65% as efficient in conducting electricity as are copper wires, but aluminum wires are significantly lighter in weight and less expensive than copper wires. Even so, aluminum wiring is not used in homes because of its high electrical resistance, which can build up heat and may cause fires.

Aluminum reacts with acids and strong alkali solutions. Once aluminum is cut, the fresh surface begins to oxidize and form a thin outer coating of aluminum oxide that protects the metal from further corrosion. This is one reason aluminum cans should not be discarded in the environment. Aluminum cans last for many centuries (though not forever) because atmospheric gases and soil acids and alkalis react slowly with it. This is also the reason aluminum is not found as a metal in its natural state.

Abundance and Source

Aluminum is the third most abundant element found in the Earth's crust. It is found in concentrations of 83,200 ppm (parts-per-million) in the crust. Only the nonmetals oxygen and silicon are found in greater abundance. Aluminum oxide (Al_2O_3) is the fourth most abundant compound found on Earth, with a weight of 69,900 ppm. Another alum-type compound is potassium aluminum sulfate [$KAl(SO_4)_2 \bullet 12H_2O$]. Although aluminum is not found in its free metallic state, it is the most widely distributed metal (in compound form) on Earth. Aluminum is also the most abundant element found on the moon.

Almost all rocks contain some aluminum in the form of aluminum silicate minerals found in clays, feldspars, and micas. Today, bauxite is the major ore for the source of aluminum metal. Bauxite was formed eons ago by the natural chemical reaction of water, which then formed aluminum hydroxides. In addition to the United States, Jamaica and other Caribbean islands are the major sources of bauxite. Bauxite deposits are found in many countries, but not all are of high concentration.

History

Beginning in the 1700s and beyond, scientists suspected that an unknown metal existed in alum. Alum is found in the form of several compounds—for example, aluminum ammonium sulfate [$AlNH_4ISO_4)_2 \bullet 12H_2O$], aluminum potassium sulfate [$AlK(SO_4)_2 \bullet 12H_2O$], or aluminum sulfate [$Al_2(SO_4)_3$]. The scientists' problem was that they had no techniques or knowledge of how to extract the metal from its ore until 1825 when the Danish chemist Hans Christian Oersted (1777–1851) isolated a very small amount of aluminum by melting alum with potassium. However, it was not enough for practical use. Napoleon considered it a precious metal. Later, a German scientist, Friederich Wohler (1800–1882), found a way to extract enough aluminum to analyze its chemical and physical properties. In the latter part of the nineteenth century, a French chemist, Henri Etienne Sainte-Claire Deville (1818–1881), produced the first commercial aluminum, which reduced the price from about $1,200 per pound to about $40 a pound. This was still too expensive for commercial use, but aluminum "silverware" and other utensils were all the rage and used by royalty during this time in history.

This picture changed in the 1886 when an American chemist, Charles Martin Hall (1863–1914), and a French chemist, Paul Louis-Toussaint Heroult (1863–1914), both discovered, at about the same time, a new process for extracting aluminum from molten aluminum oxide by electrolysis. (It might be noted that both discoverers have the same birth and death dates as well as the same date of discovery.) Hall was inspired by his teacher to find a way to inexpensively produce aluminum metal. He wired together numerous "wet cells" to form a "battery" that produced enough electricity to separate the aluminum from the melted aluminum oxide (mixed with the minerals cryolyte or fluorite), by the process known as **electrolysis.** Hall formed the Pittsburgh Reduction Co., which is now known as the Aluminum Company of America, or Alcoa. His company produced so much aluminum that the price dropped to about sixty cents per kilogram.

A few years later, an Austrian chemist, Karl Joseph Bayer, refined Hall's process, and it is now called the Hall-Heroult or Bayer process, which is the method used today for obtaining aluminum at very reasonable prices

Common Uses

Aluminum is a very versatile metal with many uses in today's economy, the most common of which are in construction, in the aviation-space industries, and in the home and automobile industries. Its natural softness is overcome by alloying it with small amounts of copper or magnesium that greatly increase its strength. It is used to make cans for food and drinks, in pyrotechnics, for protective coatings, to resist corrosion, to manufacture die-cast auto engine blocks and parts, for home cooking utensils and foil, for incendiary bombs, and for all types of alloys with other metals.

Aluminum does not conduct electricity as well as copper, but because it is much lighter in weight, it is used for transmission lines, though not in household wiring. A thin coating of aluminum is spread on glass to make noncorroding mirrors. Pure oxide crystals of aluminum are known as corundum, which is a hard, white crystal and one of the hardest substances known. Corundum finds many uses in industry as an abrasive for sandpaper and grinding wheels. This material also resists heat and is used for lining high-temperature ovens, to form the white insulating part of spark plugs, and to form a protective coating on many electronic devices such a transistors.

Aluminum oxide is used to make synthetic rubies and sapphires for lasers beams. It has many pharmaceutical uses, including ointments, toothpaste, deodorants, and shaving creams.

Aluminum scrap is one of the salvaged and recycled metals that is less expensive to reuse than it is to extract the metal from its ore. In other words, it takes much less electricity to melt scrap aluminum than it does to extract aluminum from bauxite.

Examples of Compounds

Several examples of compounds in aluminum's oxidation state of +3 follow:

Aluminum chloride: $Al^{3+} + 3Cl^{1-} \rightarrow AlCl_3$ Aluminum chloride is a crystal that vaporizes in air and is explosive in water as it forms aluminum oxide and hydrochloric acid, as follows: $2AlCl_3 + 3H_2O \rightarrow Al_2O_3 + 6HCl$. Aluminum chloride is used as a catalyst in many organic reactions.

Aluminum fluoride: $2Al^{3+} + 3F^{2-} \rightarrow Al_2F_3$. Aluminum fluoride is used to produce low-melting aluminum metal, as a flux in ceramic glazes and white enamels, and as a catalyst in chemical reactions.

Aluminum oxide: $2Al^{3+} + 3O^{2-} \rightarrow Al_2O_3$ is known as the mineral bauxite. Its main use is for the production of aluminum metal by electrolysis. It is also used in many other chemical reactions.

Aluminum sulfate: $Al_2O_3 + 3H_2SO_4 \rightarrow Al_2(SO_4)_3 + 3H_2O$. This reaction treats bauxite with sulfuric acid, resulting in aluminum sulfate plus water. Aluminum sulfate is also known as alum, which acts as an astringent to stop a minor flow of blood or dry up blisters (potassium aluminum sulfate is also an alum).

Aluminum alloys are not really compounds, but rather mixtures of other metals with aluminum to produce stronger metal. Some of the metals used as alloy metals with aluminum are copper, manganese, silicon, magnesium, zinc, chromium, zirconium, vanadium, lead, and bismuth. One of the most important alloys is known as duralumin. It is a heat-treated mixture of aluminum, copper, magnesium, and manganese. It is very strong, lightweight, and noncorrosive, making it useful in the aerospace industries.

Hazards

Aluminum dust and fine powder are highly explosive and can spontaneously burst into flames in air. When treated with acids, aluminum chips and coarse powder release hydrogen. The heat from the chemical reaction can then cause the hydrogen to burn or explode. Pure aluminum foil or sheet metal can burn in air when exposed to a hot enough flame. Fumes from aluminum welding are toxic if inhaled.

GALLIUM

SYMBOL: Ga   **PERIOD:** 4   **GROUP:** 13 (IIIA)   **ATOMIC NO:** 31
ATOMIC MASS: 69.723 amu   **VALENCE:** 2 and 3   **OXIDATION STATE:** +3
   NATURAL STATE: Solid
ORIGIN OF NAME: Latin word *Gallia,* meaning "Gaul," an early name for France.
ISOTOPES: There are 33 isotopes of gallium, two of which are stable. They are Ga-69, which makes up 60.108% of the element's presence in the Earth's crust, and Ga-71, which contributes 39.892% of the gallium found in the Earth's crust. All the other 31 isotopes are radioactive with half-lives ranging from a few nanoseconds to about 15 hours.

ELECTRON CONFIGURATION

Energy Levels/Shells/Electrons	Orbitals/Electrons
1-K = 2	s2
2-L = 8	s2, p6
3-M = 18	s2, p6, d10
4-N = 3	s2, p1

Properties

Gallium is soft and bluish off-white when solid and silvery in color as a liquid. It is soft enough to cut with knife and has an extremely low melting point. When held in the hand, it will melt from body heat as it becomes mirror-like in color. It expands when changing back from a liquid to a solid. When cold, it becomes hard and brittle. Of all the metals, gallium exhibits the largest range of temperatures from its liquid phase to its solid phase, and, like water, it expands when it freezes. Its melting point is 29.76°C, its boiling point is 2,204°C, and its density is 5.903 g/cm^3,

Characteristics

Gallium is truly an "exotic" element in that it has so many unusual characteristics. It can form **monovalent** and **divalent** as well as **trivalent** compounds. It is considered a "post-transitional metal" that is more like aluminum than the other elements in group 13. It has few similar characteristics to the two elements just below it in group 13 (In and Ti).

Gallium reacts strongly with boiling water, is slightly soluble in alkali solutions, acids, and mercury, and is used as an amalgam. It has some semiconductor properties but only if "doped" with elements in group 14, such as As, P, and Sb. It is also used as a "dope" for other semiconducting elements.

Gallium is easy to mix with several other metals to produce alloys with low melting points.

Abundance and Source

Gallium is the 34th most abundant element, but it is not widely distributed as an elemental metal. It is usually combined with other elements, particularly zinc, iron, and aluminum ores. It is found in diaspore, sphalerite, germanite, gallite, and bauxite. Although small amounts are recovered from burning coal used for heating or generation of electricity, it is mostly recovered as a by-product from the production of ores of other metals. Gallium is about as abundant as lead in the Earth's crust.

Since 1949, the Aluminum Company of America has extracted gallium metal from aluminum bauxite ore. In the past gallium had few uses. Only recently, with the development of microprocessors, chips, computer, and the like, has gallium found many profitable uses.

History

Gallium is one of the elements Mendeleev predicted to fill the space just below aluminum. He named it "eka-aluminum" and even gave it the chemical symbol Ea because, when found, it would mostly resemble aluminum. He also suggested that it would combine with oxygen with the formula Ea$_2$O$_3$.

In 1875 Paul-Emile Lecoq de Boisbaudran (1838–1912), a French chemist, used Mendeleev's clues along with the aid of a spectroscope to identify the metal missing between aluminum and indium. He found gallium's spectral lines in the mineral sphalerite, which is a zinc/sulfide mineral that at one time was thought to be "useless lead." Sometime later, the scientist used the process of electrolysis with molten gallium hydroxide [Ga(OH)$_3$] and potassium hydroxide (KOH) to produce a pure sample of gallium metal. Paul-Emile Lecoq de Boisbaudran is given credit for the discovery of gallium, which he named after his native country, France, the Latin name for which is Gaul.

Common Uses

The compound gallium arsenide (GaAs) has the ability to convert electricity directly into laser-light used as the laser beam in compact disc players. It is also used to make light-emitting diodes (LEDs) for illuminated displays of electronic devices such as watches. Gallium is also a semiconductor that when used in computer chips generates less heat than silicon chips, making it a viable option for designing supercomputers that otherwise would generate excessive heat.

The radioisotope of gallium-67 is one of the first to be used in medicine. It has the ability to locate and concentrate on malignant tissue, such as skin cancers, without harming normal tissue in the same area.

One of the more recent uses of gallium is based on the fact that normal gallium, when bombarded by neutrinos, is converted into the radioisotope germanium-71, which can be detected by sensitive instruments. Neutrinos are subatomic particles that "bathe" the Earth as a product of the sun's thermonuclear activity and, from outer space, and can easily go through miles of solid rock. Neutrinos are classed as leptons, which are somewhat like electrons, but with no electrical charge and with no or very little mass. Two large gallium detectors are buried deep underground, one in the tunnel known as Gran Sasso in Italy, and the other, named SAGE, under the Caucasus Mountains in Russia. Scientists from the United States also run this neutrino detector in cooperation with the Russians, thus the name Soviet-American Gallium Experiment (SAGE). Buried deep underground is a deposit of 250,000 pounds of gallium that has a market price of about $400 a pound. Reportedly, attempts have been made to steal this stash of gallium. The purpose of this experiment is to indirectly identify the elusive neutrinos as they convert the gallium trichloride ($GaCl_3$) to the radioactive isotope germanium-71, which is then exposed to sensitive instruments that detect radiation, thus revealing the existence of neutrino activity.

Gallium makes a safe substitute for mercury amalgams in dental fillings when it is combined with tin or silver.

Because of its high range of temperatures as a liquid (from 29.8°C to 2,403°C), it is used in special types of high-temperature thermometers. It is also alloyed with other metals to make alloys with low temperature melting points.

Because of the unique property of some of its compounds, gallium is able to translate a mechanical motion into electrical impulses. This makes it invaluable for manufacturing transistors, computer chips, semiconductors, and **rectifiers**.

A unique use of gallium metal is to "glue" gemstones to metal jewelry.

Examples of Compounds

Gallium can combine with –1, –2, and –3 ions as follows:

Gallium chloride: $Ga^{3+} + 3Cl^{1-} \rightarrow GaCl_3$.

Gallium oxide: $Ga^{3+} + 3O^{2-} \rightarrow Ga_2O_3$.

Gallium arsenide ($Ga^{3+} + As^{3-} \rightarrow GaAs$) is **electroluminscent** in infrared light and is used for telephone equipment, lasers, solar cell, and other electronic devices.

Galllium phosphide (GaP) is a light-colored highly pure crystal form used as "whiskers" and crystals in semiconductor devices.

Gallium antimonide (GaSb), when in pure form, is used in semiconductor industries.

Hazards

Most gallium compounds are toxic, particularly the metal gallium arsenide. When forms of gallium are used in the electronics industry, great care must be taken to protect workers.

INDIUM

SYMBOL: In   **PERIOD:** 5   **GROUP:** 13 (IIIA)   **ATOMIC NO:** 49
ATOMIC MASS: 114.818 amu   **VALENCE:** 1 and 3   **OXIDATION STATE:** +3
   NATURAL STATE: Solid
ORIGIN OF NAME: Indium's name is derived from the Latin word *indicum,* meaning "indigo," which is the color of its spectral line when viewed by a spectroscope.
ISOTOPES: There are a total of 73 isotopes of indium. All are radioactive with relatively short half-lives, except two that are considered stable. Isotope In-113 makes up just 4.29% of the total indium found in the Earth's crust. The isotope In-115, with a half-life of 4.41×10^{-14} years contributes the balance (95.71%) of the element's existence in the Earth's crust.

ELECTRON CONFIGURATION

Energy Levels/Shells/Electrons	Orbitals/Electrons
1-K = 2	s2
2-L = 8	s2, p6
3-M = 18	s2, p6, d10
4-N = 18	s2, p6, d10
5-O = 3	s2, p1

Properties

Indium is silvery-white and malleable and looks much like aluminum and tin. However, it is softer than lead. Indium metal is so soft that it cannot be "wiped" onto other surfaces as with a graphite pencil. Because it is noncorrosive and does not oxidize at room temperatures, it can be polished and will hold its shine better than silver. Its melting point is 156.60°C, its boiling point is 2,075°C, and its density is 7.31 g/cm^3.

Characteristics

Indium has one odd characteristic in that in the form of a sheet, like the metal tin, it will emit a shrieking sound when bent rapidly. Indium has some of the characteristics of other metals near it in the periodic table and may be thought of as an "extension" of the second series of the transition elements. Although it is corrosion-resistant at room temperature, it will oxidize at higher temperatures. It is soluble in acids, but not in alkalis or hot water.

Abundance and Source

Indium is a rather rare metal. It is the 69th most abundant element, which is about as abundant as silver at 0.05 ppm. Although it is widely spread over the Earth's crust, it is found in very small concentrations and always combined with other metal ores. It is never found in its natural metallic state.

Indium is recovered as a by-product of smelting other metal ores such as aluminum, antimony, cadmium, arsenic, and zinc. About 1,000 kg of indium is recovered each year (or a concentration of 1 part indium per 1000 parts of dust) from the flue stacks (chimneys) of zinc refineries.

Indium is found in metal ores and minerals located in Russia, Japan, Europe, Peru, and Canada, as well as in the western part of the United States.

History

While searching for traces of one element contained in some zinc ore, Ferdinand Reich (1799–1882) in 1863 accidentally produced a yellow sulfide of what he suspected was a new element. By this time in history the spectroscope was being used to identify elements by their unique color spectrum. Reich was colorblind and could not use the instrument, but he asked his assistant, Hieronymous Theodor Richter (1824–1898), to examine the new element with the spectroscope. (A spectroscope is an instrument used to analyze the wavelength of light emitted or absorbed by elements when excited. Each element has a unique line or color in the light spectrum.) Richter saw the separate deep-purple line of the new element, and together Reich and Richter became the codiscoverers of indium, which they named after the indigo plant that grew in India, as well as in Egypt and China.

The indigo plant was used as early as 3000 BCE to make purple dyes. For many years humans used insects, snails, and plants to make dyes of bright red, deep red, purple, brown, yellow, and black. The processes of extracting the dyes were expensive. Only the wealthy, and primarily royalty at this early time in history, could afford the deep purple dyes for their clothing and robes. Thus, the tradition of purple as the color of kings was originated.

By the late 1800s, many chemists were trying to synthetically produce dyes such as indigo, but it was not until 1900 that synthetic indigo production was a commercial success. The synthetic production of indigo and other synthetic dyes, most notably the color mauve, destroyed the commercial agriculture trade of the extensive indigo-growing regions of India and other countries that depended on growing the plants for their livelihood.

Common Uses

Indium's low melting point is the major factor in determining its commercial importance. This factor makes it ideal for soldering the lead wires to semiconductors and transistors in the electronics industry. The compounds of indium arsenide, indium antimonide, and indium phosphide are used to construct semiconductors that have specialized functions in the electronics industry.

Another main use is as an alloy with other metals when it will lower the melting point of the metals with which it is alloyed. Alloys of indium and silver and indium and lead have the ability to carry electricity better than pure silver and lead.

Indium is used as a coating for steel bearings to increase their resistance to wear. It also has the ability to "wet" glass, which makes it an excellent mirror surface that lasts longer than mercury mirrors. Sheets of indium foil are inserted into nuclear reactors to help control the nuclear fission reaction by absorbing some of the neutrons.

Examples of Compounds

Indium's stable oxidation state is +3 and thus, being trivalent, it forms compounds with elements with −1, −2, and −3 oxidation states as follows:

Indium trichloride ($In^{3+} + 3Cl^{1-} \rightarrow InCl_3$). Hazardous: keep container closed.

Indium trioxide ($2In^{3+} + 3O^{2-} \rightarrow In_2O_3$). Used in specialty glass production.

Indium arsenide ($In^{3+} + As^{3-} \rightarrow InAs$). Used in semiconductor devices.

There are many other compounds of indium, most of which are useful in the electronics and semiconductor industries. Some other examples are InP, $In_2(SO_4)_3$, In_2Te_3, and $InSb$.

Hazards

Indium metal dust, particles, and vapors are toxic if ingested or inhaled, as are most of the compounds of indium. This requires the semiconductor and electronics industries that use indium compounds to provide protection for their workers.

THALLIUM

SYMBOL: Tl   **PERIOD:** 6   **GROUP:** 13 (IIIA)   **ATOMIC NO:** 81

ATOMIC MASS: 204.383 amu   **VALENCE:** 1 and 3   **OXIDATION STATE:** +1 and +3   **NATURAL STATE:** Solid

ORIGIN OF NAME: From the Greek word *thallos,* meaning "young shoot" or "green twig." Named for the green spectral line produced by the light from the element in a spectroscope.

ISOTOPES: There are a total of 55 isotopes for thallium. All are radioactive with relatively short half-lives, and only two are stable. The stable ones are Tl-203, which constitutes 29.524% of the element's existence in the Earth's crust, and Tl-205, which makes up 70.476% of the element's natural abundance found in the Earth's crust.

ELECTRON CONFIGURATION

Energy Levels/Shells/Electrons	Orbitals/Electrons
1-K = 2	s2
2-L = 8	s2, p6
3-M = 18	s2, p6, d10
4-N = 32	s2, p6, d10, f14
5-O = 18	s2, p6, d10
6-P = 3	s2, p1

Properties

Thallium has much the same look (silvery) and feel as lead and is just as malleable. Unlike lead, which does not oxidize readily, thallium will oxidize in a short time, first appearing as a dull gray, then turning brown, and in just a few years or less turning into blackish corroded chunks of thallium hydroxide. This oxide coating does not protect the surface of thallium because it merely flakes off exposing the next layer to oxidation.

Thallium is just to the left of lead in period 6, and both might be considered extensions of the period 6 transition elements. Thallium's high corrosion rate makes it unsuitable for most commercial applications. Its melting point is 304°C, its boiling point is 1,473°C, and its density is 11.85 g/cm³.

Characteristics

Elemental thallium metal is rare in nature mainly because it oxidizes if exposed to air (oxygen) and water vapor, forming thallium oxide, a black powder. Although some compounds of thallium are both toxic and carcinogenic, they have some uses in the field of medicine. Some compounds have the ability to alter their electrical conductivity when exposed to infrared light.

Abundance and Source

Thallium is the 59th most abundant element found in the Earth's crust. It is widely distributed over the Earth, but in very low concentrations. It is found in the mineral/ores of crooksite (a copper ore; CuThSe), lorandite (TlAsS$_2$), and hutchinsonite (lead ore, PbTl). It is found mainly in the ores of copper, iron, sulfides, and selenium, but not in its elemental metallic state. Significant amounts of thallium are recovered from the flue dust of industrial smokestacks where zinc and lead ores are smelted.

History

Prior to its discovery, no one knew that thallium existed, and thus, no one was looking for it. Therefore, it was discovered by accident by Sir William Crookes (1832–1919) in 1861. After removing selenium from a sample of minerals, he observed the sample with his spectroscope, expecting to see some blue lines of leftover selenium and the yellow lines of tellurium, but instead he observed a never-before-seen bright green line, which indicated a new element. This green line inspired him to name the new element after the Greek word *thallos,* which stands for "green twig."

Common Uses

Thallium is used as an alloy with mercury and other metals. One main use is in photoelectric applications and for military infrared radiation transmitters.

It is also used to make artificial gemstones and special glass and to make green colors in fireworks and flares. It formerly was used as a rat poison, but is no longer used for this purpose because it is very toxic to humans.

Another main use is the radioisotope TlCl-201, with the relatively short half-life of about 73 hours, in cardiac stress tests to identify potential heart abnormalities. TlCl-201 has an ability to bind with the heart muscle, but only if the heart is receiving an adequate supply of blood. Restricted blood flow by blocked or narrow arteries in the heart limits the supply of TlCl-201 absorbed. First, a small dose of TlCl-201 is injected into the patient, and the patient then engages in a strenuous workout on a treadmill. Both before and after the test, the patient is scanned by a "gamma" detector that sends the results to a computer where the physician can compare the uptake of TlCl-201 before and after the treadmill stress test to determine the condition of the patient's heart. An area where the heart's muscle is weak and the blood flow is limited will show up as a darkish spot on the computer screen. Since the radioisotope TlCl-201 has such a short half-life, it is soon excreted from the body. Thus, there are no long-term detriments to the body.

Examples of Compounds

Two examples of compounds in thallium +1 and +3 oxidation states follow:
Thallium (I) hydroxide (Tl^{1+} + OH^{1-} → TlOH) is used in optical glass.
Thallium (III) chloride: Tl^{3+} + $3Cl^{1-}$ → $TlCl_3$.
A few other compounds include the following:
Thallium carbonate (Tl_2CO_3) is used to make artificial diamonds (along with several other thallium compounds).
Thallium sulfide (Tl_2S) is used to make infrared-sensitive photoelectric cells and as a pesticide.
Thallium-mercury is an amalgam—not really a compound, but more like an alloy mixture. It is used in low-temperature thermometers and as a substitute for mercury in low-temperature switches.

Hazards

In all forms, thallium is very toxic if inhaled, when in contact with the skin, and in particular, if ingested. Mild thallium poisoning causes loss of muscle coordination and burning of the skin, followed by weakness, tremor, mental aberration, and confusion.

Thallium disease (thallotoxicosis) results from the ingestion of relatively large doses (more than a few micrograms). The severity may vary with the age and health of the patient. Nerves become inflamed, hair is lost, the patient experiences stomach pain, cramps, hemorrhage, rapid heartbeat, delirium, coma, and respiratory paralysis. The disease has the potential to cause death in about one week. In the past thallium was one of the poisons of choice used by murderers because it acts slowly and makes victims suffer. In 1987 then Iraqi dictator Saddam Hussein's agents mixed thallium powder in orange juice or yogurt and fed it to people he perceived to be his enemies. There were at least 40 thallium poisonings, mostly of Kurdish leaders. (William Langewiesche, "The Accuser," *Atlantic,* March 2005, 56.)

The Carbon Group (Metalloids to Semiconductors): Periods 2 to 6, Group 14 (IVA)

Introduction

Elements in the carbon group are fairly well known except for germanium. They are arranged vertically in group 14 (IVA) of the periodic table with carbon (C) at the top of the group, followed by silicon (Si), germanium (Ge), tin (Sn), and then lead (Pb).

The elements Si and Ge of group 14 act as semiconductors. A semiconductor is an element that can, to some extent, conduct electricity and heat, meaning it has the properties of both metal and nonmetals. The ability of semiconductors to transmit variable electrical currents can be enhanced by controlling the type and amount of impurities. This is what makes them act as "on-off" circuits to control electrical impulses. This property is valuable in the electronics industry for the production of transistors, computer chips, integrated circuits, and so on. In other words, how well a semiconductor conducts electricity is not entirely dependent on the pure element itself, but also depends on the degree of its impurities and how they are controlled.

Tin and lead might even be thought of as extensions of the second and third series of transition elements. The elements in group 14 form many useful compounds, but are also recognizable and useful in their elemental forms. With the exception of silicon, they are not found in great abundance in the Earth's crust, but they may be found concentrated in their mineral ores. Carbon is one of the most versatile elements found on Earth. It can form a multitude of compounds, both inorganic and organic, and is the basis for all living organisms

CARBON

SYMBOL: C   PERIOD: 2   GROUP: 14 (IVA)   ATOMIC NO: 6
ATOMIC MASS: 12.01115 amu.   VALENCE: 4   OXIDATION STATE: +2, +4, and −
    4   NATURAL STATE: Solid
ORIGIN OF NAME: Carbon's name is derived from the Latin word *carbo,* which means, "charcoal."

PERIODIC TABLE OF THE ELEMENTS

GROUPS / PERIODS	1 IA	2 IIA	3 IIIB	4 IVB	5 VB	6 VIB	7 VIIB	8	9 VIII	10	11 IB	12 IIB	13 IIIA	14 IVA	15 VA	16 VIA	17 VIIA	18 VIIIA
1	1 H 1.0079																	2 He 4.00260
2	3 Li 6.941	4 Be 9.01218											5 B 10.81	6 C 12.011	7 N 14.0067	8 O 15.9994	9 F 18.9984	10 Ne 20.179
3	11 Na 22.9898	12 Mg 24.305											13 Al 26.9815	14 Si 28.0855	15 P 30.9738	16 S 32.066(6)	17 Cl 35.453	18 Ar 39.948
4	19 K 39.0983	20 Ca 40.08	21 Sc 44.9559	22 Ti 47.88	23 V 50.9415	24 Cr 51.996	25 Mn 54.9380	26 Fe 55.847	27 Co 58.9332	28 Ni 58.69	29 Cu 63.546	30 Zn 65.39	31 Ga 69.72	32 Ge 72.59	33 As 74.9216	34 Se 78.96	35 Br 79.904	36 Kr 83.80
5	37 Rb 85.4678	38 Sr 87.62	39 Y 88.9059	40 Zr 91.224	41 Nb 92.9064	42 Mo 95.94	43 Tc (98)	44 Ru 101.07	45 Rh 102.906	46 Pd 106.42	47 Ag 107.868	48 Cd 112.41	49 In 114.82	50 Sn 118.71	51 Sb 121.75	52 Te 127.60	53 I 126.905	54 Xe 131.29
6 ★	55 Cs 132.905	56 Ba 137.33	★	72 Hf 178.49	73 Ta 180.948	74 W 183.85	75 Re 186.207	76 Os 190.2	77 Ir 192.22	78 Pt 195.08	79 Au 196.967	80 Hg 200.59	81 Tl 204.383	82 Pb 207.2	83 Bi 208.980	84 Po (209)	85 At (210)	86 Rn (222)
7 ▲	87 Fr (223)	88 Ra 226.025	▲	104 Unq (261)	105 Unp (262)	106 Unh (263)	107 Uns (264)	108 Uno (265)	109 Une (266)	110 Uun (267)	111 Uuu (272)	112 Uub	113 Uut	114 Uuq	115 Uup	116 Uuh	117 Uus	118 Uuo

TRANSITION ELEMENTS

6 ★ Lanthanide Series (RARE EARTH)

57 La 138.906	58 Ce 140.12	59 Pr 140.908	60 Nd 144.24	61 Pm (145)	62 Sm 150.36	63 Eu 151.96	64 Gd 157.25	65 Tb 158.925	66 Dy 162.50	67 Ho 164.930	68 Er 167.26	69 Tm 168.934	70 Yb 173.04	71 Lu 174.967

7 ▲ Actinide Series (RARE EARTH)

89 Ac 227.028	90 Th 232.038	91 Pa 231.036	92 U 238.029	93 Np 237.048	94 Pu (244)	95 Am (243)	96 Cm (247)	97 Bk (247)	98 Cf (251)	99 Es (252)	100 Fm (257)	101 Md (258)	102 No (259)	103 Lr (260)

ISOTOPES: There are 15 isotopes of carbon, two of which are stable. Stable carbon-12 makes up 98.89% of the element's natural abundance in the Earth's crust, and carbon-13 makes up just 1.11% of carbon's abundance in the Earth's crust. All the other isotopes of carbon are radioactive with half-lives varying from 30 nanoseconds (C-21) to 5,730 years (C-14).

ELECTRON CONFIGURATION

Energy Levels/Shells/Electrons	Orbitals/Electrons
1-K = 2	s2
2-L = 4	s2, p2

Properties

All the elements in group 14 have four electrons in their outer valence shell. Carbon exhibits more nonmetallic properties than do the others in group 14 and is unique in several ways. It has four forms, called allotropes:

1. *Carbon black* is the amorphous allotrope (noncrystal form) of carbon. It is produced by heating coal at high temperatures (producing coke); burning natural gas (producing jet black); or burning vegetable or animal matter (such as wood and bone), at high temperatures with insufficient oxygen, which prevents complete combustion of the material, thus producing charcoal.

2. *Graphite* is a unique crystal structure of carbon wherein layers of carbon atoms are stacked parallel to each other and can extend indefinitely in two dimensions as in the shafts of carbon fiber golf clubs. Graphite is also one of the softest elements, making it an excellent dry lubricant.

3. *Diamonds* are another allotrope whose crystal structure is similar to graphite. Natural diamonds were formed under higher pressure and extreme temperatures. Synthetic diamonds have been artificially produced since 1955.

4. *Fullerenes* are another amorphous (no crystal structure) form of carbon that have the basic formula of $C_{60}H_{60}$ and are shaped like a soccer ball. (See the "Atomic Structure" section of the book for more on fullerenes.)

The different allotropes of carbon were formed under varying conditions in the Earth, starting with different minerals, temperature, pressure, and periods of time. Once the distinct crystal structures are formed, they are nearly impossible to change.

Carbon-12 is the basis for the average atomic mass units (amu) that is used to determine the atomic weights of the elements. Carbon is one of the few elements that can form covalent bonds with itself as well as with many metals and nonmetals.

Each allotropic form of carbon has its own melting point, boiling point, and density. For instance, the density of the amorphous allotrope is 1.9 g/cm^3, and it is 2.25 g/cm^3 for graphite and 3.52 g/cm^3 for diamonds.

Characteristics

Carbon is, without a doubt, one of the most important elements on Earth. It is the major element found in over one million organic compounds and is the minor component in minerals such as carbonates of magnesium and calcium (e.g., limestone, marble, and dolomite), coral, and shells of oysters and clams.

The carbon cycle, one of the most essential of all biological processes, involves the chemical conversion of carbon dioxide to carbohydrates in green plants by photosynthesis. Animals consume the carbohydrates and, through the metabolic process, reconvert the carbohydrates back into carbon dioxide, which is returned to the atmosphere to continue the cycle.

Abundance and Source

Carbon is the 14th most abundant element, making up about 0.048% of the Earth's crust. It is the sixth most abundant element in the universe, which contains 3.5 atoms of carbon for every atom of silicon. Carbon is a product of the cosmic nuclear process called **fusion,** through which helium nuclei are "burned" and fused together to form carbon atoms with the atomic number 12. Only five elements are more abundant in the universe than carbon: hydrogen, helium, oxygen, neon, and nitrogen.

History

Carbon was known in prehistoric times in the form of charcoal and later as peat and coal deposits. Graphite's name was derived from the Greek word *graphein,* meaning "to write." resulting from the fact that graphite was used to make dark marks on paper. At one time, early chemists confused graphite with lead and molybdenite. Diamonds were also known in ancient times. The word diamond is from the Greek word *adamas,* meaning "the invincible," which well describes this allotrope of carbon.

Antoine-Laurent Lavoisier (1743–1794) is known as the "father of modern chemistry" because he believed in weighing, measuring, observing, heating, and testing the substances with which he experimented, as well as in keeping accurate records of his findings. He was among the first to experiment with carbon chemistry, and his techniques led to the field of modern quantitative chemistry.

Lavoisier, along with a number of other chemists, pooled their funds and purchased a diamond, which they placed in a closed glass jar. Using a magnifying glass, they focused the sun's rays on the diamond. This produced enough heat to make the diamond disappear. Given that the weight of the glass container that held the diamond was unchanged, Lavoisier determined that the colorless substance in the glass container was a gas—in this case, carbon dioxide. He concluded that the no-longer-visible diamond was the carbon that combined with the oxygen in the container to form CO_2.

For years it was thought that diamonds were made of carbon atoms, just like graphite and coal, but no one could demonstrate this. In 1955 scientists were able to produce the tremendous pressure (over 100,000 times normal) and temperatures over 2,500°C to form a synthetic diamond from graphite that appears to be as real as a naturally formed diamond. However, these high-grade "real" man-made diamonds are much too expensive to mass-produce. Nevertheless, the natural supply of diamonds cannot meet the needs of industry, so low-grade industrial diamonds are produced synthetically by forcing graphite, under great pressure and temperature, to form industrial diamonds that are used as abrasives and are placed on the tips of saw blades to improve wearability.

Today, carbon chemistry is more closely related to **organic** and **hydrocarbon** chemistry than to the elemental allotropes of carbon. Over the past century organic and hydrocarbon chemistry has opened up vast areas of research and development leading to new commercial processes and products.

Common Uses

There are many uses for the very versatile element carbon. It, no doubt, forms more compounds than any other element, particularly in the world of modern carbon chemistry. Carbon's nature allows the formation-rings and straight- and branched-chains types of compounds that are capable of adding hydrogen as well as many different types of elemental atoms to these structures. (See figure 5 in the book's section titled "Atomic Structure" for a depiction of a snake eating its tail as an analogy for the carbon ring of benzene.) In addition, these ringed, straight, and branched carbon molecules can be repeated over and over to form very large molecules such as the polymers, proteins, and carbohydrates that are required for life.

Carbon is an excellent reducing agent because it readily combines with oxygen to form CO and CO_2. Thus, in the form of coke in blast furnaces, it purifies metals by removing the oxides and other impurities from iron.

Carbon, as graphite, has strong electrical conductivity properties. It is an important component in electrodes used in a variety of devices, including flashlight cells (batteries). Amorphous carbon has some superconduction capabilities.

Graphite is used for the "lead" in pencils, as a dry lubricant, and as electrodes in arc lamps. Of course, carbon is a popular jewelry item (e.g., diamonds).

Future uses of carbon in the forms of fullerenes (C_{60} up to C_{240}) and applications of nanotechnology will provide many new and improved products with unusual properties. (See the section on "Atomic Structure" for more on these topics.)

Examples of Compounds

Carbon-14 is a naturally occurring radioactive form of carbon with a half-life of 5,730 years. **Carbon dating** is used to "date" any type of substance that was at one time "living." A small amount of C-14 is always found with C-12. Because carbon-14 is radioactive, the rate of decay of carbon-14 can be calculated accurately to confirm the date when the organic substance was living. (Carbon was not "lost" over time, nor is any more added to the ancient sample, but C-14 does lose a specific amount of radiation at a given rate over time.) Therefore, calculating the ratio between C-12 and C-14 in an organic sample will tell you when that sample was alive.

Sucrose ($C_{12}H_{22}O_{11}$) is one of many forms of sugars (carbohydrates) that are important organic compounds for maintaining life.

Carbon dioxide (CO_2) is the 18th most frequently produced chemical in the United States. It has numerous uses, including in refrigeration, in the manufacture of carbonated drinks (e.g., soda pop), in fire extinguishers, in providing an inert atmosphere (unreactive environment), and as a moderator for some types of nuclear reactors.

Hydrocarbons are used as fuels and as the basic source of many other chemical compounds. The production of coke from coal also produces by-products known as coal-tars, which are used in the pharmaceutical, dye, food, and other industries. The refining of crude oil produces gasoline and many other fractions of the crude oil as well as **petrochemical** by-products. The range of useful products we derive from crude oil is very broad. These products not only power our automobiles, trucks, trains, and planes, but also provide the base for many of our medicines, foods, and numerous other essential products. (See the section of the book titled "Atomic Structure" for more on the chemistry of hydrocarbons.)

Hazards

Many compounds of carbon, particularly the hydrocarbons, are not only toxic but also carcinogenic (cancer-causing), but the elemental forms of carbon, such as diamonds and graphite, are not considered toxic.

Carbon dioxide (CO_2) in its pure form will suffocate you by preventing oxygen from entering your lungs. Carbon monoxide (CO) is deadly, even in small amounts; once breathed into the lungs, it replaces the oxygen in the bloodstream.

Carbon dioxide is the fourth most abundant gas in the atmosphere at sea level. Excess CO_2 produced by industrialized nations is blamed for a slight increase in current temperatures around the globe. CO_2 makes up only 0.03+ percent by volume of the gases in the atmosphere. However, even a small amount in the upper atmosphere seems to be responsible for some global warming. Since pre-industrial times, the concentration of CO_2 in the Earth's atmosphere has risen by approximately one-third, from 280 ppm (parts per million) to about 378 ppm. At the same time methane (CH_4) doubled its concentration over the years to about 2 ppm in the atmosphere. Methane is many times more effective as a "greenhouse" gas than is carbon dioxide, even though it breaks down in a shorter period of time. Some Scandinavian countries have experimented with pumping excess CO_2 produced by their industries deep onto the ocean floor where it will reenter the carbon cycle just as it does through trees and vegetation on the surface of the Earth. There are a number of super-computer programs attempting to predict the extent of global warming. The problem is the number of variables affecting climate change. The process is akin to trying to determine the shape of a cloud over the next hour. Unfortunately, neither well-meaning politicians nor scientists can agree on the extent of potential damage that excess carbon dioxide may do to the Earth in the future. Global warming and cooling are cyclic, which means that these processes have been alternating over eons of time.

SILICON

SYMBOL: Si   **PERIOD:** 3   **GROUP:** 14 (IVA)   **ATOMIC NO:** 14
ATOMIC MASS: 28.0855 amu   VALENCE: 4   OXIDATION STATES: +2, +4 and −4
   NATURAL STATE: Solid
ORIGIN OF NAME: Silicon was named after the Latin word silex, which means "flint."
ISOTOPES: There are 21 isotopes of silicon, three of which are stable. The isotope Si-28 makes up 92.23% of the element's natural abundance in the Earth's crust, Si-29 constitutes 4.683% of all silicon found in nature, and the natural abundance of Si-30 is merely 3.087% of the stable silicon isotopes found in the Earth's crust.

ELECTRON CONFIGURATION

Energy Levels/Shells/Electrons	Orbitals/Electrons
1-K = 2	s2
2-L = 8	s2, p6
3-M = 4	s2, p2

Properties

Silicon does not occur free in nature, but is found in most rocks, sand, and clay. Silicon is electropositive, so it acts like a metalloid or semiconductor. In some ways silicon resembles metals as well as nonmetals. In some special compounds called **polymers,** silicon will act in conjunction with oxygen. In these special cases it is acting like a nonmetal.

There are two allotropes of silicon. One is a powdery brown amorphous substance best known as sand (silicon dioxide). The other allotrope is crystalline with a metallic grayish luster best known as a semiconductor in the electronics industry. Individual crystals of silicon are grown through a method known as the Czochralski process. The crystallized silicon is enhanced by "doping" the crystals (adding some impurities) with other elements such as boron, gallium, germanium, phosphorus, or arsenic, making them particularly useful in the manufacture of solid-state microchips in electronic devices.

The melting point of silicon is 1,420°C, its boiling point is 3,265°C, and its density is 2.33 g/cm^3.

Characteristics

The characteristics of silicon in some ways resemble those of the element germanium, which is located just below it in the carbon group.

Flint is the noncrystalline form of silicon and has been known to humans since prehistoric times. When struck with a sharp blow, flint would flake off sharp-edged chips that were then used as cutting tools and weapons.

In addition to silica (silicon dioxide SiO_2), the crystal form of silicon is found in several semiprecious gemstones, including amethyst, opal, agate, and jasper, as well as quartz of varying colors. A characteristic of quartz is its piezoelectric effect. This effect occurs when the quartz crystal is compressed, producing a weak electrical charge. Just the opposite occurs when electric vibrations are fed to the crystal. These vibrations are then duplicated in the crystal. Quartz crystals are excellent timekeeping devices because of this particular characteristic.

Abundance and Source

Silicon, in the form of silicon dioxide (SiO_2), is the most abundant compound in the Earth's crust. As an element, silicon is second to oxygen in its concentration on Earth, yet it is only the seventh most abundant in the entire universe. Even so, silicon is used as the standard (Si = 1) to estimate the abundances of all other elements in the universe. For example, hydrogen equals 40,000 times the amount of silicon in the cosmos. Hydrogen is the most abundant of all elements in the universe, and carbon is just three and half times as abundant as silicon in the entire universe. On Earth silicon accounts for 28% of the crust, oxygen makes up 47% of the crust, and much of the rest of the crust is composed of aluminum.

It is believed that silicon is the product of the cosmic nuclear reaction in which alpha particles were absorbed at a temperature of 10^9 Kelvin into the nuclei of carbon-12, oxygen-16, and neon-20. Pure elemental silicon is much too reactive to be found free in nature, but it does form many compounds on Earth, mainly oxides as crystals (quartz, cristobalite, and tridymite) and amorphous minerals (agate, opal, and chalcedony). Elemental silicon is produced by reducing silica (SiO_2) in a high-temperature electric furnace, using coke as the reducing agent. It is then refined. Silicon crystals used in electronic devices are "grown" by removing starter crystals from a batch of melted silicon.

History

The practice of heating and forming glass-like figures from melted silica (sand) was known at least three or four thousand years ago. In the eighteenth century early chemists were aware of some kind of link between sand (silica) and quartz but were unaware that a new, unidentified element was involved.

Sir Humphry Davy attempted to isolate this unidentified element through electrolysis—but failed. It was not until 1824 that Jöns Jakob Berzelius (1779–1848), who had earlier discovered cerium, osmium, and iridium, became the first person to separate the element silicon from its compound molecule and then identify it as a new element. Berzelius did this by a two-step process that basically involved heating potassium metal chips with a form of silica (SiF_4 = silicon tetrafluoride) and then separating the resulting mixture of potassium fluoride and silica ($SiF_4 + 4K \rightarrow 4KF + Si$). Today, commercial production of silicon features a chemical reaction (reduction) between sand (SiO_2) and carbon at temperatures over 2,200°C ($SiO_2 + 2C + heat \rightarrow 2CO + Si$).

Silicon research carried out in the nineteenth and early twentieth centuries led to many forms and uses of silicon and its compounds, including silicone plastics, resins, greases, and **polymers.**

Common Uses

Silicon's tetravalent pyramid crystalline structure, similar to tetravalent carbon, results in a great variety of compounds with many practical uses. Crystals of silicon that have been contaminated with impurities (arsenic or boron) are used as semiconductors in the computer and electronics industries. Silicon semiconductors made possible the invention of transistors at the Bell Labs in 1947. Transistors use layers of crystals that regulate the flow of electric current. Over the past half-century, transistors have replaced the vacuum tubes in radios, TVs, and other electronic equipment that reduces both the devices' size and the heat produced by the electronic devices.

Silicon can be used to make solar cells to provide electricity for light-activated calculators and satellites. It also has the ability to convert sunlight into electricity.

When mixed with sodium carbonate (soda ash) and calcium carbonate (powdered limestone) and heated until the mixture melts, silica (sand) forms glass when cooled. Glass of all types has near limitless uses. One example is Pyrex, which is a special heat-resistant glass that is manufactured by adding boron oxide to the standard mixture of silica, soda ash, and limestone. Special glass used to make eyewear adds potassium oxide to the above standard mixture.

Silicon is also useful as an alloy when mixed with iron, steel, copper, aluminum, and bronze. When combined with steel, it makes excellent springs for all types of uses, including automobiles.

When silicon is mixed with some organic compounds, long molecular chains known as silicone polymers are formed. By altering the types of organic substances to these long silicone polymer molecules, a great variety of substances can be manufactured with varied physical properties. Silicones are produced in liquid, semisolid, and solid forms. Silicones may be rubbery, elastic, slippery, soft, hard, or gel-like. Silicone in its various forms has many commercial and industrial uses. Some examples are surgical/reconstructive implants, toys, Silly Putty, lubricants, coatings, water repellents for clothing, adhesives, cosmetics, waxes, sealants, and electrical insulation.

Examples of Compounds

Silicon dioxide (SiO_2) is the most abundant compound in the Earth's crust. Known as common sand, it also exists in the forms of quartz, rock crystal, amethyst, agate, flint, jasper, and opal. It has many industrial uses.

Sodium silicate (Na_2SiO_3), better known as water glass, is one of the few silicon compounds that dissolves in water. Produced at high temperatures (SiO_2 + 2NaOH + heat$\rightarrow$ H_2O + Na_2SiO_3), it is used in the manufacture of soaps, adhesives, and food preservatives.

Silicon nitride (Si_3N_4), resistant to oxidation, is an excellent coating for metals, as well as an adhesive and abrasive, and it is used in high-temperature crucibles. It has proven useful as heat-resistant substance for the nozzles on rocket engines.

Silicon carbide (SiC), nearly as hard as diamonds, is used as an abrasive in grinding wheels and metal-cutting tools, for lining furnaces, and as a **refractory** in producing nonferrous metals.

Silicon tetrachloride ($SiCl_4$), produced when both silicon and chlorine are combined at high temperatures, is used by the military to produce smoke screens. When released in air, it reacts with the moisture in the atmosphere to produce dense clouds of water vapor.

Hazards

The dust of silicon oxide (silicate) can burn or explode and is very harmful if inhaled. Continued exposure to silica dust causes silicosis, a form of pneumonia.

The hydrides of silicon (silicon plus hydrogen) are extremely volatile and spontaneously burst into flames in air at room temperatures. They must be kept in special vacuum chambers.

Over the past several decades, there has been some concern over the potential hazards and safety of the cosmetic use of silicone body implants—breast implants, in particular. Several manufactures have been sued over the failure of the implants, and the federal government

(FDA) withdrew its approval for their use. Congressional hearings with manufacturers in 2005 produced new information that has reversed the FDA's ban on their use—but only with certain manufacturers of implants. The debate continues.

GERMANIUM

SYMBOL: Ge **PERIOD:** 4 **GROUP:** 14 (IVA) **ATOMIC NO:** 32

ATOMIC MASS: 72.61 amu **VALENCE:** 2 and 4 **OXIDATION STATE:** +2 and +4 **NATU-RAL STATE:** Solid

ORIGIN OF NAME: Germanium's name was derived from the Latin word *Germania,* meaning "Germany."

ISOTOPES: There are a total of 38 isotopes of Germanium, five of which are stable. The stable isotopes of germanium and their natural abundance are as follows: Ge-70 = 20.37%, Ge-72 = 27.31%, Ge-73 = 7.76%, Ge-74 = 36.73%, and Ge-76 = 7.83%. Ge-76 is considered stable because it has such a long half-life ($0.8 \times 10^{+25}$ years)All the other 33 isotopes are radioactive and are produced artificially.

ELECTRON CONFIGURATION

Energy Levels/Shells/Electrons	Orbitals/Electrons
1-K = 2	s2
2-L = 8	s2, p6
3-M = 18	s2, p6, d10
4-N = 4	s2, p2

Properties

Germanium has a gray shine with a metallic silvery-white luster. It is a brittle element classed as a semimetal or metalloid, meaning it is neither a metal such as iron or copper nor a nonmetal, such as phosphorus, sulfur, or oxygen. Germanium has some properties like a metal and some like a nonmetal. It is a crystal in its pure state, somewhat like silicon. It will combine with oxygen to form germanium dioxide, which is similar to silicon dioxide (sand).

Germanium is not found in its free elemental state because it is much too reactive. For the most part, it is found combined with oxygen, either as germanium monoxide or as germanium dioxide. Also, it is recovered from the ores of zinc, copper, and arsenic and the flue deposits of burning coal.

The crystal structure of germanium is similar to that of diamonds and silicon, and its semiconducting properties are also similar to silicon.

The melting point of germanium is 938.3°C, its boiling point is 2833°C, and its density is 5.323 g/cm³.

Characteristics

Once germanium is recovered and formed into blocks, it is further refined by the manufacturer of semiconductors. It is melted, and the small amounts of impurities such as arsenic, gallium, or antimony, are added. They act as either electron donors or acceptors that are infused (doped) into the mix. Then small amounts of the molten material are removed and used to grow crystals of germanium that are formed into semiconducting transistors on a germanium chip. The device can now carry variable amounts of electricity because it can act as both an insulator and a conductor of electrons, which is the basis of modern computers.

Abundance and Source

Germanium, the 52nd most abundant element in the Earth's crust, is widely distributed, but never found in its natural elemental state. It is always combined with other elements, particularly oxygen.

Germanium's main minerals are germanite, argyrodite, renierite and canfieldite, all of which are rare. Small amounts of germanium are found in zinc ore, as well as in copper and arsenic ores. It is known to concentrate in certain plants on Earth, particularly in coal: commercial quantities are collected from the soot in the stacks where coal is burned.

History

In 1871 Dmitri Mendeleev predicted the existence of a new element that would have similar properties as silicon. He called this yet to be found element "eka-silicon," which he assigned the symbol "Es" with an atomic weight of 72 and a specific gravity of 5.5.

When Clemens Alexander Winkler (1838–1904) was analyzing silver ore (Ag_8GeS_6), he came up short on the resulting products by 7%. He kept working on this problem, and in 1886 he found the missing 7% as a newly identified element that he then named germanium after his native country, Germany.

In 1948 William Bradford Shockley (1910–1989), who is considered the inventor of the transistor, and his associates at Bell Research Laboratories, Walter Houser Brattain (1902–1987) and John Bardeen (1908–1991), discovered that a crystal of germanium could act as a semiconductor of electricity. This unique property of germanium indicated to them that it could be used as both a rectifier and an amplifier to replace the old glass vacuum tubes in radios. Their friend John Robinson Pierce (1910–2002) gave this new solid-state device the name transistor, since the device had to overcome some resistance when a current of electricity passed through it. Shockley, Brattain, and Bardeen all shared the 1956 Nobel Prize in Physics.

Common Uses

By far, the most common use for germanium is in the semiconductor and electronics industries. As a semiconductor, germanium can be used to make transistors, diodes, and numerous types of computer chips. It was the first element that could be designed to act as different types of semiconductors for a variety of applications just by adding variable amounts of impurities (doping) to the germanium crystals.

Germanium is also used as a brazing alloy, for producing infrared transmitting glass and other types of lenses, and for producing synthetic garnets (semiprecious gemstones) that have special magnetic properties.

Examples of Compounds

Germanium monoxide and germanium dioxide are examples of the element's oxidation states of +2 and +4, as follows:

Germanium (II) monoxide: $Ge^{2+} + O^{2-} \rightarrow GeO$.

Germanium (IV) dioxide: $Ge^{4+} + 2O^{2-} \rightarrow GeO_2$.

When Germanium dioxide is 99.999% pure, it is used as a semiconductor in transistors, in diodes, and to make special infrared transmitting glass. (The same types of compounds can be formed with chlorine, and Germanium IV) compounds occur mostly with the sulfide of zinc ore.)

Germanium tetrahydride (GeH_4) is used to produce crystals of germanium. It is extremely toxic.

Hazards

Many of the chemicals used in the semiconductor industries are highly toxic. For example, germanium-halogen compounds are extremely toxic, both as a powder and in a gaseous state. Precautions should be taken when working with germanium as with similar metalloids from group 14 (IVA).

TIN

SYMBOL: Sn **PERIOD:** 5 **GROUP:** 14 (IVA) **ATOMIC NO:** 50
ATOMIC MASS: 118.710 amu **VALENCE:** 2 and 4 **OXIDATION STATE:** +2 and
+4 **NATURAL STATE:** Solid
ORIGIN OF NAME: The name "tin" is thought to be related to the pre-Roman Etruscan god
Tinia, and the chemical symbol (Sn) comes from *stannum,* the Latin word for tin.
ISOTOPES: There are 49 isotopes of tin, 10 of which are stable and range from Sn-112
to Sn-124. Taken together, all 10 stable isotopes make up the natural abundance of tin
found on Earth. The remaining 39 isotopes are radioactive and are produced artificially in
nuclear reactors. Their half-lives range from 190 milliseconds to $1 \times 10^{+5}$ years.

ELECTRON CONFIGURATION

Energy Levels/Shells/Electrons	Orbitals/Electrons
1-K = 2	s2
2-L = 8	s2, p6
3-M = 18	s2, p6, d10
4-N = 18	s2, p6, d10
5-O = 4	s2, p2

Properties

Tin is a soft, silvery-white metal located in the carbon group, similar in appearance to fresh-cut aluminum. When polished, it takes on a bluish tint caused by a thin protective coating of oxidized tin. This property makes it useful as a coating for other metals. It is malleable and ductile, meaning it can be pounded, rolled, and formed into many shapes, as well as "pulled" into wires through a die.

There are two allotropes of tin. One is known as gray or alpha (α) tin, which is not very stable. The other is known as white tin or beta (β), which is the most common allotrope. The two forms (allotropes) of tin are dependent on temperature and crystalline structure. White tin is stable at about 13.2°C. Below this temperature, it turns into the unstable gray alpha form. There is also a lesser-known third allotrope of tin called "brittle tin," which exists above 161°C. Its name is derived from its main property.

Tin's melting point is 231.93°C, its boiling point is 2,602°C, and the density is 5.75 g/cm³ for the gray allotrope (alpha) and 7.287 g/cm³ for the white allotrope (beta).

Characteristics

Although tin is located in group 14 as a metalloid, it retains one of the main characteristics of metals: in reacting with other elements, it gives up electrons, forming positive ions just as do all metals.

Tin has a relatively low melting point (about 231°C or 4,715°F), and it reacts with some acids and strong alkalis, but not with hot water. Its resistance to corrosion is the main characteristic that makes it a useful metal.

There is an interesting historical event related to the two main allotropes of tin. At temperatures below 13 degrees centigrade, "white" tin is slowly transformed into "gray" tin, which is unstable at low temperatures, and during the brutally cold winter of 1850 in Russia, the tin buttons sewn on soldiers' uniforms crumbled as the tin changed forms. In the 1800s, tin was also widely used for pots, pans, drinking cups, and dinner flatware. However, at very low temperatures, these implements also disintegrated as their chemical structure was altered.

Abundance and Source

Tin is the 49th most abundant element found in the Earth's crust. Although tin is not a rare element, it accounts for about 0.001% of the Earth's crust. It is found in deposits in Malaysia, Thailand, Indonesia, Bolivia, Congo, Nigeria, and China. Today, most tin is mined as the mineral ore cassiterite (SnO_2), also known as tinstone, in Malaysia. Cassiterite is tin's main ore. There are no significant deposits found in the United States, but small deposits are found on the southeast coast of England. To extract tin from cassiterite, the ore is "roasted" in a furnace in the presence of carbon, thereby reducing the metal from the slag.

History

Tin was known and used at least 5,500 years ago, when it was described in the oldest written records of the Mediterranean region. It is included in the book of Numbers in the Old Testament, along with the other then-known seven metals (gold, silver, copper, lead, brass, and iron). Bronze is an alloy of 20% tin and 80% melted copper, resulting in a metal that is harder than either pure tin or pure copper. Bronze was an extremely important metal because of its ability to maintain a cutting edge, thus making it suitable for weapons and tools. Before tin was used to make bronze, early humans used arsenic, which was problematic given that it resulted in death for those working with the toxic element.

Another interesting bit of history regarding tin came around 500 BCE, when the tin mines of the eastern Mediterranean region were depleted. This created a problem because tin, a rather rare metal, was very necessary to make the hardened bronze alloy for weapons and tools. This also is thought to be the first time in human history that a mineral was depleted in a region due to mining.

As the story goes, the Phoenicians set sail westward through the Straits of Gibraltar and discovered what were then known as the "tin islands" somewhere in the Atlantic Ocean. Because this was a rich find of an essential metal, they kept the source a secret. It is now believed that they actually found tin on the coast of England in a section we now know as Cornwall, which still produces tin ore.

Common Uses

One of the most important uses of tin is in the coating of thin steel sheets to make "tin plate," which in turn is used to make what is known as the "tin can." The tin coating is thin, inexpensive to apply, and resistant to most foods for extended periods of time. Other inert coatings are sometimes used on the inside of the can to further protect the foods for longer periods of time.

Tin is alloyed with many metals. It is added to lead to make low-melting alloys for fire-prevention sprinkler systems and easy-melting solder.

It is used for bearings, to plate electrodes, and to make pewter, Babbitt metal, and dental amalgams.

Tin also has been mixed with other metals for making castings for letter type used in printing presses.

Some compounds of tin are used as fungicides and insecticides. Tin is also used for "weighting" silk, to give the fabric more body and heft.

Molten glass is poured over a pool of molten tin to produce smooth, solid, flat plate and window glass.

Examples of Compounds

Tin has two oxidation states, +2 and +3. Some examples of tin's two ions in compounds follow:

Tin (II) chloride: $Sn^{2+} + 2\ Cl^{1-} \rightarrow SnCl_2$ (stannous chloride).

Tin (IV) chloride: $Sn^{4+} + 4Cl^{1-} \rightarrow SnCl_4$ (stannic chloride).

Both of these previous compounds are used as electrolytes in the electrotinning process and as stabilizers in perfumes and soaps.

Stannic oxide (SnO_2) is a whitish powder used as a ceramic glaze and polishing agent.

Stannous fluoride (SnF_2) is used as a toothpaste additive to help prevent tooth decay.

Hazards

Tin, as the elemental metal, is nontoxic. Most, but not all of tin's inorganic salts and compounds are also nontoxic.

In contrast, almost all organic tin compounds (tin compounds composed of carbon and hydrocarbons) are very toxic and should be avoided. If they are used, special equipment and care must be taken in handling.

(Note: When chemical formulas use the letter "R" preceding an element's symbol, it designates some form of organic compound—for example, R_4Sn. If the letter "X" follows the element's symbol in a formula, it designates some form of inorganic compound—for example, SnX_2. Thus, a whole series of tin compounds could be designated as R_4Sn_2, R_2Sn, or SnX_4, SnX_2, and so forth.)

LEAD

SYMBOL: Pb **PERIOD:** 6 **GROUP:** 14 (IVA) **ATOMIC NO:** 82
ATOMIC MASS: 207.19 amu VALANCE: 2 and 4 **OXIDATION STATE:** +2 and +4 **NAT-URAL STATE:** Solid
ORIGIN OF NAME: The name "lead" is the old Anglo-Saxon word for this well-known element, and the symbol for lead (Pb) is derived from the Latin word *plumbum,* which is also the root word for "plumber," related to the use of lead pipes in ancient Roman plumbing systems. Some of these lead pipes can still be seen in parts of modern-day Rome.
ISOTOPES: There are 47 isotopes of lead, four of which are stable. One of these four is Pb-204, which makes up 1.4% of the natural abundance of lead found on Earth. In reality this isotope is not stable but has a half-life that is so long ($1.4 \times 10^{+17}$ years), with some of the ancient deposits still existing, that it is considered stable. The other three stable isotopes of lead and their proportion to the total natural abundance are as follows: Pb-206 = 24.1%, Pb-207 = 22.1%, and Pb-208 = 52.4%. All the other isotopes are radioactive.

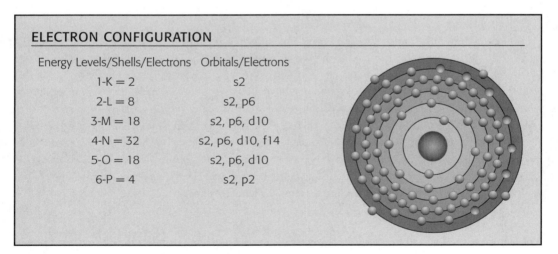

ELECTRON CONFIGURATION

Energy Levels/Shells/Electrons	Orbitals/Electrons
1-K = 2	s2
2-L = 8	s2, p6
3-M = 18	s2, p6, d10
4-N = 32	s2, p6, d10, f14
5-O = 18	s2, p6, d10
6-P = 4	s2, p2

Properties

Lead is a bluish-white, heavy metallic element with properties that are more metal-like than the properties of metalloids or nonmetals. Lead can be found in its native state, meaning that elemental metallic lead can be found in deposits in the Earth's crust. However, most lead is first mined as galena ore (lead sulfide, PbS). The galena is mixed with lead sulfate, lead sulfide, and lead oxide and is then roasted at a high temperature. The air supply is reduced, followed by an increase in heat and the vaporization of the sulfates and oxides of lead, which are drawn off as gases. The molten lead is then recovered.

Lead is only slightly soluble in water. However, it is also toxic. This is the reason lead is no longer used to pipe fresh water into homes. It does not react well with acids, with the exception of nitric acid. Lead's melting point is 327.46°C, its boiling point is 1,740°C, and its density is 11.342 g/cm^3.

Characteristics

Although lead can be found as a metal in the Earth's crust, it is usually mined and refined from minerals and ores. Lead is one of the most common and familiar metallic elements known. Although it is somewhat scarce, found at proportions of 13 ppm, it is still more prevalent than many other metals. Lead is **noncombustible.** and it resists corrosion.

When lead, which is very soft, is freshly cut, it has shiny blue-white sheen, which soon oxidizes into its familiar gray color. Lead is extremely malleable and ductile and can be worked into a variety of shapes. It can be formed into sheets, pipes, buckshot, wires, and powder. Although lead is a poor conductor of electricity, its high density makes it an excellent shield for protection from radiation, including X-rays and gamma rays.

Abundance and Source

Lead is the 35th most abundant element on Earth. Although it has been found in its free elemental metal state, it is usually obtained from a combination of the following ores: galena (PbS), anglesite (PbSO$_4$), cerussite (PbCO$_3$), and minum (Pb$_3$O$_4$). Lead ores are located in Europe (Germany, Rumania, and France), Africa, Australia, Mexico, Peru, Bolivia, and Canada. The largest deposits of lead in the United States are in the states of Missouri, Kansas, Oklahoma, Colorado, and Montana.

One of the most famous mining towns is the high-altitude western city of Leadville, Colorado. The boom started with the gold rush of the 1860s, followed by silver mining in the 1870s and 1880s. Today, this city is the site of mining operations not only for lead, but also for zinc and molybdenum. At the height of its fame, Leadville had a population of almost 50,000 people. Today the population is about 2,500.

Lead is commonly obtained by roasting galena (PbS) with carbon in an oxygen-rich environment to convert sulfide ores to oxides and by then reducing the oxide to metallic lead. Sulfur dioxide gas is produced as a waste product. Large amounts of lead are also recovered by recycling lead products, such as automobile lead-acid electric storage batteries. About one-third of all lead used in the United States has been recycled.

History

Although lead is not one of the most common metals on Earth, it is one of the best known. The metallic forms of lead, mercury, arsenic, antimony, bismuth, and zinc were not known as separate elements in ancient times until methods were developed to analyze these ores and their metals. The widespread knowledge of lead is attributed to the ancient Romans, who developed many practical uses for this heavy metal. Lead-lined pipes were used by the ancient Romans to bring water from their famous aqueducts to their homes. In addition, most of the population of Rome cooked their food in pots and pans made of lead and lead alloys. Because lead is slightly soluble in water, it is possible that much of the population was poisoned, to some degree, by lead. Although there is scant evidence that mass lead poisoning existed, it has been speculated that lead pipes and cooking utensils may have helped accelerate the decline of the Roman Empire.

Lead was also known in other regions of the ancient world. Lead sculptures, coins, and other artifacts have been found in Egyptian tombs dating back to 5000 BCE. Lead was also known in ancient biblical times and is mentioned in the books of Job and Exodus.

Common Uses

Lead has many uses and is an important commercial commodity. One of the most common uses is in the acid-lead electrical storage batteries used in automobiles. Much of the lead in these devices can be recycled and used again.

In the past, tetraethyl lead was added to gasoline to slow its burning rate in order to prevent engine "knock" and increase performance. This caused serious and harmful pollution, and lead has since been eliminated as a gasoline additive in most countries. Most exterior (and some interior) house paints once contained high levels of lead as well. Today, the amount of lead in paint is controlled, with not more than 0.05% allowed in the paint material.

Lead is used to make a number of important alloys. One is solder, an alloy of 1/2 lead and 1/2 tin. Solder is a soft, low-melting metal that, when melted, is used to join two or more other metals—particularly electrical components and pipes.

Babbitt metal is another alloy of lead that is used in the manufacture of wheel bearings that reduces friction. Lead is an ingredient in several types of glass, such as lead crystal and flint glass.

TV screens are coated with lead to absorb any radiation projected by the mechanism, and over 500,000 tons of lead is used in consumer electronics (computers, phones, games, and so on). Much of it ends up in solid waste dumps.

Many lead compounds are poisonous; thus, their uses in insecticides and house paints have been limited as other less toxic substances have been substituted. For example, lead arsenate [$Pb_3(AsO_4)$], which is very poisonous, has been replaced in insecticides by less harmful substances.

Examples of Compounds

Lead has a great range of oxides, many of which are of commercial importance. Although lead's two common oxidation states are +2 and +4, it can also combine in +1 and +3 oxidation states.

Lead (I) suboxide (Pb_2O) is the thin black film that forms naturally on a freshly cut piece of lead and that retards further oxidation.

Lead (III) oxide (Pb_2O_3) is a reddish-yellow solid used to manufacture glass, glazes, enamels, and the like. It is also used as a packing substance between pipe joints.

Some examples of lead +2 and +4 are as follows:

Lead (II) chloride ($PbCl_2$) is commonly known as the mineral cotunnite.

Lead (IV) oxide (PbO_2) is also known as lead dioxide. It is a brown substance important in the operation of the lead-acid storage battery.

Lead arsenate [$Pb_3(AsO_4)_4$] is a toxic commercial insecticide and herbicide.

Lead carbonate ($PbCO_3$) is found in nature as cerussite. It can also be produced in the laboratory by reacting sodium carbonate with chlorine. It is a crystalline poison that was, and to a lesser extent still is, used as a pigment in white house paints.

Lead chromate ($PbCrO_4$) is found in nature as yellow crystals in the mineral crocoite. It can be produced by reacting lead chloride and sodium dichromate. It is a popular and safe yellow pigment.

Lead azide [$Pb(N_3)_2$] is very unstable and must be handled with care. It is used as a detonator of explosives.

Hazards

Lead is probably one of the most widely distributed poisons in the world. Not only is the metal poisonous, but most lead compounds are also extremely toxic when inhaled or ingested. A few, such as lead alkalis, are toxic when absorbed through skin contact.

Workers in industries using lead are subject to testing of their blood and urine to determine the levels of lead in their bodies' organs. Great effort is made to keep the workers safe.

Unfortunately, many older homes (built prior to 1950) have several coats of lead-based paints that flake off, which then may be ingested by children, causing various degrees of lead poisoning, including mental retardation or even death.

Young children are more susceptible to an accumulation of lead in their systems than are adults because of their smaller body size and more rapidly growing organs, such as the kidneys, nervous system, and blood-forming organs. Symptoms may include headaches, dizziness, insomnia, and stupor, leading to coma and eventually death.

Lead poisoning can also occur from drinking tap water contained in pipes that have been soldered with lead-alloy solder. This risk can be reduced by running the tap water until it is cold, which assures a fresher supply of water.

Another hazardous source of lead is pottery that is coated with a lead glaze that is not stabilized. Acidic and hot liquids (citrus fruits, tea, and coffee) react with the lead, and each use adds a small amount of ingested lead that can be accumulative. Lead air pollution is still a problem, but not as great as before, given that tetraethyl lead is no longer used in gasoline. However, lead air pollution remains a problem for those living near lead smelting operations or in countries where leaded gasoline is still permitted.

Even though lead and many of its compounds are toxic and carcinogenic, our lives would be much less satisfying without its use in our civilization.

The Nitrogen Group (Metalloids to Nonmetals): Periods 2 to 6, Group 15 (VA)

Introduction

In addition to its most common designation, the nitrogen group also has an archaic name, the "pnictogen group" (also spelled pnicogen), which is of dubious origin. The pnictogen elements are the solid elements in the nitrogen group (P, As, Sb, and Bi). "Pnictogen" is a Greek verb meaning "to choke," and the designation of the name is based on nitrogen's being a "choking gas." The Greek "pnigo," or "pnigein," by definition means to choke, throttle, wring the neck, or strangle a person. The IUPAC (International Union of Pure and Applied Chemistry) does not recognize the terms "pnictogen" or "pnicogen," even though many chemists now use these terms to define some of the new technologies based on the unique chemistry of this group.

Some of the pnictogen elements are **diamagnetic paramagnets** at normal temperatures, and others are antiferromagnetic or **ferromagnetic** at very low temperatures. These properties are being researched for the possibility of the efficient transmission of electricity with little loss of heat. These elements are classed as either metals or nonmetals (solid-state metalloids) that act as semiconductors, ranging from novel semiconductor devices to nonlinear, light-splitting, frequency-doubling effects for optical materials, and solid-state refrigeration.

All the elements in group 15 have five electrons in their outer valence shells. They can lose, gain, or share any number of these. For example, the lightest element in group 15 is nitrogen, which, as a diatomic molecular gas (N_2), is not very reactive, yet it forms thousands of compounds. As an element, it exhibits valences of 1, 2, 3, 4, and 5. It can gain electrons, share its electrons, or give up electrons with many other elements—both metals and nonmetals. The heavier elements (the pnictogens), located in the lower section of group 15, are more likely to give up or share electrons similar to metals, as compared with the lighter elements at the top of the group 15 that can also receive electrons.

NITROGEN

SYMBOL: N **PERIOD:** 2 **GROUP:** 15 (VA) **ATOMIC NO:** 7
ATOMIC MASS: 14.0067 amu   **VALENCE:** 1, 2, 3, 4, and 5   OXIDATION STATES: −1, +1,− 2, +2, +3, +4, +5   **NATURAL STATE:** Gas

PERIODIC TABLE OF THE ELEMENTS

TRANSITION ELEMENTS

GROUPS / PERIODS	1 IA	2 IIA	3 IIIB	4 IVB	5 VB	6 VIB	7 VIIB	8 VIII	9 VIII	10 VIII	11 IB	12 IIB	13 IIIA	14 IVA	15 VA	16 VIA	17 VIIA	18 VIIIA
1	1 H 1.0079																	2 He 4.00260
2	3 Li 6.941	4 Be 9.01218											5 B 10.81	6 C 12.011	7 N 14.0067	8 O 15.9994	9 F 18.9984	10 Ne 20.179
3	11 Na 22.9898	12 Mg 24.305											13 Al 26.9815	14 Si 28.0855	15 P 30.9738	16 S 32.066(6)	17 Cl 35.453	18 Ar 39.948
4	19 K 39.0983	20 Ca 40.08	21 Sc 44.9559	22 Ti 47.88	23 V 50.9415	24 Cr 51.996	25 Mn 54.9380	26 Fe 55.847	27 Co 58.9332	28 Ni 58.69	29 Cu 63.546	30 Zn 65.39	31 Ga 69.72	32 Ge 72.59	33 As 74.9216	34 Se 78.96	35 Br 79.904	36 Kr 83.80
5	37 Rb 85.4678	38 Sr 87.62	39 Y 88.9059	40 Zr 91.224	41 Nb 92.9064	42 Mo 95.94	43 Tc (98)	44 Ru 101.07	45 Rh 102.906	46 Pd 106.42	47 Ag 107.868	48 Cd 112.41	49 In 114.82	50 Sn 118.71	51 Sb 121.75	52 Te 127.60	53 I 126.905	54 Xe 131.29
6	55 Cs 132.905	56 Ba 137.33	★	72 Hf 178.49	73 Ta 180.948	74 W 183.85	75 Re 186.207	76 Os 190.2	77 Ir 192.22	78 Pt 195.08	79 Au 196.967	80 Hg 200.59	81 Tl 204.383	82 Pb 207.2	83 Bi 208.980	84 Po (209)	85 At (210)	86 Rn (222)
7	87 Fr (223)	88 Ra 226.025	▲	104 Unq (261)	105 Unp (262)	106 Unh (263)	107 Uns (264)	108 Uno (265)	109 Une (266)	110 Uun (267)	111 Uuu (272)	112 Uub	113 Uut	114 Uuq	115 Uup	116 Uuh	117 Uus	118 Uuo

★ 6 Lanthanide Series (RARE EARTH)

▲ 7 Actinide Series (RARE EARTH)

★	57 La 138.906	58 Ce 140.12	59 Pr 140.908	60 Nd 144.24	61 Pm (145)	62 Sm 150.36	63 Eu 151.96	64 Gd 157.25	65 Tb 158.925	66 Dy 162.50	67 Ho 164.930	68 Er 167.26	69 Tm 168.934	70 Yb 173.04	71 Lu 174.967
▲	89 Ac 227.028	90 Th 232.038	91 Pa 231.036	92 U 238.029	93 Np 237.048	94 Pu (244)	95 Am (243)	96 Cm (247)	97 Bk (247)	98 Cf (251)	99 Es (252)	100 Fm (257)	101 Md (258)	102 No (259)	103 Lr (260)

ORIGIN OF NAME: From the two Greek words *nitron* and genes, which together stand for "soda or saltpeter forming."

ISOTOPES: There are 19 isotopes of nitrogen, two of which are stable. The stable ones and their proportion to the natural abundance of nitrogen on Earth follow: N-14 = 99.634% and N-15 = 0.366%. The other 17 isotopes are radioactive and man-made in nuclear reactors and have half-lives ranging from a few nanoseconds to 9.965 minutes.

ELECTRON CONFIGURATION

Energy Levels/Shells/Electrons   Orbitals/Electrons

| 1-K = 2 | s2 |
| 2-L = 5 | s2, p3 |

Properties

In its natural gaseous state, nitrogen is a relatively inert diatomic molecule (N_2) that is colorless, odorless, and tasteless, yet it is responsible for hundreds of active compounds. It makes up about 78% of the air we breathe. We are constantly taking it into our lungs with no stimulation or sensation; therefore, we really do not detect its presence. When liquefied, it is still colorless and odorless and resembles water in density. The melting point of nitrogen is −209.86°C, its boiling point is −195.8°C, and its density as a gas is 0.0012506 g/cm^3.

Characteristics

There are approximately 4,000 trillion tons of gas in the atmosphere, and nitrogen makes up about 78% of these gases. It is slightly soluble in water and alcohol. It is noncombustible and is considered an **asphyxiant** gas (i.e., breathing pure nitrogen will deprive the body of oxygen).

Although nitrogen is considered an inert element, it forms some compounds that are very active. Of the diatomic molecules, such as CO_2, it is difficult to separate the two atoms in nitrogen's molecules because of their strong binding energy. This is the reason that, along with carbon dioxide, nitrogen gas is stable. However, once separated, the individual atoms of nitrogen (N) become very reactive and do combine with hundreds of other elements.

Nitrogen can be liquefied easily, making it useful in many applications wherein sustained cooling is needed. At high temperatures, nitrogen reacts with many metals to form nitrides.

Abundance and Source

Nitrogen is the 30th most abundant element on Earth. There is an almost unlimited source of nitrogen available to us considering that our atmosphere constitutes 4/5, or over 78%, of the nitrogen by volume. Over 33 million tons of nitrogen is produced each year by liquefying air and then using **fractional distillation** to produce nitrogen as well as other gases in the atmosphere. During this process the air is cooled and then slowly warmed to fractional temperature points at which each specific gas in the air will "boil" off. (Note: Oxygen, argon, carbon dioxide, and nitrogen all have specific boiling points and these gases can be used to collect the specific gas during the fractionation process.) When the temperature –reaches –195.8°C, the nitrogen is boiled off and collected.

There is a balance of nitrogen with other gases in the atmosphere that is maintained by what is called the nitrogen cycle. This cycle includes several processes, including nitrogen fixation of bacteria in the soil by **legumes** (bean and pea plants). Lightning produces nitrogen, as do industrial waste gases and the **decomposition** products of organic material (i.e., organic proteins and amino acids in plants and animals contain nitrogen). In time, these sources replace the nitrogen in the atmosphere to complete the cycle.

Ammonia (NH_3) is the first binary molecule discovered in outer space of our galaxy, the Milky Way. It may also be the main compound that forms the rings of the planet Saturn.

History

In 1772 a student, Daniel Rutherford (1749–1819), at the suggestion of his mentor, Joseph Black (1728–1799), conducted an experiment in which he burned a candle in a closed container of air. (Joseph Black was famous for his concept of "fixed air," which was an important step in understanding gases in chemistry.) Chemists of the time already knew that air contained at least two gases, one that supported life and one that did not. Rutherford started his experiment by placing a mouse in a sealed glass jar until it suffocated and had reduced the volume of air by 1/16. He repeated this with a candle and noticed that there was still a large quantity of gas in the container after the burning candle had consumed the oxygen. There was no carbon dioxide because he had chemically removed that gas. Rutherford experimented further and found that this leftover gas could not support combustion or life, so he called it noxious air. He never did identify nitrogen but came up with some several suggestions. That led to the identification of a new element, which was later named "nitrogen" for the Greek word meaning "niter producer," after the compound, potassium nitrate (saltpeter KNO_3), which contained nitrogen.

At the same time Rutherford conducted his experiments, three other chemists—Priestley, Cavendish, and Scheele—were also investigating "fixed air" gases, including nitrogen. However, Rutherford was given credit for discovering nitrogen.

Common Uses

Nitrogen has many uses. It is the second most commonly produced chemical in the United States. Its chemical and physical properties, along with the five electrons in its outer shell, make it a versatile element that can react as a metal or nonmetal to produce numerous compounds. Some of its uses are based on its inertness as a gas (N_2) and its ability to be liquefied to provide very low temperatures.

When recovered as a gas in the atmosphere, it is used to produce anhydrous ammonia (NH_3), which is the fifth most commonly produced chemical in the United States. It is also used as the basis for making many nitrogen compounds. At one time it was believed to be impossible to combine hydrogen with nitrogen to form ammonia, a natural product of animal waste that was used as a fertilizer and textile bleach, among other things. In 1905 the German chemist Fritz Haber (1868–1934) demonstrated that it was possible to combine hydrogen with nitrogen in a process that directly produced ammonia. The Haber process requires high temperatures (500°C) and very high pressure. It is the main source of ammonia today. Millions of tons of ammonia are produced every year worldwide. Haber received the 1918 Nobel Prize in Chemistry for his work.

Most of us are familiar with the liquid form of ammonia known as ammonium hydroxide (NH_4OH), a colorless liquid that, with its strong odor, is irritating to the eyes and potentially harmful to the moist mouth and nose, throat, and lungs if its vapors are breathed. Weak solutions of NH_4OH are ingredients in household cleaning ammonia. Concentrated ammonium hydroxide has many industrial uses, including the manufacture of rayon, fertilizers, refrigerants, rubber, pharmaceuticals, soaps lubricants, inks, explosives, and household cleaners.

Many fertilizers are based on ammonia compounds. Modern agriculture requires more nitrogen in soils than is normally replaced by the nitrogen cycle, lightning, decaying plants and animals, and other natural means

Nitric acid (HNO_3) is an important commercial chemical and was manufactured commercially to produce fertilizers and explosives as well as plastics and many other products. In 1902 a German chemist, Wilhelm Ostwald (1853–1932), developed a process wherein at high temperatures he used platinum catalysts to convert ammonia into nitric acid. When nitric acid is reacted with glycerol, the result is nitroglycerine—an unstable explosive unless dissolved in inert material, such as clay. It can then be stabilized as dynamite.

Nitrates are formed when nitric acid is neutralized by a base such as sodium hydroxide (NaOH). These nitrates change form to become nitrites that are used as preservatives, particularly in canned goods and to keep meat looking fresh.

Sodium azide (NaN_3) is an explosive salt of nitrogen that produces large quantities of gas upon its explosion. This quality has made it ideal as the chemical contained in automobile air safety bags. When triggered it explodes immediately, producing the expanding gases that fill the bag.

The radioisotope nitrogen-13 has a relatively short half-life of about 10 minutes that produces a positron as it decays. This makes N-13 useful in PET (Positron Emission Tomography) scan technology, in which it is injected into the patient. The positive electrons (positrons) interact with the patient's negative electrons to produce an image similar to an X-ray.

Oil companies force nitrogen under great pressure into depleted oil wells to force residual crude oil to the surface.

By far, nitrogen compounds are of the utmost importance to the diets and welfare of both plants and animals. Nitrogen is essential to living things. Plants require nitrogen, and so do animals, which get their nitrogen from eating plants and other animals.

Examples of Compounds

Nitrogen has numerous positive and negative oxidation states. For example, it can form six different compounds with oxygen using the oxidation states of +1 through +6.

Ammonium nitrate (NH_4NO_3, also known as "Norway saltpeter") is mainly used as a fertilizer. It is also known as the chemical that was mixed with diesel fuel to create the explosion that demolished the Murrah Federal Building in Oklahoma City in 1995.

Nitrogen oxides (NO_x; the "x" represents the proportion of nitrogen to oxygen atoms in the various oxidation states of nitrogen) have many uses, including the production of nitric acid. Most of the oxides of nitrogen, especially NO_2, are toxic if inhaled. On the other hand, nitrous oxide (N_2O), although explosive in air, is known as "laughing gas" and is used as an anesthetic in dentistry and surgery.

Nitroglycerine, as mentioned, is an unstable explosive. It is also used as a **vasodilator** to reduce high blood pressure and angina pectoris by dilating the blood vessels of heart patients whose hearts are not receiving an adequate blood supply.

Recently a team of chemists discovered a new allotrope of nitrogen that consisted of five nitrogen atoms in the form of a V. This surprised them because it was thought that any structure with three or more nitrogen atoms would be unstable—and indeed, the V form turned out to be so unstable that it created an extreme explosion that destroyed the equipment being used to analyze it.

Hazards

Nitrogen is nontoxic, but it is an asphyxiate gas that cannot, by itself, support oxidation (combustion) or support life. If you breathe pure nitrogen for any period of time, you will die—not because the nitrogen gas is a poison, but because your body will be deprived of oxygen.

Nitrogen oxides are formed under certain conditions when nitrogen combines with oxygen, thus contributing to pollution. One source is from the internal combustion engine that produces NO similar to lightning. Once released, it combines with more oxygen to form NO_2, which is a very reactive polluting gas. Nitrogen dioxide NO_2 is the main cause of "brown" smog over some cities and is harmful to plants, animals, and humans. To make matter worse, if there is adequate sunlight at the time of the smog, the ultraviolet light of the sun will break down the N and O of the NO_2 to form free radicals of oxygen that are reactive, forming ozone (O_3), which is itself a strong oxidizing agent that adds to pollution.

Several of the oxygen, hydrogen, and halogen compounds of nitrogen are toxic when inhaled. A common error made in using household cleaners is to mix or use together ammonia cleaning fluids (containing nitrogen) and Clorox-type cleaning fluids (containing chlorine). The combined fumes can be deadly in any confined area. NEVER mix Clorox with ammonia-type cleaning fluids.

PHOSPHORUS

SYMBOL: P   **PERIOD:** 3   **GROUP:** 15 (VA)   **ATOMIC NO:** 15
ATOMIC MASS: 30.97376 amu.   **VALENCE:** 1, 3, 4, and 5   **OXIDATION STATE:** +3, −3, and +5   **NATURAL STATE:** Solid
ORIGIN OF NAME: Its name is derived from the Greek word *phosphoros,* which means "bringer of light" or "light bearing."
ISOTOPES: There are a 23 isotopes of phosphorus, ranging from P-24 to P-46, with half-lives that range from a few nanoseconds to about two and half minutes. The one stable isotope is phosphorus-31, which accounts for 100% of the natural phosphorus on Earth.

ELECTRON CONFIGURATION

Energy Levels/Shells/Electrons	Orbitals/Electrons
1-K = 2	s2
2-L = 8	s2, p6
3-M = 5	s2, p3

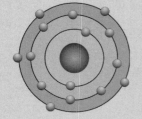

Properties

Although phosphorus is in group 15 with some other metalloids, it is usually classed as a nonmetal since it resembles nitrogen somewhat, the element above it in group 15. Both are essential to the biochemical field as vital elements to support life. Phosphorus has 10 known allotropic forms. This is an unusually high number for any element. A system of categorizing the allotropes by three colors has made it easier to keep track of them. These three colors are white, red, and black phosphorus.

White phosphorus has a white waxy appearance that turns slightly yellow with age and impurities. There are two allotropic forms of white phosphorus. The alpha (α) form has a cubic crystal structure, and the beta (β) form has a hexagonal crystalline structure. White phosphorus is extremely reactive and will spontaneously burst into flame when exposed to air at a temperature of about 35°C. It must be kept under water. But this property of spontaneous combustion has made it useful for military applications.

White phosphorus is the most useful version of the three allotropes, and it is used in processes to manufacture the other two versions of phosphorus. White phosphorus's melting point 44.15°C, its boiling point is 280.5°C, and its density is 1.82 c/cm^3.

Exposing white phosphorus to a process of heat produces red phosphorus. Red phosphorus has a density of 2.34 g/cm^3.

Black phosphorus also starts with heating white phosphorus. The difference is that the white phosphorus is heated in the presence of a mercury catalyst and a small amount of already-formed black phosphorus. Its density is 2.4 g/cm^3.

Characteristics

White phosphorus occurs in nature in phosphate rock. It is insoluble in water and alcohol and will ignite spontaneously in air. It exhibits what is known as **phosphorescence;** that is, it glows in the dark at room temperature. White phosphorus is poisonous and must be stored under water.

Red phosphorus is less reactive than the white variety. It is not poisonous, but large amounts can explode. It is used in fireworks and matches.

Black phosphorus is the only one of the three that will conduct electricity; white and red are poor conductors. Black phosphorus has no significant commercial uses.

Abundance and Source

Phosphorus is the 12th most abundant element. It makes up about 0.1% of the Earth's crust.

Phosphorous occurs in nature in several forms, mostly as phosphates. The most common source is phosphate rock [$Ca_3(PO_4)_2$] and a mineral called "apatite." Phosphorus is found in all animal bones and teeth and in most living tissue. Phosphorous nodules are found on the ocean floor along with manganese nodules.

Most commercial phosphorus is produced in electric furnaces where the phosphate-rich minerals are heated to drive off the phosphorus as a gas, which is then condensed under water. Another process uses sulfuric acid to remove the phosphorus.

History

In 1669 a German physician, Hennig Brand (1630–1692), a proponent of ancient alchemy, attempted to extract gold from urine. He collected a bucketful of urine that he allowed to evaporate. He soon expanded his experiment to 60 buckets of urine placed all around the laboratory until they evaporated. He then proceeded to boil down the urine, filter it, and perform some other processes on it. He ended up with a distilled residue that glowed in the dark and would burst into flames. He had discovered white phosphorus. He named it "phosphors" (Greek for "light bearer") after the bright early morning star that had already been given that name. Brand's slow, smelly, ineffective procedure led to a different process that soon supplanted the old alchemist's practice. Today, phosphorus is removed from phosphate rock, $Ca_3(PO_4)_2$. In retrospect, it is amazing that Brand was able to isolate even small amounts of white phosphorus using his simple techniques.

In 1841 Jöns Jakob Berzellius (1779–1884), who introduced the term "allotropy," transformed white phosphorus to red phosphorous. In 1865 Johann Wilhelm Hittorf (1824–1914) was the first to produce metallic phosphorus. Brand, however, was given credit for the discovery of phosphorus.

Common Uses

The allotropes and compounds of phosphorus have many important uses and are an essential commercial commodity. Phosphorus is essential to all living tissue, both plant and animal. It is the main element in the compound adenosine triphosphate (ATP), the main energy source for living things.

Red phosphors are formed either by heating white phosphorus or by exposing white phosphorus to sunlight. It is quite different from the explosive white phosphorus. For instance, when scratched on a surface, the heads of safety matches made of red phosphorus convert back to white phosphorus and ignite due to the heat of the slight friction of the match on a rough surface. Red phosphorus is also used in fireworks, smoke bombs, and pesticides and to make phosphoric acid, electroluminescent paints, and fertilizers.

Most elemental phosphorus is used to manufacture phosphoric acid, a solid that is used to produce triple-phosphate fertilizers. Some soils require large amounts of phosphorus to produce a viable crop.

Sodium tripolyphosphate is the main phosphate found in detergents. It acts as a water softener and counteracts the elements that are responsible for "hard water" while at the same time making the detergent a more effective cleaner.

Some phosphorus compounds glow in the dark as they emit light after absorbing radiation. This makes them useful in special fluorescent lights and the color screens of television sets and computers.

Examples of Compounds

Phosphorus-32, the most important radioisotope of phosphorus, has a half-life of 14 days. It provides beta radiation (high-speed electrons) and is made by inserting phosphorus into nuclear reactor piles. P-32 is used as a "tag" to trace biochemical reactions in patients. It is also used to treat leukemia and skin and thyroid diseases.

Phosphorus pentasulfide (phosphoric sulfide, P_2S_5) is an insecticide. It is also an additive to oils and a component of safety matches.

Phosphorus trichloride (PCl_3) is used in the manufacture of other phosphorus compounds. It is also an insecticide, a gasoline additive, an ingredient in dyes, and a "finisher" for the surface of textiles.

Phosphoric acid (H_4PO_4). Although phosphoric acid is derived from phosphate rocks, it is the 7th highest in terms of volume of chemical produced and one of the most widely used compounds of phosphorus wherein the PO_4^{3-} ion is required in the chemical reactions. It might be best known as a "flavor" in soft drinks, to make solvents, and as a component of fertilizers. It has many uses, including the manufacture of soaps and detergents, sugar refining, water treatment, animal feeds, electroplating, gasoline additive, and binder for foods, to name a few.

Hazards

Many of the compounds of phosphorus are extremely dangerous, both as fire hazards and as deadly poisons to the nervous system of humans and animals. Some of the poisonous compounds (PCl_x) can be absorbed by the skin as well as inhaled or ingested. Flushing with water is the only way to stop the burning of white phosphorus on the skin, but water does not affect the combustion of some phosphorus compounds. Although red phosphorus is not as dangerous or poisonous as white phosphorus, merely applying some frictional heating will induce the red allotrope to change back to the explosive white allotrope (the striking of a safety match is an example).

Some of the main types of poisonous gases used in warfare have a phosphorus base. Many countries stockpile these gases, but, by agreement, the supplies are being reduced.

ARSENIC

SYMBOL: As   **PERIOD:** 4   **GROUP:** 15 (VA)   **ATOMIC NO:** 33
ATOMIC MASS: 74.92158 amu   **VALENCE:** 2, 3, and 5   **OXIDATION STATE:** +3, −3, and +5   **NATURAL STATE:** Solid
ORIGIN OF NAME: Derived either from the Latin word *arsenicum* or the Greek word arse-nikon, both meaning a yellow pigment. It is possible that the Arabic word azzernikh was also an ancient name for arsenic.
ISOTOPES: There are a total of 35 isotopes of arsenic, ranging from As-60 to As-92, with half-lives spanning from a few nanoseconds to 80 days. Although some references claim there are no stable isotopes of arsenic, arsenic-75 is classed as a stable isotope that makes up 100% of arsenic found in the Earth's crust.

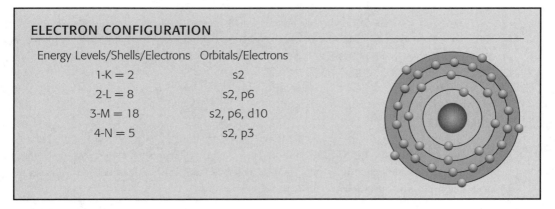

ELECTRON CONFIGURATION

Energy Levels/Shells/Electrons	Orbitals/Electrons
1-K = 2	s2
2-L = 8	s2, p6
3-M = 18	s2, p6, d10
4-N = 5	s2, p3

Properties

Arsenic is classed as a semimetal, meaning that it is neither a metal like aluminum or lead, nor quite a nonmetal such as oxygen, sulfur, or chlorine. Arsenic's main allotrope is a silvery-gray, brittle, metal-like substance. Its other two isotopes are unstable crystalline substances.

Gray arsenic exhibits an unusual property in that its boiling point (614°C) is lower than its melting point (817°C). As its temperature changes, it sublimates, which means it goes from the solid state, skipping the liquid state, into a vapor state. Cooling the vapor of **sublimation,** the black allotrope condenses out and in turn changes from the black to the gray allotrope. If yellow arsenic is rapidly cooled from its sublimation point, yellow arsenic will condense out and will not revert back to gray arsenic upon cooling.

The following information is for the gray semimetal form of arsenic only. Its melting point is 817°C, its sublimation point varies between 613°C and 814°C depending on the atmospheric pressure, and its density is 5.776 g/cm^3.

Characteristics

Arsenic in the elemental form is a brittle, grayish crystal that becomes darker when exposed to air. It is seldom found in the pure elemental form but rather in minerals (compounds). It has a long history of use as a poison, and many alchemists were poisoned when using it in their attempts to produce gold from base metals.

Arsenic has limited commercial use.

Abundance and Source

Arsenic is the 53rd most abundant element and is widely distributed in the Earth's crust. It occurs naturally in several minerals, but high-grade deposits are rare. Most of the minerals and ores that contain arsenic also contain other metals. Some major sources of arsenic are the minerals orpiment, scherbenkobalt, arsenopyrite, niccolite, realgar, gersdorffite, and smaltite. In addition, most sulfide ores of other metals also contain some arsenic. The three major minerals that produce arsenic are: realgar (arsenic monosulfide, AsS), orpiment (arsenic trisulfide, As$_2$S$_2$), and arsenopyrite (iron arsenosulfide, FeAsS).

Today, most arsenic is recovered as a by-product from the smelting of nickel, copper, iron, and tin. It is also recovered from the flue dust of copper- and lead-smelting furnaces.

History

In about 3600 BCE, ores containing both arsenic and copper were known and mined by the early Greeks and Romans, as well as by Chinese alchemists. This is about the time when copper was smelted and alloyed to make bronze. Some ores of copper produced harder metals than others because of impurities. One of these impurities was arsenic. Because the workers were becoming ill when smelting these types of ore, the process was abandoned, and tin was added to copper to form bronze. "Bronze" may have been the Persian (Iranian) word for "copper."

The honor for discovering the element arsenic in or about 1250 goes to a German alchemist, Albertus Magnus (1193–1280). He was also the first to propose the concepts of *affinitas,* which explained how chemicals were held together. In addition, he learned how to separate several compounds into their constituent elements.

Common Uses

Over the years a number of practical uses for arsenic developed, particularly related to its poisonous nature. Today, it is not of great commercial value except as an insecticide and herbicide.

It is used in the semiconductor industry to coat solid-state devices. Some compounds are used in paints and fireworks. The major uses are in medicine, where its toxic properties are important for the treatment of diseases.

At the beginning of the twentieth century, arsenic compounds were used to kill spirochete bacteria, which cause the sexually transmitted disease syphilis. After this amphetamine compound was used to treat syphilis in Europe, the disease rate was reduced by more than half. The antibiotic penicillin has replaced arsenic for most medical purposes

The compound arsphenamine is also called "compound 606" because it was the 606th arsenic compound that Paul Ehrlich (1854–1915) had **synthesized** for use in treating diseases. His assistant found this compound to be effective as a treatment for syphilis. Ehrlich also coined the word "chemotherapy."

Historically, arsenic was one of the "poisons of choice" because it killed slowly and mimicked many other common ailments, such as gastrointestinal diseases, that could not easily be diagnosed and treated. Once autopsies started being generally performed and it became possible to identify the presence of arsenic as the culprit in the death of the patient, arsenic lost its "luster" as the ideal poison.

Examples of Compounds

Arsenic disulfide (AsS) is also known as ruby arsenic because it is a reddish-orange powder. It is used as a depilatory agent, a paint pigment, and a rat poison and to make red glass and fireworks.

Arsenic trichloride ($AsCl_3$), also known as arsenic chloride, is used in the pharmaceutical industry and to make insecticides and ceramics.

Arsine (AsH_3), as a colorless gas, is also known as arsenic hydride. It is used to synthesize organic compounds and as the major ingredient of several military poisons, including the wartime gas lewisite.

There are two important oxides of arsenic:

Arsenic trioxide (As_2O_3) is found in nature as arsenolite and claudetite. It is extremely poisonous.

Arsenic pentoxide (As_2O_5) is manufactured by adding oxygen to the trioxide form and is used to form many arsenic compounds.

Hazards

Most of the compounds of arsenic are toxic when in contact with the skin, when inhaled, or when ingested. As with arsenic's cousin phosphorus above it in group 15 of the periodic table, care must be taken when using arsenic. The compound arsenic trioxide (As_2O_3), an excellent weed-killer, is also carcinogenic. Copper acetoarsenite, known as Paris green, is used to spray cotton for boll weevils. A poisonous dose of arsenic as small as 60 milligrams can be detected within the body by using the Marsh test.

ANTIMONY

SYMBOL: Sb   **PERIOD:** 5   **GROUP:** 15 (VA)   **ATOMIC NO:** 51
ATOMIC MASS: 121.760 amu   **VALENCE:** 3, 4, and 5   **OXIDATION STATE:** +3, −3, and
  +5   **NATURAL STATE:** Solid

ORIGIN OF NAME: The element's name comes from the Greek words *anti* and minos, which mean "not alone," and antimony's symbol (Sb) is derived from the name for its ancient source mineral, stibnium.

ISOTOPES: There are 53 isotopes of antimony. They range from Sb-103 to Sb-139 (a few have two forms). Their half-lives range from 150 nanoseconds to 2.7 years. The two stable isotopes of antimony and their contribution to the natural abundance of antimony on Earth are as follows: Sb-121 = 57.21% and Sb-123 = 42.79%.

ELECTRON CONFIGURATION

Energy Levels/Shells/Electrons	Orbitals/Electrons
1-K = 2	s2
2-L = 8	s2, p6
3-M = 18	s2, p6, d10
4-N = 18	s2, p6, d10
5-O = 5	s2, p3

Properties

Physically, antimony's properties are related to sulfur and some of the nonmetals, but chemically, its properties are related to metals. It behaves like a metal and is often found in nature along with other metals. In its pure form it is rather hard and brittle with a grayish crystal structure

Characteristics

There are two allotropes of antimony. The native metallic form is one allotrope, and the other allotrope is an amorphous grayish form. Antimony is a true metalloid that is brittle with a low melting point. And similar to nonmetals, it is a poor conductor of heat and electricity.

Antimony is unique in that when it solidifies from a molten liquid state to a solid state, it expands, which is just the opposite of most metals. This is useful in making some typesetting castings in which the expansion assures an accurate reproduction of the letter mold.

Abundance and Source

Although antimony is not a rare metal, it is not well known, despite having been known and used for many centuries. It is the 63rd most abundant element on Earth, and it occurs mainly as sulfide ores or in combination with the ores of other metals. The ore that is the primary source of antimony is the mineral stibnite (antimony sulfide, Sb_2S_3). Antimony is also found in copper, silver, and lead ores. Breithauptite (NiSb) and ullmanite (NiSbS) are two ores containing nickel. Dicrasite (Ag_2Sb) and pyrargyrite (Ag_3SbS_3) are silver ores containing some antimony.

History

Antimony in its black form (Stibnite, Sb_2S_2) was known at least 4,000 to 5,000 years ago. Egyptian women used it as an eyeliner for both themselves and their children. It was known during biblical times and was referred to as "stick-stone"; it was also used during this time as black eyeliner for wealthy women.

Antimony was known in the days of alchemy (500 BCE to 1600 CE) when it was associated with other metals and minerals such as arsenic, sulfides, and lead used as medications. It is possible that an alchemist, Basilus Valentinus (fl. 1450), knew about antimony and some of its minerals and compounds sometime around the mid-fifteenth century CE. Physicians of this period—and earlier periods—used elements such as mercury and antimony to cure diseases, although they knew that these elements were toxic in larger doses. Antimony was used to treat **depression,** as a laxative, and as an **emetic** for over two thousand years. Despite the elements' poisonous nature, physicians of that early era considered both mercury and antimony good medicines.

Nicolas Lemery (1645–1715), a French chemist, studied and analyzed antimony and its compounds in detail. He published his findings in 1707. Antimony was used primarily as a treatment for diseases, but as a chemical element it was neglected it until modern days, when it was used to produce low-melting-point alloys.

Common Uses

Today the most common use of antimony is as an alloy metal with lead to make the lead harder. This lead–antimony alloy is used for electrical storage batteries, for sheathing for electrical and TV cables, in the making of wheel bearings, and as **solder.**

Although it is a brittle metal, it has found a use in the semiconductor industry in the production of diodes and infrared devices. It is also used to flameproof material and **vulcanize** rubber and can also be a component in paints, ceramic enamels, glass, pottery, and fireworks.

The head of a safety match is a mixture of antimony trisulfide (Sb_2S_3) and an oxidizing agent (potassium chlorate ($KClO_3$). Red phosphorous is placed on the tip so that when it is struck against a rough surface, it ignites with enough flame to ignite the other chemicals in the head and then burn the wood match.

Both antimony-124 and antimony-125 radioisotopes are used as industrial and metallurgical **tracers.** A unique use is as a partition or separator between different types of fluids flowing through the same pipeline. Its radioactivity can mark where one type of fluid ends and a different fluid begins.

Antimony has a long history of medical use. Some scholars believe that Mozart died after being given antimony by the physicians who were treating his depression and who were unaware of just how poisonous antimony was. The evidence for this story is scant. It is also known that around 870 BCE, Queen Jezebel and her contemporaries used the mineral or ore antimony sulfide as a cosmetic to darken their eyelashes and as an eyeliner. It is still used for this purpose in many countries.

Antimony has few other uses except as an alloy to harden other materials. One recent development was to add antimony oxide to polyvinyl chloride (PVC pipe) to act as a flame retardant.

Examples of Compounds

Common oxidation states for antimony are +3, +5, and –3. A number of compounds are formed with these states. For instance, both chlorine and sulfur antimony compounds are common.

Antimony chloride: $Sb^{3+} + 3Cl^{1-} \rightarrow SbCl_3$. This compound is sometimes called "butter of antimony."

Antimony trisulfide: $2Sb^{3+} + 3S^{2-}$. Antimony trisulfide is better known as the mineral stibnite and is used as a yellow paint pigment and in the manufacture of ruby glass, fireworks, and matches. It is also used to make percussion caps that set off explosives.

Antimony pentoxide (Sb_2O_3) is used as a flame retardant for textile materials and as a source to prepare other antimony compounds.

A unique use of antimony is to make glass that reflects infrared radiation; thus, what is behind the glass cannot be detected by the infrared ray source.

Hazards

The powder and dust of antimony are toxic and can cause damage to the lungs. The fumes of antimony halogens (chlorides and fluorides) are especially dangerous when inhaled or in contact the skin.

Many of the salts of antimony are carcinogenic and can cause lung cancer if inhaled, as well as other cancers if ingested. This is a major hazard with the radioisotopes of antimony used in industry. Some of its sulfide compounds are explosive.

BISMUTH

SYMBOL: Bi   **PERIOD:** 6   **GROUP:** 15 (VA)   **ATOMIC NO:** 83
ATOMIC MASS: 208.98038 amu   **VALENCE:** 3 and 5   **OXIDIATION STATE:** +3 and
   +5   **NATURAL STATE:** Solid
ORIGIN OF NAME: Bismuth was known and used by the ancient alchemists along with
   other metals both for chemical reactions and for medical purposes. The name comes
   from the German *bismu,* which had been changed from wismu, meaning "white."
ISOTOPES: There are a total of 59 radioactive isotopes for bismuth, ranging in half-lives
   from a few milliseconds to thousands of years. At one time it was thought that there was
   just one stable isotope (Bi-209), but it was later found that Bi-209 is radioactive with
   a half-life of 19,000,000,000,000,000,000 years. Such a long half-life means that Bi-
   209 has not completely disintegrated and is still found in nature, and is thus considered
   stable. In this case, Bi-209 makes up 100% of Bismuth's natural abundance.

ELECTRON CONFIGURATION

Energy Levels/Shells/Electrons	Orbitals/Electrons
1-K = 2	s2
2-L = 8	s2, p6
3-M = 18	s2, p6, d10
4-N = 32	s2, p6, d10, f14
5-O = 18	s2, p6, d10
6-P = 5	s2, p3

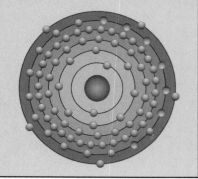

Properties

Bismuth is the fifth element in the nitrogen group, and its properties are the most metal-like of the five. Elemental bismuth is a heavy, brittle, hard metal that can be polished to a bright gray-white coat with a pinkish hue. It is not found in this state very often because it is more likely to be combined with other metals and minerals, such as tin, lead, iron and cadmium. These are mixtures with low melting points, making them useful in fire-detection devices.

When heated in air, bismuth burns with a blue flame, giving off clouds of its yellow oxide. Bismuth's melting point is 271.40°C, its boiling point is 1,564°C, and its density is 9.807 g/cm³.

Characteristics

Bismuth is more resistant to electrical current in its solid state than it is in its liquid form. Its thermal conductivity is the lowest of all metals, except mercury. Even though it is considered a metal-like element, it is a very poor conductor of heat and electricity.

Bismuth has a characteristic similar to water. It expands when changing from the liquid phase to the solid phase. This factor makes it useful as an alloy in metals that are used to fill molds, given that it will expand to the cast's dimensions.

Abundance and Source

Bismuth is the 70th most abundant element, and it is widely spread over the Earth's crust, but in very small amounts. There are no major concentrated sources. It occurs both in the free elemental state and in several ores. The major ore, bismuthinite (B_2S_3), is found in South America.

The United States gets most of its bismuth as a by-product from smelting ores of lead, silver, copper, and gold. It is also recovered from the refining of tin and tungsten ores.

History

Bismuth, arsenic, antimony, and zinc were known as metals in the days of alchemy (500 BCE to 1600 CE). Philippus Aureolus Theophrastus bombastus von Hohenheim (1493–1541), better known as Paracelsus, mentioned bismuth as well as otherthen-known metals in his writings. In the middle of the fifteenth century, bismuth was described by a German monk, Basilus Valentinus (fl. 1640). He called it *wismut masse,* which is the German term for "white stuff or white mass." One problem was distinguishing bismuth from lead and tin. Georgius Agricola (1494–1555) worked on this problem at the same time as Paracelsus, but more as a scientist than as an alchemist. He tentatively separated the metals as *bismutum* (bismuth), *plumbum candidum* (tin), and *nigrum* (lead). Bismuth was not definitively identified as a separate element until 1659 when Johann Glauber used glass beads to demonstrate the importance of color in flames of different substances when trying to identify them. In 1753 Claude J. Geoffroy (1729–1753) separated bismuth from its ore.

Not much is known of the element's use until the knowledge of **metallurgy** developed in the late eighteenth century, when arsenic, antimony, and bismuth were used to form alloys with other metals.

Common Uses

Bismuth is used to make the drugs such as Pepto-Bismol for upset stomachs and diarrhea and in medicine to treat intestinal infections. Bismuth is used in the cosmetics industry to provide the "shine" for lipsticks, eye shadow, and other products.

It is added to steel and other metals as an alloy to make the metals easier to roll, press, pull into wires, and turn on a lathe. It is also used in the semiconductor industry and to make permanent magnets.

Bismuth is similar to antimony in that it expands from the molten liquid state to the solid state. This property makes it an excellent material to pour into molds and can produce fine details in whatever is being molded, such as metallic printing type and similar fine castings.

Examples of Compounds

Bismuth has a +3 oxidation state that in some ways limits the type of compounds it forms.

Bismuth chloride ($Bi^{3+} + 3Cl^{1-} \rightarrow BiCl_3$) reacts with water to produce what is known as "bismuth white" pigment.

Bismuth antimonide (BiSb) is not really a compound but single crystals of the alloy of bismuth and antimony. The crystals are used as semiconductors in the electronics industry and to produce type for printing presses and low-melting-point electrical fuses.

Bismuth subcarbonate [$(BiO)_2CO_3$] is used to make other bismuth compounds, cosmetics, enamel, and ceramic glazes. Its major use is as an opaque substance placed in the digestive tract to show up on X-rays. The bismuth blocks X-rays, and thus the physician can see patterns inside the stomach and intestines.

Bismuth subnitrate ($BiNO_3$) is used in the cosmetics industry and to make ceramic and enamel glazes.

Hazards

Bismuth is flammable as a powder. The halogen compounds of bismuth are toxic when inhaled or ingested. Some of the salts of bismuth can cause metallic poisoning in a manner similar to mercury and lead.

At the beginning of the twentieth century, before penicillin, bismuth compounds were used to treat some venereal diseases. However, the treatment was generally unsuccessful.

The Oxygen Group (Oxidizers and Nonmetals): Periods 2 to 6, Group 16 (VIA)

Introduction

The oxygen group (group 16; VIA) is at times also referred to as the "chalcogen group." The word "chalcogen" means "ore-forming" and is derived from the Greek *chalcos* (meaning ore) and -*gen* (meaning formation). The term was popular in the 1930s in Germany because it was analogous to the useful term for the "halogens" in group 17 (meaning salt-formers), and the chalcogenides were ores and common minerals such as iron pyrite (FeS_2). The compounds containing oxygen are extensive, and the minerals and ore are found worldwide. The oxidation number of the chalcogens is generally −2, but may be as high as +6 in some sulfate compounds. As oxidizers they "collect" electrons from metals to form ionic bonds or share electrons to form covalent bonds with many elements.

All five elements in the oxygen group have six electrons in their outer orbits. They are all oxidizers (they accept electrons), but they are not all alike. They range from a nonmetal gas (oxygen) to a nonmetal solid (sulfur) to a nonmetallic semiconductor (selenium) to a semi-metal (tellurium) and finally to a radioactive metal (polonium).

OXYGEN

SYMBOL: O   **PERIOD:** 2   **GROUP:** 16 (VIA)   **ATOMIC NO:** 8

ATOMIC MASS: 15.9994 amu   **VALENCE:** 2   **OXIDATION STATE:** −2   **NATURAL STATE:** Gas

ORIGIN OF NAME: From the Greek words *oxys* (which means sharp or acid) and gen (which means forming); together they stand for "acid-forming." In the eighteenth century, it was believed that all acids contained oxygen.

ISOTOPES: There are a total of 15 isotopes of oxygen, three of which are stable. The stable ones are O-16, which accounts for 99.762% of all the oxygen on Earth; O-17, which contributes only 0.038% of the Earth's oxygen; and O-18, which makes up just 0.200% of Earth's oxygen.

PERIODIC TABLE OF THE ELEMENTS

Group headers (top): GROUPS / PERIODS

Group	1	2	3	4	5	6	7	8	9	10	11	12	13	14	15	16	17	18
	IA	IIA	IIIB	IVB	VB	VIB	VIIB	VIII	VIII	VIII	IB	IIB	IIIA	IVA	VA	VIA	VIIA	VIIIA
1	H 1 1.0079																	He 2 4.00260
2	Li 3 6.941	Be 4 9.01218											B 5 10.81	C 6 12.011	N 7 14.0067	O 8 15.9994	F 9 18.9984	Ne 10 20.179
3	Na 11 22.9898	Mg 12 24.305											Al 13 26.9815	Si 14 28.0855	P 15 30.9738	S 16 32.066(6)	Cl 17 35.453	Ar 18 39.948
4	K 19 39.0983	Ca 20 40.08	Sc 21 44.9559	Ti 22 47.88	V 23 50.9415	Cr 24 51.996	Mn 25 54.9380	Fe 26 55.847	Co 27 58.9332	Ni 28 58.69	Cu 29 63.546	Zn 30 65.39	Ga 31 69.72	Ge 32 72.59	As 33 74.9216	Se 34 78.96	Br 35 79.904	Kr 36 83.80
5	Rb 37 85.4678	Sr 38 87.62	Y 39 88.9059	Zr 40 91.224	Nb 41 92.9064	Mo 42 95.94	Tc 43 (98)	Ru 44 101.07	Rh 45 102.906	Pd 46 106.42	Ag 47 107.868	Cd 48 112.41	In 49 114.82	Sn 50 118.71	Sb 51 121.75	Te 52 127.60	I 53 126.905	Xe 54 131.29
6	Cs 55 132.905	Ba 56 137.33	★ La 57 138.906	Hf 72 178.49	Ta 73 180.948	W 74 183.85	Re 75 186.207	Os 76 190.2	Ir 77 192.22	Pt 78 195.08	Au 79 196.967	Hg 80 200.59	Tl 81 204.383	Pb 82 207.2	Bi 83 208.980	Po 84 (209)	At 85 (210)	Rn 86 (222)
7	Fr 87 (223)	Ra 88 226.025	▲ Ac 89 227.028	Unq 104 (261)	Unp 105 (262)	Unh 106 (263)	Uns 107 (264)	Uno 108 (265)	Une 109 (266)	Uun 110 (267)	Uuu 111 (272)	Uub 112	Uut 113	Uuq 114	Uup 115	Uuh 116	Uus 117	Uuo 118

TRANSITION ELEMENTS (groups 3–12, IIIB → IIB)

★ Lanthanide Series (RARE EARTH) — Period 6

La 57 138.906	Ce 58 140.12	Pr 59 140.908	Nd 60 144.24	Pm 61 (145)	Sm 62 150.36	Eu 63 151.96	Gd 64 157.25	Tb 65 158.925	Dy 66 162.50	Ho 67 164.930	Er 68 167.26	Tm 69 168.934	Yb 70 173.04	Lu 71 174.967

▲ Actinide Series (RARE EARTH) — Period 7

Ac 89 227.028	Th 90 232.038	Pa 91 231.036	U 92 238.029	Np 93 237.048	Pu 94 (244)	Am 95 (243)	Cm 96 (247)	Bk 97 (247)	Cf 98 (251)	Es 99 (252)	Fm 100 (257)	Md 101 (258)	No 102 (259)	Lr 103 (260)

ELECTRON CONFIGURATION

Energy Levels/Shells/Electrons	Orbitals/Electrons
1-K = 2	s2
2-L = 6	s2, p4

Properties

There are three allotropes (different forms) of oxygen: (1) atomic oxygen (O), sometimes referred to as nascent or "newborn" oxygen; (2) diatomic oxygen (O_2), or molecular oxygen (gas); and (3) ozone (O_3), also a gas.

The atmospheric oxygen that we breathe is a very reactive nonmetal and is colorless, odorless, and tasteless, but it is essential to all living organisms. It readily forms compounds with most other elements. With six electrons in its outer valence shell, it easily gains two more electrons to form a negative (–2) ion; or as covalent, it can share electrons with other elements to complete its outer shell.

Almost all the oxygen in the atmosphere (˜21%) is the allotropic form of molecular oxygen (O_2). This essential gas we breathe is the result of photosynthesis, which is how green plants (with chlorophyll) use the energy of the sun to convert carbon dioxide (CO_2) and water to starches and sugars with molecular oxygen as the by-product.

Liquid oxygen has a slightly bluish cast to it. As it boils, pure oxygen gas is released. The melting point for oxygen is –218.79°C, its boiling point is –182.95°C, and its density is 0.001429 g/cm³.

Characteristics

Oxygen is, without a doubt, the most essential element on Earth. It is required to support all plant and animal life, and it forms more compounds with other elements than any other element.

Oxygen is soluble in both water and alcohol. Contrary to what many people believe, oxygen is NOT combustible (it will not burn), but rather it actively supports the combustion of many other substances. After all, if oxygen burned, every time a fire was lit, all the O_2 in the atmosphere would be consumed!

Burning is a form of oxidation wherein oxygen chemically combines with a substance rapidly enough to produce adequate heat to cause fire and light, or to maintain a fire once started.

The oxidation of iron is called rusting. Rusting in an example of "slow oxidation," which is the reaction of O_2 with Fe to form Fe_2O_3 or Fe_3O_4. This chemical reaction is so slow that the heat it produces is dissipated; thus, there is no fire.

Recently a new allotrope of oxygen was discovered. When O_2 is subjected to great pressure, it is converted into O_4, which is a deep red solid that is a much more powerful **oxidizer** than the other forms of oxygen.

Abundance and Source

Oxygen is the third most abundant element in the universe, making up nearly half the mass of the Earth's crust and nine-tenths of the total mass of water. Even the mass of our bodies consists of two-thirds oxygen. Oxygen is also the most abundant element in the Earth's atmosphere at 20.947% by volume.

Oxygen is produced commercially by liquefying air under reduced temperatures and increased pressure. Then oxygen (and other gases) can be collected as the temperature rises in the liquid air, allowing the various gases to boil off at their specific boiling points. This process is known as **fractional distillation.** Liquid air can be transported in vacuum vessels in the liquid form as long as there is a small vent to allow the escape of some of the gas that boils away as temperatures rise above the boiling point.

Fractional distillation is based on the principle that each element has its own temperature at which it changes from a liquid to a gas. Thus, any gas can be separated from other liquefied components of air and then collected. The same process is used in the petroleum industry to separate various fractions from the crude oil.

There are several methods of producing oxygen gas in the laboratory.

Oxygen can be produced by electrolysis of water using a salt as an electrolyte that produces hydrogen at the opposite electrode. When potassium chlorate ($KClO_3$) is heated in a test tube with a small amount of manganese dioxide (MnO_2) as a catalyst, the chemical reaction that releases the oxygen from potassium chlorate will be accelerated. Use of potassium nitrate (KNO_3) will also produce small amounts of oxygen.

A recent, and more productive, method is to pass air through fine molecular-size sieves of material that will absorb the nitrogen gas of air, which then allows the oxygen gas to pass through the sieve to be collected.

History

At one time in history it was thought that air was an elemental substance. Leonardo da Vinci (1452–1519) believed that air was composed of at least two gases. He was the first to report that air was a mixture of two gases and that one of them supported life and combustion.

Alchemists of the seventeenth century were aware that when metals were heated in air, they gained weight; when heated in the absence of air, metals lost weight. We now know that oxides of metals are heavier than the metals themselves.

In the eighteenth century, several chemists worked with the concept that air consists of more than one gas. Some of them produced oxygen without realizing it. They all knew that air supported life. For example, Joseph Priestley (1733–1804) produced a special gas by heating oxides of mercury and lead. He called it "dephlogisticated air." In 1772 Carl Wilhelm Scheele (1742–1786) published information explaining how air is composed of two different gases and how one supported fire whereas the other did not. He called one of his gases "fire air" (also "empyreal air") and the other "vitiated air." Daniel Rutherford (1749–1819) called the gas he isolated "phlogisticated air."

Antoine-Laurent Lavoisier (1743–1794) followed up Priestley's work by making quantitative measurements of the ratio of oxygen to nitrogen in air. At first he named the new gas "highly respirable air" and later, "vital air." Lavoisier is often considered the "father of modern chemistry" for the experimental procedures he used and for making precise measurements and recording his data when conducting investigations. Credit for naming the element was given to Lavoisier. He and other scientists of his time believed that all acids contained oxygen. Because acids smell "sharp," Lavoisier named the element oxygen for *oxys,* the Greek word for "sharp." Later it was discovered that not all acids contained oxygen (e.g., hydrochloric acid [HCl]).

Common Uses

Oxygen has many uses due to its high electronegativity with the ability to oxidize many other substances. Only fluorine has higher electronegativity and is thus a stronger oxidizer. Besides the essential use to support life, oxygen has many other uses.

It is used in the smelting process to free metals from their ores. It is particularly important in the oxygen-converter process in the production of steel from iron ore.

Oxygen is used in making several important **synthetic** gases and in the production of ammonia, methyl alcohol, and so on.

It is the oxidizer for liquid rocket fuels, and as a gas, oxygen is used in a mixture with helium to support the breathing of astronauts and divers and to aid patients who have difficulty breathing. It is use to treat (oxidize) sewage and industrial organic wastes.

Oxygen has many uses because of its ability to accept electrons from other elements to form ionic bonds or to share electrons with other elements to form covalent bonds.

As previously mentioned, oxygen is the by-product of photosynthesis, wherein carbon dioxide and water are transformed into food by the chlorophyll of green plants, using sunlight as energy. Carbon dioxide is produced by the decay of organic material and the burning of fossil fuels in homes, in industry, and for transportation. There is much concern about the increase of carbon dioxide in our atmosphere. There is about 0.03+% carbon dioxide in the atmosphere, and this amount has increased over the past century. It is claimed that excess carbon dioxide produced by modern society is responsible for global warming, a claim with which not all scientists agree. At the same time, an atmosphere enriched with CO_2 enhances the growth of plants, which are ultimately the source of all our food. The respiration (breathing) of plants and animals also involves the intake of oxygen, followed by the process of oxidation, which converts food into energy that is required to sustain all life.

Examples of Compounds

The 10 most common compounds found in the Earth's crust are oxides. Silicone dioxide (SiO_2), or common sand, makes up about half of the oxides in the crust.

Oxygen, with an −2 oxidation state, reacts vigorously with group 1 (1A) metals, which have a +1 oxidation state, and is somewhat less reactive with group 2 (2A) metals, which have a +2 oxidation state. A general formula is used for the reactions of oxygen with the group 1 elements (Li, Na, K, Rb, Cs, and Fr). An upper case "M" represents the metallic element in oxidation reactions: $4M^{1+} + O_2^{2-} \rightarrow 2M_2O$. Similarly, a common formula for +2 group 2 metals (Be, Mg, Ca, Sr, Ba, and Ra) uses the "M" to represent the metals: $2M^{2+} + O_2^{2-} \rightarrow 2MO$.

One of the most important oxides is *dihydrogen oxide,* or rather water (H_2O). There are numerous oxygen compounds on Earth, many of them with more that two elements. They include the silicates, which make up rocks and soil, as well as limestone (calcium carbonate), gypsum (calcium sulfate), bauxite (aluminum oxide), and many iron oxides.

The lanthanides, sometimes referred to as "rare-earths," are another important group of oxides. Most exist in nature as their oxides. (See the section titled "Lanthanide Series.")

There are general formulas for other oxygen compounds. For example, alcohols can be generalized as R-OH, aldehydes as R-CHO, and carboxylic acids as R-COOH.

Hazards

Although oxygen itself is not flammable or explosive, as is sometimes believed, its main hazard is that, in high concentrations, oxygen can cause other materials to burn much more rapidly.

Oxygen is toxic and deadly to breathe when in a pure state at elevated pressures. In addition, such pure oxygen promotes rapid combustion and can produce devastating fires, such as the fire that killed the Apollo 1 crew on a test launch pad in 1967. It spread rapidly because the pure oxygen was at normal pressure rather than the one-third pressure used during flight.

Oxygen used for therapeutic purposes in adults can cause **convulsions** if the concentration is too high. At one time, high levels of oxygen were given to premature infants to assist their breathing. It was soon discovered that a high concentration of O_2 caused blindness in some of the infants. This practice has been abandoned, or the oxygen levels have since been reduced, and this is no longer a medical problem.

Oxygen involved in metabolic processes are prone to form "**free radicals**," which are thought to cause damage to cells and possibly be associated with cancer and aging.

Ozone (Also Group 16)

This section on ozone is included under oxygen in group 16 because of its importance today in the lives of citizens and its effect on the environment. It is treated as another element with its own properties and characteristics, uses, and hazards.

Properties

Ozone is an allotropic molecular form of oxygen containing three atoms of oxygen (O_3). It is a much more powerful oxidizing agent than diatomic oxygen (O_2) or monatomic oxygen (O). It is the second most powerful oxidizer of all the elements. Only fluorine is a stronger oxidizer. It is not colorless as is oxygen gas. Rather, ozone is bluish in the gaseous state, but blackish-blue in the liquid and solid states (similar to the color of ink).

Ozone's boiling point is –112°C, and its freezing point is –192°C.

Characteristics

Ozone has a very distinctive pungent odor. It exists in our lower atmosphere in very small trace amounts. In higher concentrations it is irritating and even poisonous. Ozone is in relatively low concentrations at sea level. In the upper atmosphere, where it is more concentrated, it absorbs ultraviolet radiation, which protects the Earth and us from excessive exposure to ultraviolet radiation.

Electrical discharges in the atmosphere produce small amounts of ozone. You can recognize the odor when running electrical equipment that gives off sparks. Even toy electric trains can produce ozone as they spark along the track. Ozone can be produced by passing dry air between two electrodes that are connected to alternating electric current with high voltage. Such a system is sometimes used to purify the air in buildings or provide ozone for commercial uses. Ozone is produced during the electrical discharges of lightning during storms. This is what makes the air seem so fresh after a thunderstorm or electrical storm. Besides being produced by electrical discharges, ozone is produced in the upper atmosphere or stratosphere by ultraviolet (UV) radiation from the sun striking O_2 molecules, breaking them down and reforming them as O_3 molecules. The vast majority of ozone is produced in the atmosphere over the tropical latitudes because this area gets most of the radiation from more direct sunlight. Normal wind currents carry the ozone to the polar regions of the Earth where it is thickest.

Following are the chemical reactions for the production of ozone at ground level. Ozone can be formed when a mixture of O_2 and NO_2 is exposed to sunlight. Given that this reaction is very slow at normal temperatures, it is not a problem until hot gases in the cylinders of automobiles' internal combustion engines cause the following more rapid reactions.

N_2 (gas) + O_2 (gas) → + heat 2NO (gas). The NO formed inside the automobile engine reacts spontaneously with O_2 to form NO_2: for example, 2NO (gas) + O_2 (gas) → 2NO_2 (gas). This nitrogen dioxide is a reddish-brown color seen in smog that dissociates when exposed to strong sunlight: NO_2 (gas) + sunlight → NO (gas) + O (gas). In the final step, the monoatomic oxygen form is very reactive and combines with O_2 to form ozone: O (gas) + O_2 (gas) → O_3 (ozone gas). On sunny days in high-traffic areas, the concentration of ozone can reach levels that are harmful to plants and animals.

History

It was once believed that air was a single element, but by the fifteenth century **ce,** scientists began to question whether it was possibly at least two separate gases. Leonardo da Vinci was one of the first to suggest the air consisted of at least two gases. He even determined that one of them would support life and fire.

In 1839 Christian Friedrich Schonbein (1799–1868) discovered a gas with an unusual odor coming from some electrical equipment. He did not know what it was, but because it had an odd smell, he called it "ozone," after the Greek word for "I smell." Although he knew that it was a chemical substance, he mistakenly associated ozone with the halogens (group 17). Others before Schonbein had smelled the gas but had not recognized its importance. Thomas Andrews (1813–1885) and several other scientists, through different experiments, identified ozone as a form of oxygen (an allotrope). It was not until 1868 that J. Louis Soret established the formula to be O_3.

Common Uses

Ozone is much more reactive than O_2, which makes it a very powerful oxidizing agent. Only fluorine is more reactive. It has many commercial uses. It is a strong oxidizer, particularly of organic compounds, it is a strong bleaching agent for textiles, oils, and waxes, and it is a powerful germicide. It is also used in the manufacture of paper, steroid hormones, waxes, and cyanide and in the processing of acids.

Ozone produced by electrical discharge is used to purify drinking water and to treat industrial wastes and sewage. It is also use to deodorize air and kill bacteria by passing dry air through special ozone-producing electronic devices.

Hazards

High concentrations of ozone are a fire and explosion hazard when in contact with any organic substance that can be oxidized.

In moderately high concentrations ozone is very toxic when inhaled, and in lesser concentrations, it is irritating to the nose and eyes. Ozone in the lower atmosphere contributes to air pollution and smog. It can cause damage to rubber, plastics, and paints. These low concentrations can cause headaches, burning eyes, and respiratory irritation. It is particular harmful to asthmatics and the elderly with respiratory problems.

The U.S. EPA defines as "unhealthy" ambient air containing 125 parts per billion (ppb) of ozone. (This means the ordinary air we breathe.) In different regions of the United States, ozone alerts are issued when the ozone level is both higher and lower than this standard. Most regions have their own standards that range from 100 ppb to 350 ppb.

Ozone Controversy

The effects of natural and man-made chemicals on the ozone layer of the upper atmosphere have been suspected and controversial for over 40 years. In 1974 F. Sherwood Rowland, from the University of California at Irving, and Mario Molina, a professor of environmental studies at MIT suggested that the problem chemicals were **chlorofluorocarbons** (CFCs) used in refrigeration and as a propellant in spray cans. As early as the 1960s, several chemists suspected that the effects of CFCs, which are extremely stable on Earth, were negatively impacting the atmospheric ozone layer. They had experimental laboratory evidence of the chemical reactions involved, but no actual data that related to the atmospheric ozone layer. Since the mid-1970s, both the scientific facts and the arguments about potential problems have become ensnared in political and environmental ideologies.

The ozone issue, as well as the possibly related issue of global warming and other environmental concerns, is often replete with invectives due to a lack of understanding of both the science and the political arguments involved.

In 1913 Charles Fabry (1867–1945) identified large quantities of ozone in the upper atmosphere found at a range of about 10 to 30 miles above the Earth's surface. This layer is call the **ozonosphere,** or more commonly, the ozone layer. The ozone gas layer of the ozonosphere, although thin, is massive and covers most of the Earth at high altitudes. It is essential because it partially blocks and absorbs the UV rays of the sun, preventing them from reaching the Earth, particularly in the tropical and subtropical areas. Not all the UV is absorbed by the ozone, but a large portion of the stronger, shorter-wavelength rays, which are the most harmful to living organisms, are absorbed and do not reach the Earth. Some UV radiation is needed to activate vitamin D, but excessive ultraviolet light can cause skin cancer, cataracts, mutations, and death in plants and animal cells. One theory is that early in formation of life and in the following ages, the lack of an ozone layer permitted short UV rays to reach Earth, and this many have been partially responsible for the proliferation of the species, because of genetic **mutations.**

Through the natural processes of the UV rays of the sun passing through this layer, the O_3 absorbs the rays and is broken down to O_2 molecules and O atoms. This process is reversible, and ozone (O_3) is constantly being reformed from UV effects on O_2. However, the separation can be accelerated faster than the reformation of new O_3 by induction of other chemical gases into the ozone layer. Of particular concern is that chlorine from CFCs and from other sources, such as the ocean and volcanic eruptions, combines with atomic oxygen that is broken down from O_3 by UV radiation. It then forms chlorine monoxide (ClO), which means the atomic oxygen is not available for reformation into O_3 by UV radiation. Herein lie the potential problem and the controversy.

At issue is a group of man-made chemicals called chlorofluorocarbons (CFCs), which are used as the refrigeration fluids (Freon) in the air-conditioning and refrigeration units of automobiles, homes, and industry and which, in the past, were used as the pressure gas in spray cans of paint, deodorants, and so forth. There are a number of compounds containing chlorine, fluorine, and carbon that are inert, as are the chlorofluorocarbons, and they remain that way until they reach the upper atmosphere, where intense UV radiation from the sun liberates the chlorine atoms for the compound. These chlorine atoms catalytically destroy ozone. An example follows:

$$O_3 + UV \text{ light} \rightarrow O_2 + O$$

$$Cl + O_3 \rightarrow ClO + O_2$$

$$ClO + O \rightarrow Cl + O_2$$

Results: $2O_3 \rightarrow 3O_2$

When CFCs slowly rise in the atmosphere and reach the ozone layer, they are broken down into component molecular compounds and atoms by the UV rays of the sun. Some of these chemicals then react with ozone to break it down, thus reducing the amount of O_3. Further, some chlorine (also from the oceans) and some other elements combine with the O and O_3 to form other chemicals. This also contributes to the reduction of ozone faster than natural processes can reform it. Ozone is a renewable resource. The issue is this: can a balance be obtained between the destruction of ozone in the atmosphere, by both natural and man-made causes, and its natural regeneration?

Also of concern because of their possible connection with global warming are the hydrocarbon gases, such as methane (CH_4), which is produced naturally in large quantities, as the gas from the digestive process of cows, from decaying of organic matter, and from petroleum refining; and the gas carbon dioxide (CO_2), which is produced in nature by respiration of plants and animals, by volcanic action and forest fires, and by humans in the burning of wood, coal, gas, and oil products. Nitrogen-based compounds from automobile exhausts also contribute to the problem, but automobile exhaust gases do not directly produce ozone.

There are parallel policy issues for global warming and the periodic increase and decrease of the ozone layer, but the science is not parallel. The following are some factors to keep in mind when examining these issues.

1. There is an agreement that a "hole" (really a thin area) in the ozone layer over the Earth's polar regions, particularly Antarctica, changes in size over periods of time. The ozone is produced over the tropics and spreads to the polar areas. But not all scientists agree on the associated causes, seriousness, dangers involved, and remedies.

2. The ozone layer is dynamic and unpredictable, which means it is constantly changing and seems to change in ways we cannot yet understand. The large thin area (hole) over Antarctica seems not only to move but also to become larger and then smaller. The ozone layer is thickest over the poles of the Earth, yet it is mostly produced over the equator. The ozone layer seems to follow some cyclic pattern that may affect the size of the thin layer over time.

3. Much of the data supporting claims on both sides of the controversy of ozone depletion and global warming need to be analyzed with great care because the issues have gone beyond the science involved. Our ability to predict long-term weather and atmospheric conditions, even with computer modeling, is limited due to the large number of variables involved. A new science called "chaoplexity," which combines chaos theory with related

problems of complexity, may lead to a useful method of predicting "unpredictable" events, However, it also may not be useful.

4. Global warming may or may not be a problem related to ozone depletion. There has been a 30% increase in the amount of carbon dioxide produced by humans since the early 1700s. Carbon dioxide makes up about 0.03+% of the atmosphere. Only a small fraction of CO_2 remains in the atmosphere. Some is used by plants to make food (photosynthesis), some is dissolved in the oceans, and some combines with other elements. The argument is that CO_2 and some other gases from the burning of hydrocarbon fuels may form a "greenhouse" effect in the upper atmosphere to hold in the heat of the Earth and have some effect on the ozone layer.

According to the Council on Environmental Quality, there was a warming trend of the Earth from 1870 to 1940. This trend was reversed from 1940 to 1960 when the Earth became cooler. Several days in the 1990s were considered the warmest days on record until the year 2005, which was the warmest on record in the northern hemisphere, but not the southern hemisphere. There has been a general (average) temperature increase of about 1°F of the Earth's atmosphere over the last 100 years. Some of this increase is a result of natural atmospheric conditions that tend to be cyclic, and some is a result of increased concentrations of trace gases, such as carbon dioxide, methane, CFCs, sulfur, and nitrogen compounds. Studies indicate that methane, which is an important greenhouse gas, is produced in the wetlands and the sediment of lakes in Siberia, as well as other sources such as vents in the ocean floor. Much of the controversy about proving the reality of climate change is because the wrong effects are being measured and analyzed by inadequate computer programs that do not consider all the physical, chemical, and biological variables that may affect climate change.

Today, there are over 30% more trees growing in the United States than there were 100 years ago, and we plant four trees for every three harvested, although the growth rate of these new trees may not equal the harvest rate of more mature trees. Sadly, this is not true for all countries, some of which have depleted not only their forests but also most natural vegetation, causing the formation of deserts. One way of reducing the amount of carbon dioxide in the atmosphere, besides reducing the burning of hydrocarbon fuels, is to greatly increase the amount of plant life on Earth. We need to reestablish forests worldwide.

5. The U.S. government banned the use of chlorofluorocarbon aerosols in paint and spray cans in 1978. In 1986 an international agreement was adopted that required all industries to reduce the manufacture and use of CFCs by 50% by the year 2000. More recently, an outright ban on the use of Freon by industrialized countries has been agreed to. Research is being conducted to find suitable substitutes. Materials in one group of promising compounds are called HFCs and HCFCs, which are not as stable as CFCs. They break down before reaching the ozone level of the stratosphere and thus do not react with the ozone. The problem is that some of these new HCFCs are flammable and possibly carcinogenic.

6. It is doubtful that any international agreement to control the so-called greenhouse gases by all nations can be effective. First, many developing nations will continue to develop industries that produce these gases. Second, the cost of developing alternate, nonpolluting energy sources is tremendous. And finally, Earth, through many natural systems and

processes, will continue to produce variable amounts of gases, such as carbon dioxide, chlorine, methane, and ozone, in the future, and the Earth's tendency is to establish her own balance, as she always has.

There is a need for much more scientific study of the systems and processes related to the atmosphere and the changing cyclic conditions that affect the Earth. Understanding the cyclic nature and systems of the Earth is required, as are the objective generation and interpretation of reliable data.

SULFUR

SYMBOL: S   **PERIOD:** 3   **GROUP:** 16 (VIA)   **ATOMIC NO:** 16
ATOMIC MASS: 32.065 amu   **VALENCE:** 2, 4, and 6   **OXIDATION STATE:** −2, +4, and +6   **NATURAL STATE:** Solid
ORIGIN OF NAME: From the Sanskrit word *sulvere* and the Latin word sulphurim.
ISOTOPES: There are a total of 24 isotopes of sulfur; all but four of these are radioactive. The four stable isotopes and their contribution to sulfur's total abundance on Earth are as follows: S-32 contributes 95.02% to the abundance of sulfur; S-33, just 0.75%; S-34, 4.21%; and S-36, 0.02%.

ELECTRON CONFIGURATION

Energy Levels/Shells/Electrons	Orbitals/Electrons
1-K = 2	s2
2-L = 8	s2, p6
3-M = 6	s2, p4

Properties

Sulfur is considered a nonmetallic solid. It is found in three allotropic crystal forms:

1. Orthorhombic (or rhombic) octahedral lemon-yellow crystals, which are also called "brimstone" and referred to as "alpha" sulfur. The density of this form of sulfur is 2.06 g/cm^3, with a melting point of 95.5°C.
2. Monoclinic, prismatic crystals, which are light-yellow in color. This allotrope is referred to as "beta" sulfur. Its density is 1.96 g/cm^3, with a melting point of 119.3°C.
3. Amorphous sulfur is formed when molten sulfur is quickly cooled. Amorphous sulfur is soft and elastic, and as it cools, it reverts back to the orthorhombic allotropic form.

Sulfur, in its elemental form, is rather common and does not have a taste or odor except when in contact with oxygen, when it forms small amounts of sulfur dioxide.

Characteristics

Sulfur exhibits a remarkable array of unique characteristics. Today, there are chemists devoting large portions of their careers to studying this unusual element. For example, when sulfur is melted, its **viscosity** increases, and it turns reddish-black as it is heated. Beyond 200°C, the color begins to lighten, and it flows as a thinner liquid.

Sulfur burns with a beautiful subdued blue flame. The old English name for sulfur was "brimstone," which means "a stone that burns." This is the origin of the term "fire and brimstone" when referring to great heat. Above 445°C, sulfur turns to a gas, which is dark orange-yellow but which becomes lighter in color as the temperature rises.

Sulfur is an oxidizing agent and has the ability to combine with most other elements to form compounds.

Abundance and Source

Sulfur has been known since ancient times primarily because it is a rather common substance. It is the 15th most common element in the universe, and though it is not found in all regions of the Earth, there are significant deposits in south Texas and Louisiana, as well in all volcanoes. Sulfur makes up about 1% of the Earth's crust.

Sulfur is an element found in many common minerals, such as galena (PbS), pyrite (fool's gold, FeS_2), sphalerite (ZnS), cinnabar (HgS), and celestite ($SrSO_4$), among others. About 1/4 of all sulfur procured today is recovered from petroleum production. The majority of sulfur is the result of or a by-product of mining other minerals from the ores containing sulfur.

Sulfur is mined by the recovery method known as the Frasch process, which was invented by Herman Frasch in Germany in the early 1900s. This process forces superheated water, under pressure, into deep underground sulfur deposits. Compressed air then forces the molten sulfur to the surface, where it is cooled. There are other methods for mining sulfur, but the Frasch process is the most important and most economical.

Sulfur is found in Sicily, Canada, Central Europe, and the Arabian oil states, as well as in the southern United States in Texas and Louisiana and offshore beneath the Gulf of Mexico.

History

Sulfur was known in the days of early humans. No single person can take credit for its discovery. It was probably one of the first "free" elements ancient humans tried to use and understand besides the "always known" air, fire, and water.

Sometime around 1300 BCE, an unknown alchemist described sulfuric acid. Not much is known about the early use of sulfur or sulfuric acid. In 1579 an alchemist named Andreas Libavius described the progress of alchemy. In his book he described how hydrochloric and sulfuric acids are produced and mentioned the formation of **aqua regia,** which is a mixture of acids that is strong enough to dissolve gold—the royal metal.

Most chemists considered sulfur to be a mixture of elements until 1777, when Antoine Lavoisier (1743–1794), the "father of modern chemistry," convinced them that it was a true chemical element.

Common Uses

Sulfur is one of the four major commodities of the chemical industry. The other three are limestone, coal, and salt. Most sulfur that is produced is used to manufacture sulfuric acid (H_2SO_4). Forty million tons are produced each year in the manufacture of fertilizers, lead-acid batteries, gunpowder, desiccants (drying agent), matches, soaps, plastics, bleaching agents, rubber, road asphalt binders, insecticides, paint, dyes, medical ointment, and other pharmaceutical products, among many, many other uses. Sulfur is essential to life.

Examples of Common Compounds

Hydrogen sulfide (H_2S) is one of the most important compounds of sulfur. It is a colorless gas with a foul, rotten-egg odor. It is well known in school laboratories when sulfur is being studied. It is produced by the reaction of hydrochloric acid with iron sulfide ($2HCl + FeS \rightarrow FeCl_2 + H_2S$).

Sulfur dioxide (SO_2) has many uses, including as a bleaching agent, solvent, disinfectant, and refrigerant. It is mostly used in the production of sulfuric acid. Sulfur dioxide is generated by burning sulfur in air, resulting in the following chemical reaction: $S + O_2 \rightarrow SO_2$. Sulfur dioxide can also be produced by roasting metal sulfides and by the reaction of acids with metallic sulfides. Sulfur dioxide combines with water to form a weak acid known as sulfurous acid ($SO_2 + H_2O \rightarrow H_2SO_3$).

Sulfur chloride (S_2Cl_2) is combustible and will react when in contact with water. It is used to produce carbon tetrachloride, to purify sugar juices, to extract gold from its ore, in insecticides, and as a poisonous gas for military/combat purposes.

Sulfuric acid (H_2SO_4), also known at battery acid, is the leading chemical manufactured in the United States, with an annual production of 40 million tons per year. Most of it is used in the manufacture of fertilizers, explosives, pigments, and dyes. It has the nasty attribute of being able to extract hydrogen and oxygen from organic substances, which can cause serious burns. It is strongly corrosive and, in both concentrated and weak solutions with water, will react with most metals.

Several steps are required in the production of sulfuric acid, as follows:

$$S + O_2 \rightarrow SO_2$$

$$2SO_2 + O_2 \rightarrow 2SO_3$$

$$SO_3 + H_2SO_4 \rightarrow H_2S_2SO_7$$

$$H_2S_2O_7 + H_2O \rightarrow 2H_2SO_4$$

Because there are so many metal sulfides, chemists usually use the letter "M" in the formula to indicate that sulfur can combine with just about any metal (e.g., MHS, M_2S, M_2S_3, and so on). Sulfate ions (SO_4^{2-}) also combine with many different metal atoms to form common compounds, such as copper sulfate ($CuSO_4$) and magnesium sulfate, calcium sulfate, lead sulfate, zinc sulfate, and barium sulfate.

There are four major phosphorous–sulfur compounds, ranging from P_4S_3 to P_4S_{10}, and there are six important oxygen–sulfur compounds, ranging from SO to S_2O_7, the most important being sulfur dioxide (SO_2), which can react as both an oxidizing and a reducing agent.

Hazards

Many of the sulfur compounds are toxic but essential for life. The gas from elemental sulfur and from most of the compounds of sulfur is poisonous when inhaled and deadly when ingested. This is the reason that sulfur compounds are effective for rat and mice extermination as well an ingredient of insecticides. Sulfa drugs (sulfanilamide and sufadiazine), although toxic, were used as medical antibiotics during World War II before the development of penicillin. They are still used today in veterinary medicine.

SELENIUM

SYMBOL: Se **PERIOD:** 4 **GROUP:** 16 (VIA) **ATOMIC NO:** 34
ATOMIC MASS: 79.96 amu **VALENCE:** 2, 4, and 6 **OXIDATION STATE:** −2, +4, and
  +6 **NATURAL STATE:** Solid.
ORIGIN OF NAME: Named for the Greek word *selene,* meaning "moon." Jons Jacob Berzelius (1779–1848) discovered selenium and named it after the mineral called "eucairite," which in Greek means "just in time."
ISOTOPES: There are a total of 35 isotopes of selenium. Five of these are stable, and a sixth isotope has such a long half-life that it is also considered stable: Se-82 = $0.83 \times 10^{+20}$ years. This sixth isotope constitutes 8.73% of selenium's abundance in the Earth's crust, and the other five stable isotopes make up the rest of selenium's abundance on Earth.

ELECTRON CONFIGURATION

Energy Levels/Shells/Electrons	Orbitals/Electrons
1-K = 2	s2
2-L = 8	s2, p6
3-M = 18	s2, p6, d10
4-N = 6	s2, p4

Properties

Selenium is a soft metalloid or semimetal that is similar to tellurium, located just below it in the oxygen group, and sulfur, which is just above it in the same group. Selenium has several allotropic forms that range from a gray metallic appearance to a red glassy appearance. These allotropic forms also have different properties of heat, conductivity, and density. In its amorphous state, it is a red powder that turns black and becomes crystalline when heated. Crystalline selenium has a melting point of 220°C, a boiling point of 685°C, and a density of 4.809 g/cm³.

Characteristics

Crystalline selenium is a p-type semiconductor. It acts as a rectifier that can change electric current from alternating current (AC) to direct current (to DC). It has photovoltaic properties, meaning it is able to convert light (radiant) energy that strikes it into electrical energy. Selenium's resistance to the flow of electricity is influenced by the amount of light shining on it. The brighter the light, the better the electrical conductivity.

Selenium burns with a blue flame that produces selenium dioxide (SeO_2). Selenium will react with most metals as well as with nonmetals, including the elements in the halogen group 17.

Abundance and Source

Selenium is the 67th most abundant element in Earth's crust. It is widely spread over the Earth, but does not exist in large quantities. As a free element it is often found with the element sulfur.

There is only one mineral ore that contains selenium: eucairite (CuAgSe). Although rich in selenium, it is too scarce to be of commercial use. Almost all selenium is recovered from the processing of copper and the manufacturing of sulfuric acid as a leftover sludge by-product. This makes selenium's recovery profitable. Recovering it from eucairite is not profitable.

Selenium is found in Mexico, Bosnia, Japan, and Canada. It can be found in recoverable quantities in some soils in many countries.

History

Selenium was discovered in the early 1800s by Jons Jakob Berzelius (1779–1848) along with a friend Wilhelm Hisinger (1766–1852), who was a mineralogist. Berzelius believed tellurium was contaminating the product in a sulfuric acid factory. Later, he found it to be another element similar to tellurium. After isolating it, he identified it as element 34, which turned out to be selenium.

Common Uses

The photosensitive nature of selenium makes it useful in devices that respond to the intensity of light, such as photocells, light meters for cameras, xerography, and electric "eyes." Selenium also has the ability to produce electricity directly from sunlight, making it ideal for use in solar cells. Selenium possesses semiconductor properties that make it useful in the electronics industry, where it is a component in some types of solid-state electronics and rectifiers. It is also used in the production of ruby-red glass and enamels and as an additive to improve the quality of steel and copper. Additionally, it is a catalyst (to speed up chemical reactions) in the manufacture of rubber.

Selenium is an essential trace element for both plants and animals, and it is a diet supplement in animal feed as well as for humans.

Examples of Compounds

Selenium forms a few important inorganic compounds. Some examples follow:

Selenium dioxide (SeO_2) is used as an oxidizing agent, as a catalyst, and as an antioxidant for lubricating oils and grease.

Selenium sulfide (SeS_2) is used for some medicines, as an additive for medicated shampoos to control dandruff and scalp itching, and in treatment products for acne and eczema.

There are several different compounds of selenium and the halogen chlorine that range from *selenium dichloride* ($SeCl_2$) to *selenium oxychloride* ($SeOCl_2$), which are used as solvents.

Hazards

The fumes and gases of most selenium compounds are very toxic when inhaled. SeO_2 and SeS_2 are toxic if ingested and very irritating to the skin. They are also carcinogenic.

Although some compounds of selenium are poisonous, as an element it is essential in trace amounts for humans. It is recommended that 1.1 to 5 milligrams of selenium be included in the daily diet. This amount can be maintained by eating seafood, egg yokes, chicken, milk, and whole grain cereals. Selenium assists vitamin E in preventing the breakdown of cells and some chemicals in the human body.

TELLURIUM

SYMBOL: Te **PERIOD:** 5 **GROUP:** 16 **ATOMIC NO:** 52
ATOMIC MASS: 127.60 amu **VALENCE:** 2, 4, and 6 **OXIDATION STATE:** +6, +4, and
  −2 **NATURAL STATE:** Solid
ORIGIN OF NAME: The name "tellurium" is derived from the Latin word for Earth, *tellus.*
ISOTOPES: There are a total of 48 isotopes of tellurium. Eight of these are considered
  stable. Three of the stable ones are actually radioactive but have such long half-lives
  that they still contribute to the natural abundance of tellurium in the crust of the Earth.
  The isotope Te-123 (half-life of $6\times10^{+14}$ years) contributes 0.89% of the total tellurium
  found on Earth, Te-128 (half-life of $7.7\times10^{+24}$ years) contributes 31.74% to the natural
  abundance, and Te-130 (half-life of $0.79\times10^{+21}$ years) contributes 34.08% to the tellurium in the Earth's crust. The other five stable isotopes and the percentage of their
  natural abundance are as follows: Te-120 = 0.09%, Te-122 = 2.55%, Te-124 = 4.74%,
  Te-125 = 7.07%, and Te-126 = 18.84%. The other 40 isotopes are all radioactive with
  short half-lives.

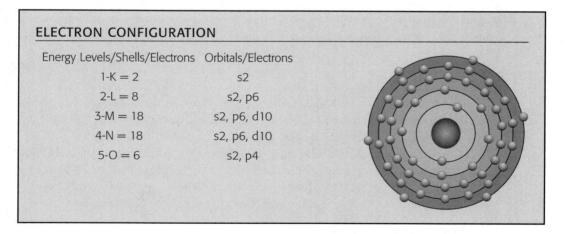

ELECTRON CONFIGURATION

Energy Levels/Shells/Electrons	Orbitals/Electrons
1-K = 2	s2
2-L = 8	s2, p6
3-M = 18	s2, p6, d10
4-N = 18	s2, p6, d10
5-O = 6	s2, p4

Properties

Tellurium is a silver-white, brittle crystal with a metallic luster and has semiconductor characteristics. It is a metalloid that shares properties with both metals and nonmetals, and it has some properties similar to selenium and sulfur, located just above it in group 16 of the periodic table.

There are two allotropic forms of tellurium: (1) the crystalline form that has a silvery metallic appearance and a density of 6.24 g/cm^3, a melting point of 499.51°C, and a boiling point of 988°C; and (2) the amorphous allotrope that is brown in color and has a density of 6.015g/cm^3 and ranges for the melting and boiling point temperatures similar to the crystalline form.

Characteristics

The pure form of tellurium burns with a blue flame and forms tellurium dioxide (TeO_2). It is brittle and is a poor conductor of electricity. It reacts with the halogens of group 17, but not with many metals. When it reacts with gold, it forms gold telluride. Tellurium is insoluble in water but readily reacts with nitric acid to produce tellurous acid. If inhaled, it produces a garlic-like odor on one's breath.

Abundance and Source

Tellurium is the 71st most abundant element on Earth. It makes up a small portion of igneous rocks and is sometimes found as a free element, but is more often recovered from several ores. Its major ores are sylvanite ($AgAuTe_4$), also known as graphic tellurium, calaverite, sylvanite, and krennerite, all with the same general formula ($AuTe_2$). Other minor ores are nagyagite, black tellurium, hessite, altaite, and coloradoite. In addition, it is recovered from gold telluride ($AuTe_2$). Significant quantities are also recovered from the anode "slime" of the electrolytic refining process of copper production.

History

Two people are responsible for the discovery of tellurium. First, Franz Joseph Muller von Reichenstein (1743–1825), chief inspector of a gold mine in Transylvania (part of Romania), experimented with the ores in his mine between 1782 and 1783. From an ore known as aurum album, he extracted an element that, at first, was thought to be antimony. He sent a sample to Martin Heinrich Klaproth (1743–1817), who 16 years later correctly identified it as a new element and named it tellurium. However, Klaproth gave Franz Joseph Muller credit for the discovery.

Common Uses

Tellurium's major use is as an alloy with copper and stainless steel. It makes these metals easier to machine and mill (cut on a lathe). It is also used as a vulcanizing agent in the production of rubber, as a coloring agent for glass and ceramics, and for **thermoelectrical** devices.

Along with lithium, it is used to make special batteries for spacecraft and infrared lamps. Tellurium can be used as a **p-type semiconductor,** but more efficient elements can do a better job. It is also used as a depilatory, which removes hair from skin.

Although tellurium forms many compounds, most of them have little commercial value.

Examples of Compounds

Following are examples of compounds associated with the three ions of tellurium, +2, +4 and +6.

Tellurium (II) dichloride ($TeCl_2$): $Te^{2+} + 2Cl^{1-} \rightarrow TeCl_2$.
Tellurium (IV) tetrachloride ($TeCl_4$): $Te^{4+} + 4Cl^{1-} \rightarrow TeCl_4$.

Tellurium (VI) trioxide (TeO_3): $Te^{6+} + 3O^{2-} \rightarrow TeO_3$.

Other compounds include the following:

Tellurium dibromide ($TeBr_2$) forms blackish-green needles that are very **hygroscopic** (readily absorbs water). It is toxic when inhaled.

Tellurium dichloride ($TeCl_2$) is similar to $TeBr_2$, but in powder form it is greenish-yellow. It is also toxic when inhaled.

Tellurium dioxide (TeO_2) is a whitish crystalline powder that is slightly soluble in water. It is also toxic when inhaled.

Hazards

All forms of tellurium are toxic in gas form. The vapors of all the compounds of the dust and powder forms of the element should not be inhaled or ingested. When a person is poisoned with tellurium, even in small amounts, the breath will smell like garlic.

POLONIUM

SYMBOL: Po   **PERIOD:** 6   **GROUP:** 16   **ATOMIC NO:** 84

ATOMIC MASS: 210 amu   **VALENCE:** 2, 4, and 6   **OXIDATION STATE:** +2 and +4   **NATURAL STATE:** Solid

ORIGIN OF NAME: Named for Poland, the native country of Marie Curie, who discovered the element.

ISOTOPES: There are 41 isotopes of polonium. They range from Po-188 to Po-219. All of them are radioactive with half-lives ranging from a few milliseconds to 102 years, the latter for its most stable isotope Po-209. Polonium is involved with several radioactive decay series, including the actinium series, Po-211 and Po-215; the thorium series, Po-212 and Po-216; and the uranium decay series, Po-210, Po-214, and Po-218.

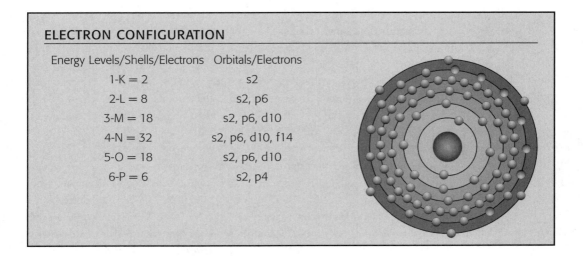

ELECTRON CONFIGURATION

Energy Levels/Shells/Electrons	Orbitals/Electrons
1-K = 2	s2
2-L = 8	s2, p6
3-M = 18	s2, p6, d10
4-N = 32	s2, p6, d10, f14
5-O = 18	s2, p6, d10
6-P = 6	s2, p4

Properties

Most of the known chemistry of polonium is based on the naturally occurring radioactive isotope polonium-210, which is a natural radioactive decay by-product of the uranium decay series. Its melting point is 254°C, its boiling point is 962°C, and its density is 9.32g/cm^3.

Po-210 is a strong emitter of alpha particles. One gram of Po-210 will produce about 140 watts of energy, making it ideal as a lightweight thermoelectric power source for space vehicles. It has a half-life of 138.39 days.

Po-209, the most stable isotope of polonium, decays into lead-205 by alpha decay. It costs about $3,000 per microcurie, which is a very small amount of polonium.

Characteristics

Polonium is more metallic in its properties than the elements above it in group 16. It is the only element in group 16 that is naturally radioactive. It is in a position on the periodic table of elements where it can be a metal, metalloid, or nonmetal. It is more often considered a metal because of its electrical conductivity decreases with an increase in temperature.

As an oxide, it is similar to the reddish color of tellurium oxide. Chemically, it behaves similar to tellurium, lead, and bismuth.

Abundance and Source

Polonium is found only in trace amounts in the Earth's crust. In nature it is found in **pitchblende** (uranium ore) as a decay product of uranium. Because it is so scarce, it is usually artificially produced by bombarding bismuth-209 with neutrons in a nuclear (atomic) reactor, resulting in bismuth-210, which has a half-life of five days. Bi-210 subsequently decays into Po-210 through beta decay. The reaction for this process is $^{209}Bi(_{n,\gamma})^{210}Bi \rightarrow ^{210}Po + \beta$-. Only small commercial milligram amounts are produced by this procedure.

History

Marie Sklodowska Curie (1867–1934) and Pierre Curie (1859–1906) are credited with discovering polonium as they sought the source of radiation in pitchblende after they removed the uranium from its ore. Their discovery in 1898 led to the modern concepts of the nucleus of the atom, its structure, and how it reacts.

They knew there must be another radioactive element in the pitchblende after the uranium was removed. Marie Curie painstakingly processed a ton of pitchblende to recover only a small amount of uranium. Even so, there was still something radioactive in all that processed pitchblende. As it turned out, there were two radioactive elements that she was able to isolate. One was radium, and the other polonium. They were identified by using **piezoelectricity,** discovered by her husband Pierre Curie, which could measure the strength of radiation given off by the radioactive elements with which Marie Curie was working.

Marie Curie named polonium after her native country of Poland. She is also given credit for coining the world "radioactivity." She is one of only two chemists to receive two Nobel Prizes. In 1903 both the Curies and Antoine-Henri Becquerel (1852–1908) shared the Nobel Prize for Physics for their work on radioactivity; in 1911 Madame Curie received the prize for chemistry for the discovery of radium and plonium. (The other scientist who received two Nobel Prizes was Linus Pauling [1901–1994], one for chemistry in 1954, and a Nobel Peace

Prize in 1962.) Madame Curie died from radiation poisoning that resulted from her work with radioactive elements.

Common Uses

There are not many uses for polonium. Probably the most important is as a source of alpha particles (nuclei of helium atoms) and high-energy neutrons for research and radiation studies. It is also used to calibrate radiation-detection devices.

Polonium is used to eliminate static electricity in industrial processes, such as rolling out paper, wire, or sheet metal in mills. Polonium is also sometimes used in "brushes" to remove dust from photographic film and in the manufacturing of spark plugs that make ignition systems in automobiles more efficient, particularly in extremely cold temperatures. It can also be used as a portable, low-level power source and, since polonium is fissionable, used in nuclear weapons and nuclear electric power plants.

Examples of Compounds

Following are two examples of polonium's +2 and +4 oxidation states, in which the element \mostly forms compounds with nonmetals.

Polonium (II) oxide (PoO): $Po^{2+} + O^{2-} \rightarrow PoO$.

Polonium (IV) tetrachloride (PoCl$_4$): $Po^{4+} + 4Cl^{1-} \rightarrow PoCl_4$.

Both the compounds $SPoO_3$ and $SePoO_3$ are similar to tellurium compounds in that they are bright red in color.

Hazards

Even though polonium is a rare element, it is a very dangerous radiation source and should be avoided.

Cigarette smoke contains a minute amount of polonium, along with many other carcinogenic chemicals, many of which can cause lung cancer. Over one hundred trace elements and compounds have been identified in cigarette smoke besides polonium. Some examples are nicotine, cresol, carbon monoxide, pyridine, and the carcinogenic compound benzopyrene.

PERIODIC TABLE OF THE ELEMENTS

GROUPS

PERIODS

Period	1 IA	2 IIA	3 IIIB	4 IVB	5 VB	6 VIB	7 VIIB	8	9 VIII	10	11 IB	12 IIB	13 IIIA	14 IVA	15 VA	16 VIA	17 VIIA	18 VIIIA
1	1 H 1.0079																	2 He 4.00260
2	3 Li 6.941	4 Be 9.01218											5 B 10.81	6 C 12.011	7 N 14.0067	8 O 15.9994	9 F 18.9984	10 Ne 20.179
3	11 Na 22.9898	12 Mg 24.305											13 Al 26.9815	14 Si 28.0855	15 P 30.9738	16 S 32.066(6)	17 Cl 35.453	18 Ar 39.948
4	19 K 39.0983	20 Ca 40.08	21 Sc 44.9559	22 Ti 47.88	23 V 50.9415	24 Cr 51.996	25 Mn 54.9380	26 Fe 55.847	27 Co 58.9332	28 Ni 58.69	29 Cu 63.546	30 Zn 65.39	31 Ga 69.72	32 Ge 72.59	33 As 74.9216	34 Se 78.96	35 Br 79.904	36 Kr 83.80
5	37 Rb 85.4678	38 Sr 87.62	39 Y 88.9059	40 Zr 91.224	41 Nb 92.9064	42 Mo 95.94	43 Tc (98)	44 Ru 101.07	45 Rh 102.906	46 Pd 106.42	47 Ag 107.868	48 Cd 112.41	49 In 114.82	50 Sn 118.71	51 Sb 121.75	52 Te 127.60	53 I 126.905	54 Xe 131.29
6	55 Cs 132.905	56 Ba 137.33	★	72 Hf 178.49	73 Ta 180.948	74 W 183.85	75 Re 186.207	76 Os 190.2	77 Ir 192.22	78 Pt 195.08	79 Au 196.967	80 Hg 200.59	81 Tl 204.383	82 Pb 207.2	83 Bi 208.980	84 Po (209)	85 At (210)	86 Rn (222)
7	87 Fr (223)	88 Ra 226.025	▲	104 Unq (261)	105 Unp (262)	106 Unh (263)	107 Uns (264)	108 Uno (265)	109 Une (266)	110 Uun (267)	111 Uuu (272)	112 Uub	113 Uut	114 Uuq	115 Uup	116 Uuh	117 Uus	118 Uuo

TRANSITION ELEMENTS

6 ★ Lanthanide Series (RARE EARTH)

57 La 138.906	58 Ce 140.12	59 Pr 140.908	60 Nd 144.24	61 Pm (145)	62 Sm 150.36	63 Eu 151.96	64 Gd 157.25	65 Tb 158.925	66 Dy 162.50	67 Ho 164.930	68 Er 167.26	69 Tm 168.934	70 Yb 173.04	71 Lu 174.967

7 ▲ Actinide Series (RARE EARTH)

89 Ac 227.028	90 Th 232.038	91 Pa 231.036	92 U 238.029	93 Np 237.048	94 Pu (244)	95 Am (243)	96 Cm (247)	97 Bk (247)	98 Cf (251)	99 Es (252)	100 Fm (257)	101 Md (258)	102 No (259)	103 Lr (260)

©1996 R.E. KREBS

The Halogen Group (Nonmetal Oxidizers):
Periods 2 to 6, Group 17 (VIIA)

Introduction

The halogens are the family of nonmetal elements in group 17 (VIIA) that are located just to the right of the oxygen group 16 on the periodic table of chemical elements. They are fluorine (F), chlorine (Cl), bromine (Br), iodine (I), and astatine (At).

Their unique characteristics are a result of their outer shells having seven electrons, and thus requiring only one electron to become complete. This −1 oxidation state makes them extremely reactive with both metals and some nonmetal elements that form negative ions, and they may form either ionic or covalent bonds. They can also form compounds with each other; these binary compounds of the halogens are called "halides."

When they form ionic bonds, they gain one electron in their outer shell to form monovalent negatively charged ions, halide ions. Halogens are excellent oxidizing agents. As their atoms become larger (i.e., as you go down in the group), they are less effective as oxidizing agents Fluorine, at the top of group 17 with the smallest atoms of the group, is also the least dense and most reactive of the five, whereas iodine, near the bottom, is the least active (excluding astatine, which is the most dense.) However, for each period (horizontal position on the periodic table) in which they are found, they are the best oxidizing agents within the period.

When the halogens are in a gaseous state, they occur as diatomic molecules (e.g., Cl_2). However, only two of the halogens are gases at room temperature: fluorine (F_2) and chlorine (Cl_2). Bromine is a liquid and iodine is a solid at room temperatures. Astatine is the only halogen that is radioactive and is not very important as a representative of the halogens.

FLUORINE

SYMBOL: F   **PERIOD:** 2   **GROUP:** 17 (VIIA)   **ATOMIC NO:** 9
ATOMIC MASS: 18.99840amu   **VALENCE:** 1   **OXIDATION STATE:** −2
   NATURAL STATE: Gas
ORIGIN OF NAME: From the Latin and French words for "flow," *fluere.*
ISOTOPES: There are a total of 16 isotopes of fluorine. Only one, F-19, is stable. It makes up 100% of the fluorine found on Earth. All the others are radioactive with half-lives ranging from 2.5 milliseconds to 4.57100×10^{-22} years.

ELECTRON CONFIGURATION

Energy Levels/Shells/Electrons   Orbitals/Electrons

1-K = 2	s2
2-L = 7	s2, p5

Properties

Fluorine does not occur in a free state in nature, and because fluorine is one of the most reactive elements, no chemical can free it from any of its many compounds. The reason for this is that fluorine atoms are the smallest of the halogens, meaning the electron donated by a metal (or some nonmetals) are closer to fluorine's nucleus and thus exert a great force between the fluorine nuclei and the elements giving up one electron. The positive nuclei of fluorine have a strong tendency to gain electrons to complete the outer shell, which makes it a strong oxidizer.

Because the fluorine atom has only nine electrons, which are close to the nucleus, the positive nucleus has a strong tendency to gain electrons to complete its outer shell. As a gas its density (specific gravity) is 1.695, and as a liquid, its density is 1.108. Its freezing point is −219.61°C, and its boiling point is −188°C. Fluorine, as a diatomic gas molecule (F_2), is pale yellow in color. Fluorine is the most electronegative nonmetallic element known (wants to gain electrons) and is, therefore, the strongest oxidizing agent known.

Characteristics

Fluorine reacts violently with hydrogen compounds, including water and ammonia. It also reacts with metals, such as aluminum, zinc, and magnesium, sometimes bursting into flames, and with all organic compounds, in some cases resulting in such complex fluoride compounds as fluorocarbon molecules. It is an extremely active, gaseous element that combines spontaneously and explosively with hydrogen, producing hydrogen fluoride acid (HF), which is used to etch glass. It reacts with most metals except helium, neon, and argon. It forms many different types of "salts" when combining with a variety of metals. Fluorine, as a diatomic gas, is extremely poisonous and irritating to the skin and lungs, as are many fluoride compounds. Fluorine and its compounds are also corrosive.

Abundance and Source

Fluorine is the 13th most abundant element on the Earth. It makes up about 0.06% of the Earth's crust. Fluorine is widely distributed in many types of rocks and minerals, but never found in its pure form. Fluorine is as plentiful as nitrogen, chlorine, and copper, but less plentiful than aluminum or iron.

The most abundant fluorine mineral is fluorite—calcium fluoride (CaF_2)—which is often found with other minerals, such as quartz, barite, calcite, sphalerite, and galena. It is mined in

Cumberland, England, and in Illinois in the United States. Other minerals from which fluorine is recovered are fluorapatite, cryolite, and fluorspar, which are found in many countries but mainly in Mexico and Africa.

Today fluorine is produced by the electrolysis of potassium fluoride (KF), hydrofluoric acid (HF), and molten potassium acid fluoride (KHF_2).

History

Fluorine was mentioned first in history in 1670 when instructions were written regarding its use to etch glass, using green fluorspar (fluorite), which is calcium fluoride (CaF_2). In the early 1700s chemists tried to identify the material that etched glass. Although Carl Wilhelm Scheele first "discovered" fluorine in 1771, he was not given credit because the element was not yet isolated and correctly identified. In 1869 George Gore produced a small amount of fluorine through an electrolytic process. He was unaware that fluorine gas would react with the hydrogen produced at the other electrode and would become extremely explosive (his apparatus exploded). In 1886 a French chemist, Ferdinand Frederick Henri Moissan (1852–1907), used platinum electrodes to produce fluorine from the electrolysis of potassium fluoride (KF) and hydrofluoric acid (HF). He was able to contain each of the gasses separately, thus preventing an explosion. Moissan was credited with the discovery of fluorine, partly because of his unique way of producing and identifying the element. He was awarded the 1906 Nobel Prize for Chemistry.

Common Uses

Probably the most common use of fluorine is its addition to municipal water supplies to help prevent tooth decay. Stannous (II) fluoride (SnF_2) is added to the water in proportions of about one part per million (1 ppm). In addition, many brands of toothpaste add stannous fluoride or other fluoride compounds to their product to help prevent tooth decay. Tooth enamel degenerates overtime. Fluorine promotes remineralization, essentially making a form of new enamel called "fluorapatite," which is resistant to decay.

Another popular use for the element fluorine is the plastic called Teflon. This is a fluoropolymer consisting of long chainlike inert molecules of carbon linked chemically to fluorine. Teflon is useful as a coating for nonstick surfaces in cookware, ironing board covers, razor blades, and so forth.

Of great importance are the inert fluorocarbons, such as dichlorodifluoromethane (CF_2Cl_2) and chlorofluorocarbon compounds (CFCs) and their usage as gas propellants in spray cans (e.g., hair spray, deodorants, and paint). They are also used as coolants in air conditioning and refrigeration (freon). The use of fluorinated carbon gases, known as fluorocarbons, in aerosol cans and refrigerants has been banned in the United States since 1978 because these gases diffuse into the upper atmosphere and react to destroy the ozone gases found in the ozone layer. A reduced ozonosphere layer allows more ultraviolet radiation to filter to the Earth's surface. Excessive strong ultraviolet radiation from the sun can be harmful to both plants and animals. The ozone layer filters out most of the harmful ultraviolet radiation

When hydrogen and fluorine gases meet, they explode spontaneously and form hydrogen fluoride (HF), which, when dissolved in water, becomes hydrofluoric acid that is strong enough to dissolve glass. It is used to etch glass and to produce "frosted" light bulbs.

The artificial radioactive fluorine isotope F-18 emits positrons (positive electrons) that, when injected into the body, interact with regular negative electrons, and they annihilate each other, producing X-ray-like radiation. This medical procedure is performed in Positron Emission Topography (PET), in which the produced radiation generates a picture of the body part being examined. Since F-18 has a short half-life of about 110 minutes, there is little chance of radiation damage to the patient.

Fluorine compounds are also used to reduce the viscosity of molten metals and slag by-products so that they will flow more easily. In addition, fluorine is a component of therapeutic chemotherapy drugs used to treat a number of different types of cancer.

Examples of Compounds

Hydrogen fluoride (hydrofluoric acid) is commercially prepared by distilling a mixture of calcium fluoride (feldspar) with concentrated sulfuric acid, as follows:

$$CaF_2 + H_2SO_4 \rightarrow 2HF + CaSO_4.$$

Fluorine nitrate (FNO_3) is a strong oxidizing gas or liquid. In the liquid state it explodes by shock or friction. It is used as an oxidizer for rocket propellant fuels.

Fluoroacetic acid (CH_2FCOOH) is very poisonous. It is used to kill rats and mice.

Chlorofluorocarbons (CFCs) come in many forms, including those used as propellants for spray cans and for refrigeration (freon). They were banned as being potentially harmful to the ozone layer of the atmosphere. In 1987 an international agreement was signed by about 90 nations to reduce the use of CFCs by 50% by the year 2000. This did not seem adequate, so in 1990 a new treaty called for the elimination of the use of all CFCs by industrial nations. Some third world countries (e.g., China, India, Russia, and Mexico) still make and sell CFCs, some of which are smuggled into the United States.

Sodium fluoride (NaF), in the concentration of one ppm, is added to municipal drinking water to help reduce tooth decay. It is also used as an insecticide, fungicide, and rodenticide, as well as in the manufacture of adhesives, disinfectants, and dental products.

Hazards

Many of the fluorine compounds, such as CFCs, are inert and nontoxic to humans. But many other types of compounds, particularly the salts and acids of fluorine, are very toxic when either inhaled or ingested. They are also strong irritants to the skin.

There is also danger of fire and explosion when fluorine combines with several elements and organic compounds.

Poisonous fluoride salts are not toxic to the human body at the very low concentration levels used in drinking water and toothpaste to prevent dental decay.

CHLORINE

SYMBOL: Cl   **PERIOD:** 3   **GROUP:** 17 (VIIA)   **ATOMIC NO:** 17
ATOMIC MASS: 35.453 amu   **VALENCE:** 1, 3, 4, 5, and 7   **OXIDATION STATE:** −1   **NAT-URAL STATE:** Gas
ORIGIN OF NAME: From the Greek word *khlôros,* meaning "greenish yellow."
ISOTOPES: There are a total of 25 isotopes of chlorine. Of these, only two are stable and contribute to the natural abundance on Earth as follows: Cl-35 = 75.77% and Cl-37 = 24.23%. All the other 23 isotopes are produced artificially, are radioactive, and have half-lives ranging from 20 nanoseconds to $3.01 \times 10^{+5}$ years.

ELECTRON CONFIGURATION

Energy Levels/Shells/Electrons	Orbitals/Electrons
1-K = 2	s2
2-L = 8	s2, p6
3-M = 7	s2, p5

Properties

As a nonmetal, chlorine exists as a greenish-yellow gas that is corrosive and toxic at room temperatures. As a halogen, chlorine is not found in the elemental (atomic) state but forms diatomic gas molecules (Cl_2). As a very active negative ion with the oxidation state of −1, chlorine forms bonds with most metals found in groups I and II.

Chlorine is noncombustible but will support combustion. It is extremely electronegative and a strong oxidizing agent. It is not as strong as fluorine, which is just above it in group 17, but is stronger than the other halogens.

As a gas, its **specific gravity** (density) is 3.214g/l or 0.003214g/cm^3. As a liquid, it is a clear amber color with a density of 1.56g/cm^3. Its melting point is −101.5°C, and its boiling point is −34.04°C.

Characteristics

Chlorine's best-known characteristic is its smell. It can be detected when used as household bleach or as an antiseptic in swimming pools. As an antiseptic, it is added to municipal drinking water supplies. Chlorine gas has a very pungent odor that is suffocating when inhaled. In a more concentrated form, Cl_2 was also a deadly poisonous gas used in combat during World War I. Because it combines with so many other elements, particularly metals, chlorine is fundamental to many industries, particularly the plastics industry.

Laboratory amounts of chlorine (Cl_2) are produced by combining hydrochloric acid (HCl) with manganese dioxide (MnO_2). The HCl provides the Cl^{-1} ion.

Abundance and Source

Chlorine is the 20th most abundant element on the Earth. It is not found as a free element (atoms) except as a diatomic gas escaping from very hot active volcanoes. It has been known for thousands of years as rock salt (halite). It is also found in sylvite and carnallite and as a chloride in seawater. In nature, it is mostly found in dissolved salts in seawater and deposits in salt mines. Its best-known compound is sodium chloride (NaCl), which is common table salt. Chlorine is important for the chemical industry. Numerically, it is the 12th most produced chemical in the United States and ranks ninth in volume of chemicals produced in the United States.

Chlorine is produced commercially by the electrolysis of a liquid solution of sodium chloride (or seawater), through which process an electric current is passed though the solution (electrolyte).

History

In 1774 Carl Wilhelm Scheele experimented with the mineral pyrolusite (manganese dioxide), which was the major source of manganese metal. He mixed pyrolusite powder with what he called muriatic acid (HCl), which is a form of hydrochloric acid. A pungent, greenish-yellow gas was produced, but Scheele did not know what it was since most of the gases he worked with were colorless. Scheele is credited with naming the gas chlorine after the Greek word *khlôros,* which means "green." Scheele isolated several other elements and compounds, including oxygen, hydrogen fluoride, hydrogen sulfide, and hydrogen cyanide.

Sometime later, Sir Humphry Davy believed that when acid reacted with a metal, the acid was the source of this unknown gas. This belief was at odds with the positions of most of the chemists of the day who believed that the source of the gas was the metals themselves—not the acid. In 1810 Davy declared that the new element was chlorine. He is generally recognized as the discoverer of chlorine because he correctly identified it as a new element. Some scientists of the day claimed that Davy believed the new element was a compound of oxygen and thus misdiagnosed the new element. Most accepted his identification of the new element and used the name proposed by Scheele.

Common Uses

In addition to the use of chlorine as an antiseptic for swimming pools and drinking water, large amounts are used during industrial processes that produce paper, plastics, textiles, dyes, medicines, insecticides, solvents, and some paints. Following are some of the more important compounds of chlorine used in industries: hydrochloric acid ($HCl + H_2O$), table salt (NaCl), chloroform ($CHCL_3$), carbon tetrachloride (CCl_4), magnesium chloride ($MgCl_2$), chlorine dioxide (ClO_2), potassium chloride (KCl), and lithium chloride (LiCl).

Chlorine is used to make plastics such as neoprene and polyvinyl chloride (vinyl). It is used to make insecticides, fireworks, explosives, and paint pigments; pharmaceuticals, chloroform, and chlorofluorocarbons (ClFCs); and **chlorohydrocarbons** (ClHCs).

Examples of Compounds

Chlorine forms ionic bonds with almost all the metals and molecular bonds with the semimetals and nonmetals. With group 1 metals it produces well-known salts when chlorine's −1 ion combines with this group's +1 ions (e.g, NaCl, LiCl, and KCl). Group 2 metals have +2 ions and thus, when combined with −1 ions of chlorine, form salts such as magnesium chloride ($MgCl_2$), calcium chloride ($CaCl_2$), and barium chloride ($BaCl_2$).

Chlorine combines with oxygen, producing four different types of oxides:

1. *Dichlorine monoxide:* Cl_2O is a reddish gas that, at subzero temperatures, is a dark brown liquid.
2. *Chlorine dioxide:* ClO_2 is a very explosive yellowish gas. It is so unstable that it must be diluted with a noble gas such as argon. It is used to bleach foods and as antiseptic for swimming pools.

3. *Dichlorine hexoxide:* Cl_2O_6 is an unstable reddish liquid that decomposes into chlorine and chlorine dioxide at room temperatures.
4. *Dichlorine heptoxide:* Cl_2O_7 is a colorless liquid that is an excellent oxidizer. It is perchloric acid with the water removed.

Chlorine combines with numerous other elements as well. Hydrogen and chlorine gases are extremely explosive when mixed, and they produce *hydrogen chloride* (HCl) that, when dissolved in water, results in hydrochloric acid: $H_2 + Cl_2 \rightarrow 2HCl$.

Carbon tetrachloride (CCl_4): Decades ago, this compound was mixed with ether and sold as Carbona, a dry-cleaning fluid for clothes. It is no longer permissible to sell or buy CCL_4 for household use. It is classed as a carcinogen by the U.S. government and is toxic if ingested, inhaled, or absorbed by the skin. Carbon tetrachloride is used to manufacture CFHCs, to fumigate grains to kill insects, and in the production of semi-conductors.

Sodium hypochlorite: NaOCl is a strong oxidizer used in swimming pools, and when diluted to 5.25%, it is known as the laundry bleach Clorox.

Phosgene: $COCl_2$ is a very poisonous gas that was used in combat in the early twentieth century. When not concentrated, it smells like newly cut hay or grass.

Chlorinated hydrocarbons: One example, DDT, is an insecticide. It was extensively used in World War II to delouse personnel and to prevent the spread of plague and other insect-borne diseases. Today, its use is restricted because of its toxicity and its very long life. Even though DDT is extremely effective, it is difficult to eliminate in nature. Its restricted use in some third-world countries has resulted in large increases in deaths due to malaria and other insect-borne diseases.

Chloroform ($CHCl_3$): Toxic and carcinogenic if ingested or inhaled over a long period of time. Formerly used as an **anesthetic** during surgical procedures, it is currently banned for use in cosmetics and items such as toothpaste and cough syrup.

Hazards

A series of chlorofluorohydrocarbons that are used as refrigerants are being phased out of manufacture and use, because of their possible deleterious effects on the ozone layer of the atmosphere. (See the entry on oxygen for more on the ozone layer.)

From time to time, railroad tank cars are involved in accidents that will leak liquid or gaseous chlorine that, when escaping into the air, forms toxic chlorine compounds. This is extremely dangerous, both as a fire hazard and for human health. When water is used to flush away the escaping chlorine, it may end up as hydrochloric acid, which can be hazardous to the water supply and to aquatic life.

Concentrated chlorine gas and many chlorine compounds will oxidize powdered metals, hydrogen, and numerous organic materials and release enough heat to generate fires or explosions. Chlorine is constantly evaporating from the oceans and drifting into the atmosphere where it causes a natural depletion of the ozone.

Warning: One should never mix, or use together, chlorine cleaners, such as Clorox, with other cleaning substances containing ammonia. It is a deadly mixture.

BROMINE

SYMBOL: Br **PERIOD:** 4 **GROUP:** 17 (VIIA) **ATOMIC NO:** 35
ATOMIC MASS: 79.904 amu **VALENCE:** 1, 3, 5, and 7 **OXIDATION STATE:** +1, −1, and
+5 **NATURAL STATE:** Liquid
ORIGIN OF NAME: Named for the Greek word *bromos,* which means "stench."
ISOTOPES: There are a total of 40 isotopes of bromine. Of these, only two are stable: Br-79
constitutes 50.69% of the stable bromine found on Earth, and Br-81 makes up 49.31%
of the naturally occurring abundance. All the other isotopes of bromine are radioactive
with half-lives ranging from 1.2 nanoseconds to 16.2 hours.

ELECTRON CONFIGURATION

Energy Levels/Shells/Electrons	Orbitals/Electrons
1-K = 2	s2
2-L = 8	s2, p6
3-M = 18	s2, p6, d10
4-N = 7	s2, p5

Properties

Bromine is a thick, dark-red liquid with a high density. It is the only nonmetallic element
that is a liquid at normal room temperatures. (The other element that is liquid at room temperatures is the metal mercury.) Bromine's density is $3.12g/cm^3$, which is three times the density of
water. Its vapor is much denser than air, and when it is poured into a beaker, the fumes hug the
bottom of the container. Bromine's melting point is −72°C, and its boiling point is 58.8°C.

Characteristics

Bromine is a very reactive nonmetallic element, located between chlorine and iodine in the
periodic table. Bromine gas fumes are very irritating and toxic and will cause severe burns if
spilled on the skin.

Bromine is soluble in most organic solvents and only slightly soluble in H_2O. Liquid bromine will attack most metals, even platinum.

Abundance and Source

Bromine is the 62nd most abundant element found on Earth. Although it is not found
uncombined in nature, it is widely distributed over the Earth in low concentrations. It is
found in seawater at a concentration of 65 ppm. This concentration is too low for the bromine
to be extracted directly, so the salt water must be concentrated, along with chlorine and other
salts, by **solar evaporation, distillation,** or both.

Most of the commercial bromine that is recovered comes from underground salt mines
and deep brine wells. A major source is the deep brine wells found in the state of Arkansas

and Great Salt Lake of Utah in the United States. This brine contains about one-half percent bromine. Chlorine gas is added to hot brine that oxidizes the bromine ions in solution, which is then collected as elemental bromine. It is also commercially produced, along with potash, from evaporation of the high-salt-content water of the Dead Sea, which is 1290 feet below sea level and is located on the borders of the Middle Eastern countries of Israel and Jordan.

History

Bromine was used centuries before it was identified. A sea mussel known as the murex secretes a liquid that was made into an expensive dye known as Tyrian purple. However, the dyemakers were unaware that a compound of bromine was the main ingredient of the dye until the 1900s.

Carl Lowig (1803–1890), a freshman chemistry student in 1825, produced a smelly, dark-reddish liquid substance. He was encouraged to produce more of this substance and to study it. However, his studies delayed his ability to do so. In the meantime, another young chemist, Antoine-Jerome Balard (1802–1876), wrote and presented a paper on his work with bromine in 1826. Therefore, Balard is credited with the discovery of bromine and was given the right to name the new element. Since his new discovery had such a strong odor, he named it "bromine" from the Greek word *bromos,* which means "stench or smell." Both young chemists went on to distinguished careers in science.

Common Uses

A major use is as a gasoline additive called ethylene dibromide, an **antiknock agent** that removes lead additives from automobile engines after the combustion of leaded gasoline, thus preventing the lead from forming deposits in the engine. The lead prevents premature burning of the gas/air mixture in the engine's cylinders that causes the "knocking" sound in the engine. The lead combines with the bromine to form lead bromide, a volatile gas, which is expelled through the car's exhaust system. Ethylene dibromide is a potent carcinogen. Since the introduction of lead free gasoline, this use has declined in importance; recently, other chemicals have been substituted for the lead as an additive in gasoline.

Methyl bromide is used as a pesticide, which is very effective against parasitic nematodes (e.g., hookworms). Silver bromide is used in photography. Compounds of bromide are used as flame-retardants, water purifiers, dyes, and pharmaceuticals. Bromine has many applications when it is combined with organic compounds. For example, it is used as a reagent to study the organic reactions of many other compounds. It is also used as a disinfectant, a fumigant, and a sedative. In the nineteenth century, many people took a "bromide" to ease tension.

Examples of Compounds

The main oxidation state of bromine is –1, and thus it reacts with group 1 alkali metals, which have an oxidation state of +1 to form common salt compounds as follows:

Sodium (I) bromide (NaBr): $Na^{1+} + Br^{1-} \rightarrow NaBr$.

Potassium(I) bromide (KBr): $K^{1+} + Br^{1-} \rightarrow KBr$.

If a metal has more than one oxidation state, bromine tends to react with the lowest value, as in the following example:

Manganese(II)bromide ($MnBr_2$): $Mm^{2+} + 2Br^{1-} \rightarrow MnBr_2$

Bromine combines with nonmetals according to the lowest positive oxidation state of the nonmetal. For example,

Phosphorus (III)bromide: (PBr_3): $P^{3+} + 3Br^{1-} \rightarrow PBr_3$.

Some other common compounds of bromine are the following:

Bromine azide (BrN_3) is a strong oxidizing agent that will explode when shocked or heated. It is used to make detonators for dynamite and other explosives.

Bromine pentafluoride (BrF_5) is very **corrosive** to the skin and explodes when in contact with water. It is used as an oxidizer in rocket fuel.

Bromophosgene (carbon oxybromide, $COBr_2$) is toxic if inhaled or ingested. It was used as a poisonous gas in military combat.

Bromine chloride ($BrCl$) is an irritating red, smelly liquid used to treat industrial and sewage wastes.

Hazards

Bromine and many bromine compounds in liquid form are very difficult to remove from the skin and produce deep burns that take a long time to heal.

Almost all bromine compounds are toxic if inhaled or ingested. Many are extremely explosive. When working with bromine, it is important to avoid breathing the vapors, as well as to avoid contact with the skin.

IODINE

SYMBOL: I **PERIOD:** 5 **GROUP:** 17 **ATOMIC NO:** 53
ATOMIC MASS: 126.9044 amu **VALENCE:** 1, 3, 5, and 7 **OXIDATION STATE:** +1, −1, +5, and +7 **NATURAL STATE:** Solid
ORIGIN OF NAME: The name originates from the Greek word *iodes*, meaning "violet-colored," which is the color of iodine's vapor.
ISOTOPES: There are a total of 145 isotopes of iodine. Only one (I-127) is stable and accounts for 100% of iodine's natural abundance on Earth. All the other 146 isotopes are radioactive with half-lives ranging from a 150 nanoseconds to $1.57\times10^{+7}$ years.

ELECTRON CONFIGURATION

Energy Levels/Shells/Electrons	Orbitals/Electrons
1-K = 2	s2
2-L = 8	s2, p6
3-M = 18	s2, p6, d10
4-N = 18	s2, p6, d10
5-O = 7	s2, p5

Properties

Iodine in its pure state is a black solid that sublimates (changes from a solid to a gas without going through a liquid state) at room temperature. It produces a deep purple vapor that is irritating to the eyes, nose, and throat. Iodine tends to form nonmetallic diatomic molecules (I_2). It is the heaviest of the naturally occurring halogens. (Although astatine, the fifth element in group 17, is heavier than iodine, it is a synthetic element and does not occur in nature except as a very small trace.) Iodine is the least reactive of the five halogens.

Iodine's melting point is 113.7°C, its boiling point is 184.4°C, and its density is 4.93g/cm^3.

Characteristics

Iodine is the least reactive of the elements in the halogen group 17. Most people associate iodine with the dark-brown color of the **tincture** of iodine used as an antiseptic for minor skin abrasions and cuts. A tincture is a 50% solution of iodine in alcohol. Although it is still used, iodine is no longer the antibiotic of choice for small skin wounds. Since iodine is a poison that kills bacteria, iodine tablets are often used by campers and others to purify water that is taken from outdoor streams.

Abundance and Source

Iodine is the 64th most abundant element on Earth. It occurs widely over the Earth, but never in the elemental form and never in high concentrations.

It occurs in seawater where some species of seaweed and kelp accumulate the element in their cells. It is also recovered from deep brine wells found in Chile, Indonesia, Japan, and Michigan, Arkansas, and Oklahoma in the United States. The iodine is recovered from cremated ashes of seaweed. The ashes are leached with water to remove the unwanted salts. Finally, manganese dioxide (MnO_2) is added to oxidize the iodine ions (I^{1-}) to produce elemental diatomic iodine (I_2). The following reaction takes place: $4I^{1-} + MnO_2 \rightarrow MnI_2 + I_2 + 2O^{2-}$.

Chilean saltpeter [potassium nitrate (KNO_3)] has a number of impurities, including sodium and calcium iodate. Iodine is separated from the impurities and, after being treated chemically, finally produces diatomic iodine. Today, iodine is mostly recovered from sodium iodate ($NaIO_3$) and sodium periodate ($NaIO_4$) obtained from Chile and Bolivia.

History

Similar to the history of many other elements, iodine's discovery was serendipitous in the sense that no one was looking for it specifically. In 1811 Bernard Courtois (1777–1838), a French chemist, attempted to remove sodium and potassium compounds from the ash of burned seaweed in order to make gunpowder. After removing these chemicals from the ash, he added sulfuric acid (H_2SO_4) to the remaining ash. However, he mistakenly added too much acid, which produced a violet-colored vapor cloud that erupted from the mixture. This violet vapor condensed on all the metallic objects in the room, leaving a layer of solid black iodine crystals. Sir Humphry Davy (1778–1829) confirmed this discovery of a new element and named it iodine after the Greek word *iodes,* which means "violet," but it was Courtois who was given credit for the discovery of iodine.

Common Uses

One of the most important uses of iodine is in the treatment of hypothyroidism, a condition in which the thyroid gland is deficient in iodine. Iodine deficiency may lead to the formation of a goiter, wherein the gland that surrounds the windpipe in the neck becomes enlarged. There are other causes of goiter, including cancer of the thyroid gland. A deficiency of iodine can also cause cretinism (infant hypothyroidism) in newborn babies, which can result in mental retardation unless the subject takes thyroid hormones for a lifetime. Green leafy foods, among other foods, contain iodine that when taken into the human body ends up in the thyroid gland. Some food grown in iodine-deficient soils do not contain adequate iodine for our diets. This is why iodine was added to table salt (about 0.01% potassium iodide) decades ago, specifically for people who live in regions with iodine-poor soils. The area around the Great Lakes in the United States is one region with soil that is deficient in iodine. A healthy diet requires 90 to 150 micrograms of iodine each day that, in addition to being available in iodized salt, can be obtained from eating a balanced diet, including seafood.

The isotope iodine-131 is an artificial radioisotope of iodine used as a tracer in biomedical research and as a treatment for thyroid disease. I-131 has a half-life of about eight days, which means it will be eliminated from the body in several weeks.

In industry, iodine is used for dyes, antiseptics, germicides, X-ray contrast medium, food and feed additives, pharmaceuticals, medical soaps, and photographic film emulsions and as a laboratory catalyst to either speed up or slow down chemical reactions.

Iodine is also used as a test for starch. When placed on starch (a potato for example), iodine turns the starch a dark blue color. Silver iodide is used in the manufacture of photographic film and paper. It is also used to "seed" clouds because of its ability to form a large number of crystals that act as nuclei upon which moisture in the clouds condenses, forming raindrops that may result in rain.

I-125, another artificially radioactive isotope of iodine, is used in pellets to treat prostate cancer. Pellets, the size of rice grains, are inserted into the cancerous prostate gland. Since the half-life of I-125 is about 17 days, the grains of I-125 deliver a strong dose of radiation for a week or so. They are either removed or become less radioactive in a few weeks. This treatment is localized and thus spares other parts of the body exposure to potentially harmful doses of radiation.

Trace amounts of iodine are required for a healthy body. Iodine is part of the hormone thyroxin produced by the thyroid gland. Thyroid secretions control the physical and mental development of the human body. A goiter, a swelling of the thyroid gland, is caused by the lack of iodine. Adding thyroid medication and iodized salt to the diet helps prevent this disease. Radioactive iodine (I-131), with a half-life of eight days, is used to treat some diseases of the thyroid gland.

Potassium iodide (KI) is used to manufacture photographic film, and when mixed with alcohol, it is used as an **antiseptic** to kill bacteria.

Examples of Compounds

Like all the lighter halogens, iodine, even though it is the heaviest halogen, easily combines with group 1 and group 2 metals to form iodide salts. Following are examples of the −1 oxidation state of iodine forming compounds:

Sodium iodide (NaI): $Na^{1+} + I^{1-} \rightarrow NaI$.

Potassium iodide (KI): $K^{1+} + I^{1-} \rightarrow KI$.

Iodine also has an oxidation state of +5, when forming the *iodate* −1 ion:

$I^{5+} + 3O^{2-} \rightarrow IO_3^{1-}$.

And it exhibits a +7 oxidation state in the following reaction:

Potassium metaperiodate (KIO_4): $K^{1+} + I^{7-} \rightarrow 4O^{2-} + KIO_4$.

Hazards

Iodine is a poison, and as such, care must be taken when handling and using it. Even in less than pure form, it can damage the skin, eyes, and mucous membranes. Both the elemental form and its compounds (gases, liquids, or solids) are toxic if inhaled or ingested. Even in diluted form (e.g., a tincture of iodine to treat minor skin wounds), it should be used with care.

Although a poison in high concentrations, iodine is required as a trace element in our diets to prevent thyroid problems and mental retardation in the very young.

ASTATINE

SYMBOL: At   **PERIOD:** 6   **GROUP:** 17 (VIIA)   **ATOMIC NO:** 85
ATOMIC MASS: 210 (most stable isotope)   **VALENCE:** 1, 3, 5, and 7   **OXIDATION STATE:** varies with isotope   **NATURAL STATE:** Solid
ORIGIN OF NAME: From the Greek word *astatos,* which means "unstable." All of its isotopes are unstable.
ISOTOPES: All 41 isotopes of astatine are radioactive, with half-lives ranging from 125 nanoseconds to 8.1 hours. The isotope As-210, the most stable isotope with an 8.1-hour half-life, is used to determine the atomic weight of astatine. As-210 decays by alpha decay into bismuth-206 or by **electron capture** into polonium-210.

ELECTRON CONFIGURATION

Energy Levels/Shells/Electrons	Orbitals/Electrons
1-K = 2	s2
2-L = 8	s2, p6
3-M = 18	s2, p6, d10
4-N = 32	s2, p6, d10, f14
5-O = 18	s2, p6, d10
6-P = 7	s2, p5

Properties

Astatine is located just below iodine, which suggests that it should have some of the same chemical properties as iodine, even though it also acts more like a metal or semimetal than does iodine. It is a fairly heavy element with an odd atomic number, which assisted chemists in learning more about this extremely rare element. The 41 isotopes are man-made in atomic reactors, and most exist for fractions of a second. The element's melting point is about 302°C, its boiling point is approximately 337°C, and its density is about 7g/cm^3.

Characteristics

Astatine is the heaviest and densest of the elements in group 17 (VIIA). It is difficult to determine the chemical and physical properties and characteristics of astatine because it is present in such small quantities that exist for extremely short periods of time. Many of its characteristics are inferred through experiments rather than by direct observations.

Abundance and Source

Chemists of the early twentieth century tried to find the existence of element 85, which was given the name "eka-iodine" by Mendeleev in order to fill the space for the missing element in the periodic table. Astatine is the rarest of all elements on Earth and is found in only trace amounts. Less than one ounce of natural astatine exists on the Earth at any one time. There would be no astatine on Earth if it were not for the small amounts that are replenished by the radioactive decay process of uranium ore. Astatine produced by this uranium radioactive decay process soon decays, so there is no long-term build up of astatine on Earth. The isotopes of astatine have very short half-lives, and less than a gram has ever been produced for laboratory study.

History

Early twentieth-century chemists were able to ascertain some of the properties of astatine from its position in the periodic table and from the fact that is was heavy and has an odd atomic number. At the beginning of World War II in 1940, Dale Raymond Corson (1914–1995), K. R. Mackenzie (1912–1995), and Emilio Gino Segre (1905–1989) produced a new element with 85 protons by using a cyclotron. Although the war interrupted their work, it was continued and confirmed in 1945 when they created astatine in the laboratory by firing high-energy alpha particles (helium nuclei) at a target of bismuth-209. This method is still used today for the production of small portions of astatine-211 plus two neutrons.

There is some question as to who discovered astatine. Some authorities state that Fred Allison and E. J. Murphy discovered astatine in 1931, but most give Corson, Mackenzie, and Segre the credit.

Astatine filled the next-to-last gap in the periodic table; at the time, element 61, promethium, had not yet been discovered.

Common Uses

In water solution astatine resembles iodine in some of its chemical and physical properties. Both are powerful oxidizing agents.

It has limited use in medicine as a radioactive source. It concentrates in the thyroid gland just like iodine, which makes it a useful radioisotope tracer.

Because of its isotopes' short half-lives and its scarcity, astatine has few practical uses outside of the laboratory for research, where less than a gram has ever been produced.

Examples of Compounds

Astatine does not have stable or useful compounds. Like a halogen, it will form halogen salts with a few other elements. No significant astatine commercial compounds have been produced with the exception of astatine-211, which has a half-life of just over seven hours and is used as a radioactive tracer for thyroid diseases.

Hazards

The major hazard is from the radiation of astatine's isotopes. However, given that these isotopes have very short half-lives, they do not pose a great long-term danger. Even so, astatine is considered a dangerous element that is a radioactive poison and carcinogen. It has been demonstrated that astatine causes cancer in laboratory animals.

PERIODIC TABLE OF THE ELEMENTS

GROUPS	1 IA	2 IIA	3 IIIB	4 IVB	5 VB	6 VIB	7 VIIB	8	9 VIII	10	11 IB	12 IIB	13 IIIA	14 IVA	15 VA	16 VIA	17 VIIA	18 VIIIA
PERIODS 1	1 H 1.0079																	2 He 4.00260
2	3 Li 6.941	4 Be 9.01218											5 B 10.81	6 C 12.011	7 N 14.0067	8 O 15.9994	9 F 18.9984	10 Ne 20.179
3	11 Na 22.9898	12 Mg 24.305											13 Al 26.9815	14 Si 28.0855	15 P 30.9738	16 S 32.066(6)	17 Cl 35.453	18 Ar 39.948
4	19 K 39.0983	20 Ca 40.08	21 Sc 44.9559	22 Ti 47.88	23 V 50.9415	24 Cr 51.996	25 Mn 54.9380	26 Fe 55.847	27 Co 58.9332	28 Ni 58.69	29 Cu 63.546	30 Zn 65.39	31 Ga 69.72	32 Ge 72.59	33 As 74.9216	34 Se 78.96	35 Br 79.904	36 Kr 83.80
5	37 Rb 85.4678	38 Sr 87.62	39 Y 88.9059	40 Zr 91.224	41 Nb 92.9064	42 Mo 95.94	43 Tc (98)	44 Ru 101.07	45 Rh 102.906	46 Pd 106.42	47 Ag 107.868	48 Cd 112.41	49 In 114.82	50 Sn 118.71	51 Sb 121.75	52 Te 127.60	53 I 126.905	54 Xe 131.29
6	55 Cs 132.905	56 Ba 137.33	★	72 Hf 178.49	73 Ta 180.948	74 W 183.85	75 Re 186.207	76 Os 190.2	77 Ir 192.22	78 Pt 195.08	79 Au 196.967	80 Hg 200.59	81 Tl 204.383	82 Pb 207.2	83 Bi 208.980	84 Po (209)	85 At (210)	86 Rn (222)
7	87 Fr (223)	88 Ra 226.025	▲	104 Unq (261)	105 Unp (262)	106 Unh (263)	107 Uns (264)	108 Uno (265)	109 Une (266)	110 Uun (267)	111 Uuu (272)	112 Uub	113 Uut	114 Uuq	115 Uup	116 Uuh	117 Uus	118 Uuo

TRANSITION ELEMENTS

6 ★ Lanthanide Series (RARE EARTH)

57 La 138.906	58 Ce 140.12	59 Pr 140.908	60 Nd 144.24	61 Pm (145)	62 Sm 150.36	63 Eu 151.96	64 Gd 157.25	65 Tb 158.925	66 Dy 162.50	67 Ho 164.930	68 Er 167.26	69 Tm 168.934	70 Yb 173.04	71 Lu 174.967

7 ▲ Actinide Series (RARE EARTH)

89 Ac 227.028	90 Th 232.038	91 Pa 231.036	92 U 238.029	93 Np 237.048	94 Pu (244)	95 Am (243)	96 Cm (247)	97 Bk (247)	98 Cf (251)	99 Es (252)	100 Fm (257)	101 Md (258)	102 No (259)	103 Lr (260)

The Noble Gases (Inert Gas Elements):
Periods 2 to 6, Group 18 (VIIIA)

Introduction

The **noble gases** are found in group 18 (VIIIA), which according to some older versions of the periodic table is called group 0. These six elements (He, Ne, Ar, Kr, Xe, and Rn) are inert and have a zero oxidation state. They also have full outer valence shells and represent the end of each period of the periodic table. Helium is placed at the beginning of group 18 (VIIIA) because its outer valence shell is completed and it is inert.

The noble gases are colorless, tasteless, and odorless. They glow in glass tubes when electric currents are passed through them. Each inert gas glows with its own distinctive color, and the color of the glowing light is dependent on its unique spectral line and the mixture of the gases.

The noble gases are sometimes referred to as "rare gases," which is not exactly accurate considering that argon makes up almost 1% of our atmosphere. Sometimes they are called "inert gases," which is also not exactly correct since, under very specific conditions of temperature and pressure, some of them will combine with oxygen or the more reactive halogens to form compounds.

The reason for the lack of chemical activity of the noble gases is that their outer valence shells have a full complement of eight electrons. They do not need to give up, receive, or share electrons with other elements. Their electron configuration is the most stable of all the elements.

Helium (H_2) is not really one of the noble gases, but because it is inactive and has a completed first shell (K = 2 electrons), it is placed at the top of group 18 (VIIIA). Radon, which has the largest molecular size of the noble gases, is the only radioactive noble gas. Its outer electron shell is furthest from the nucleus.

HELIUM

SYMBOL: He **PERIOD:** 1 **GROUP:** 18 (VIIIA) **ATOMIC NO:** 2
ATOMIC MASS: 4.002602 amu **VALENCE:** 0 **OXIDATION STATE:** 0
NATURAL STATE: Gas

ORIGIN OF NAME: From the Greek word *helios,* meaning the "sun." Through the process of spectrometry, it was discovered on the sun before it was found on Earth in 1868.

ISOTOPES: There are eight isotopes of helium. Two of these are stable. They are He-3, which makes up just 0.000137% of natural helium found on Earth, and He-4, which accounts for 99.999863% of the natural abundance of helium on Earth. Another isotope, He-5, is an extremely rare radioisotope that decays by emitting beta particles to form lithium-6 and lithium-8.

ELECTRON CONFIGURATION

Energy Levels/Shells/Electrons	Orbitals/Electrons
1-K = 2	s2

Properties

Helium is a colorless, odorless, and tasteless inert gas that is noncombustible and is the least soluble of any gas in water and alcohol. As a gas, it diffuses well in solids. Helium's freezing point is –272.2°C, and its boiling point is –268.93°C. Both temperatures are near absolute zero (–273.13°C, or –459.4°F), where all molecular and thermal motion ceases. Liquid helium has the lowest temperature of any known substance. Helium's density is 0.0001785g/cm^3.

Helium is the only element that cannot be converted into a solid by lowering the temperature. At normal pressure it remains a liquid at near absolute zero, but if the pressure is increased, it then turns into a solid.

Characteristics

When a second proton and two neutrons are added to a hydrogen nucleus, a helium atom can form after it collects two electrons. Helium is the most inert of all the noble group 18 gases. It is so inactive that it does not even combine with itself. As a gas, helium remains as a single atom. The nuclei of helium are called **alpha particles,** each of which has a charge of +2 and an atomic mass of 4.

No stable compound of helium has ever been found. However, it is possible for an atom of hydrogen to combine with helium (and other light noble elements) under special conditions to form HeH^+, an unstable ion.

Helium is not plentiful on Earth and is only the sixth most abundant gas in the atmosphere. It does not accumulate in the atmosphere because it is lighter than air. Some amount of helium continually escapes into space from the outer atmosphere of the Earth.

Liquid helium exhibits some unusual characteristics when supercooled. First, it is the only element that will not turn into a solid by just using pressure. Heat must be removed as the pressure is increased, but helium will freeze at –272.2°C, which is the lowest temperature scientists have ever achieved. Second, it is an excellent conductor of heat. As a supercold liquid, it will move toward heat—even flow up the sides and over the top of a container.

Abundance and Source

Helium is the 73rd most abundant element on Earth, but it is the second most abundant element in the universe, after hydrogen. Together, helium and hydrogen make up 99.9% of all the elements in the universe, but helium makes up only a small trace of the elements on Earth.

Most likely, helium was the first element to be formed after hydrogen during the Big Bang formation of the universe. The theory is that hydrogen atoms combined under great heat and pressure to form helium atoms. The Earth's current helium originally came from the natural decay of radioactive elements deep in the Earth. Much of it seeps up to the surface and escapes into the atmosphere, or it mixes with natural gas deposits deep in the Earth. Like hydrogen, it is a very light gas that escapes through cracks in the Earth's crust and sooner or later escapes from Earth's gravity into the atmosphere.

Helium can be obtained from the atmosphere by lowering the temperature of air until it liquefies. All the other gases in air will turn to a liquid except helium, because it has the lowest boiling point. Since helium, at this stage of cooling, will be the only vapor left, it can be removed as a pure gas. It is commercially more profitable to produce helium by separating it from a mix of natural underground gases, where its concentration is greater that in the atmosphere. Raw natural gas is a mixture of methane, nitrogen, and helium, with traces of other gases. The nitrogen and helium are separated from the methane, which is used as a fuel. This separation is accomplished by fractional distillation wherein the temperature is reduced and the gases are liquefied. As the temperature is reduced, methane is liquefied first, then nitrogen, leaving helium to be collected and sold commercially. Helium is produced in Oklahoma, Kansas, New Mexico, Arizona, and Canada by liquefying natural gas and "boiling off" the other gases. Helium is then purified to 99.995%. Amarillo, Texas, is one of the major centers for the production of helium in the United States. Over three billion cubic feet of helium is produced in the United States, where there is a military-reserve "storage" supply in natural caverns of over 32 billion cubic feet of helium. Most of the world's supply of helium comes from the United States.

History

Two astronomers identified helium independently as a part of the sun's atmosphere before it was found on Earth. Using a prism, Pierre-Jules-César Janssen (1824–1907) of France observed a yellow line in the sun's spectrum during a total solar eclipse in India in 1868. (Spectral emission lines refer to the wavelengths of light given off by each element when it is heated to a high temperature. Each element gives off a unique colored line that can be used to identify individual elements—somewhat like fingerprints.) At about the same time, an English astronomer Sir Norman Joseph Lockyer (1836–1920) realized that the wavelength of 587.49 nanometers was not related to any known element at that time. He identified this as a new element on the sun and named it helium.

It was not until 1895 that Scottish chemist Sir William Ramsay (1852–1916) first found helium on Earth when he experimented with uranium and subsequently collected the gases that were produced when he treated his samples with acid. He sent the gases to Sir William Crookes (1832–1919), who identified one gas as helium. Two Swedish chemists, Per Theodor Cleve

(1840–1905) and Niles Langlet (1868–1936), independently found helium in the decay gases of uranium at the same time. Nevertheless, Ramsay is credited with the discovery of helium.

Helium in the Earth is replaced by the decay of radioactive elements in the Earth's crust. Alpha decay produces particles (^{4}He^{++}) known as alpha particles, which can become helium atoms after they capture two electrons. This new helium works its way to the surface of the Earth and escapes into the atmosphere where, in time, it escapes into space.

Common Uses

Helium has many uses.

As an inert gas, it is used as the atmosphere in which to "grow" silicon crystals (computer chips).

As a lifting gas, it is used to inflate weather balloons and lighter-than-air ships (blimps) similar to the ones seen taking TV pictures above football games. Even though helium has less lifting power than hydrogen, it is used for all lighter-than-air ships because it is noncombustible and thus safer than hydrogen. In addition to blimps, toy balloons are filled with helium.

In arc welding, it is used as an inert gas shield that releases great heat for very long and heavy welds. Helium prevents oxidation of the metal being welded, thus preventing burning and corrosion of the metal. This is one of the major uses of helium.

Helium is used for low-temperature research (–272.2°C or –434°F). It has become important as a coolant for superconducting electrical systems that, when cooled, offer little resistance to the electrons passing through a conductor (wire or magnet). When the electrons are "stripped" from the helium atom, a positive He^{++} ion results. The positive helium ions (nuclei) occur in both natural and man-made radioactive emissions and are referred to as alpha particles. Helium ions (alpha particles) are used in high-energy physics to study the nature of matter.

In gas discharge **lasers,** helium transfers the energy to the laser gas such as carbon dioxide or another inert gas.

As an inert gas with heat-transfer capability, helium is used in gas-cooled nuclear power reactors, which operate at a higher efficiency than liquid-cooled nuclear reactors. The world's largest particle accelerators use liquid helium to cool their superconducting magnets. Astronomers use liquid helium to cool their detecting instruments. If this equipment is kept cool, the "thermal noise" produced at higher temperatures is reduced.

Helium is mixed with oxygen in air tanks for scuba diving and deep-sea diving because it is less soluble in divers' blood than is nitrogen. Divers have a greater chance of experiencing "nitrogen narcosis" and becoming disoriented and also of getting "the bends" when using compressed air (nitrogen–oxygen mixture), a condition in which the nitrogen forms bubbles in the blood as divers ascend. This condition is not only painful but even life-threatening, particularly if divers become so disoriented that they do not know where they are, or if they rise to the surface too quickly. Because helium is less soluble in the blood than nitrogen, the chances of deep-sea divers experiencing the bends and becoming disoriented are lessened when they are breathing a helium–oxygen mixture.

An interesting effect caused by breathing helium is the change in one's voice when speaking. Because helium is less dense than air, the vocal cords produce sounds at a higher pitch than normal and the speaker sounds like Donald Duck.

Examples of Compounds

The noble elements were always thought to be inert and not very reactive because their outer shells have full complements of electrons. Under special circumstances some of these inert elements can be forced to combine with a few other elements, particularly the more active halogens.

Compounds of the three heavier noble gases, krypton (Kr), xenon (Xe), and radon (Rn), have been made, but the formation of stable compounds of the lighter noble gases, helium (He), neon (Ne), and argon (Ar), has been more difficult. Recently a positive ion has been formed by combining hydrogen with helium (HeH^+).

Helium is also the result of **fusion** reactions wherein the nuclei of heavy hydrogen are "fused" to form atoms of helium. The result is the release of great amounts of energy. Fusion is the physical or nuclear reaction (not chemical reaction) that takes place in the sun and in thermonuclear weapons (e.g., the hydrogen bomb).

Hazards

Being inert, the noble gases are nontoxic. However, they can act as **asphyxiant** gases that can kill because of oxygen deprivations.

A possible hazard is when He^{++} nuclei, as alpha particles, are accelerated to high speeds and bombard a target. Alpha particles can be stopped by several inches of air or a piece of cardboard. As high-energy, charged particles generated from man-made or natural radioactivity, alpha particles can cause damage, but they are not as damaging to our bodies as are very short wavelength **gamma rays,** which can only be stopped by lead shielding.

NEON

SYMBOL: Ne **PERIOD:** 2 **GROUP:** 18 (VIIIA) **ATOMIC NO:** 10
ATOMIC MASS: 20.179 amu **VALENCE:** 0 **OXIDATION STATE:** 0 **NATURAL STATE:** Gas
ORIGIN OF NAME: The word "neon" was derived from the Greek word *neos,* meaning "new."
ISOTOPES: There are a total of 11 isotopes of neon, three of which are stable. They are Ne-20, which makes up 90.48% of the natural abundance of neon on Earth; Ne-21, which contributes just 0.27% to all the neon found in nature; and Ne-22, which contributes 9.25% to the natural abundance of neon. All the other isotopes have half-lives ranging from 3.746×10^{-21} seconds to 3.38 minutes.

ELECTRON CONFIGURATION

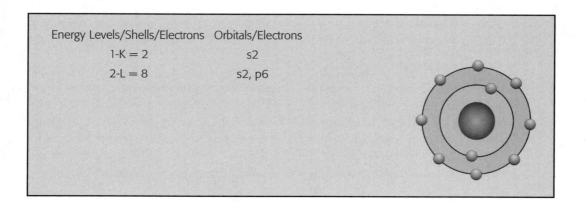

Energy Levels/Shells/Electrons	Orbitals/Electrons
1-K = 2	s2
2-L = 8	s2, p6

Properties

Neon is a monatomic atom that is considered relatively inert. It does not even combine with itself to form a diatomic molecule, as do some other gases (e.g., H_2 and O_2). During the 1960s it was discovered that the noble gases are not really inert. Neon and the heavier noble gases (Kr, Xe, and Rn) can form compounds when in an ionized state with some other elements. For example, neon can form a two-atom ionized molecule of NeH^+. Neon has also been forced to form a compound with fluorine.

Neon's melting point is $-248.59°C$, its boiling point is $-246.08°C$, and its density is 0.0008999 g/cm3.

Characteristics

As with the other noble gases, neon is colorless, tasteless, and odorless. It glows bright red when electricity is passed through it in an enclosed glass tube. It will turn from a gas to a liquid at $-245.92°C$, and only under great pressure will it become solid. It is noncombustible and lighter than air, but not as light as helium.

Abundance and Source

Neon is the fourth most abundant element in the universe, but it makes up only 18.18 ppm of the Earth's atmosphere. It is the 82nd most abundant element on Earth.

Neon is believed to be produced by radioactive decay deep in the Earth. As it rises to the surface, it escapes into the atmosphere and is soon dissipated. Some neon is found mixed with natural gas and several minerals

Neon is produced as a secondary product of the fractional distillation of liquid nitrogen and oxygen. Air is liquefied, and as it warms, nitrogen and oxygen boil off, leaving behind several other colder gases, including about 75% neon, which is then passed through activated charcoal to remove hydrogen and other gases.

History

Sir William Ramsay (1852–1916) and Morris W. Travers (1872–1961), the discoverers of krypton, experimented with liquid air, leading them to determine that there were other inert gases in the cold residues of the liquid air after oxygen and nitrogen were collected. They improved their equipment, allowing the gases to be collected separately as they boil off at their own temperatures. They then, in 1898, identified neon as one of these residue gases using a **spectrometer,** an instrument that identifies each element by its unique lines in the spectrum of light as it is excited, heated, or burned.

Common Uses

The most common use of neon is in the manufacture of **luminescent** electric tubes and specialty high-voltage indicators. Neon is placed in enclosed glass tubes of various shapes with an electrode at each end. When neon gas is ionized by the passing of a high-voltage, low-amps electrical current through it, a bright red color is produced. Other noble gases produce different colors, and they can be mixed. Unlike incandescent lamps, neon tubes can be bent and formed into unique shapes, including forming words and images for commercial advertising and signage.

Examples of Compounds

Neon is basically inert and does not normally form compounds. However, it was recently discovered that, under certain conditions of ionization, it could form the ionized two-atom NeH^+. Neon is still considered a "noble gas" that is nonreactive.

Hazards

Neon is nontoxic. As an asphyxiate gas, it can smother by removing oxygen from the lungs.

ARGON

SYMBOL: Ar **PERIOD:** 3 **GROUP:** 18 (VIIIA) **ATOMIC NO:** 18
ATOMIC MASS: 39.948 amu **VALENCE:** 0 **OXIDATION STATE:** 0 **NATURAL STATE:** Gas
ORIGIN OF NAME: The name "argon" is derived from the Greek word *argos,* meaning "inactive."
ISOTOPES: There are a total of 24 isotopes of argon, three of which are stable. They are Ar-36, which constitutes just 0.3365% of the natural amount of argon; Ar-38, which contributes just 0.0632% to the amount of argon on Earth; and Ar-40, which, by far, constitutes the most argon on Earth, 99.6003% of its natural abundance.

ELECTRON CONFIGURATION

Energy Levels/Shells/Electrons	Orbitals/Electrons
1-K = 2	s2
2-L = 8	s2, p6
3-M = 8	s2, p6

Properties

Argon is a colorless, odorless, tasteless, chemically inert noble gas that makes up about 0.93% of the Earth's atmosphere. It is the third most abundant gas in the atmosphere, meaning it is more common than carbon dioxide, helium, methane, and hydrogen.

Its melting point is −189.35°C, its boiling point is −185.85°C, and its density is 0.0017837g/cm³.

Characteristics

Although argon is considered chemically inert, at low temperatures it is possible to combine argon with other atoms to form very fragile compounds, which exist only at those very low temperatures. For instance, it can combine with fluorine and hydrogen to form argon fluorohydride (HArF). It is only slightly soluble in water.

Abundance and Source

Argon is the 56th most abundant element on Earth. It is the most abundant of all the noble gases found in the atmosphere. In fact, the only source of argon *is* the atmosphere, where it is found at just under 1% of air by volume.

There are several methods of producing argon. The most common is by fractional distillation of liquid air. Argon is collected as a by-product of this large-scale commercial process. During fractional distillation, argon boils off at its own unique temperature. It is then collected and purified by passing it through charcoal to filter out helium and other gases, producing significant amounts of argon.

History

John William Strutt, commonly referred to as Lord Rayleigh (1842–1919), isolated a gas that would not combine with oxygen and could not be identified. With the assistance of William Ramsay (1852–1916), Rayleigh separated the gas and named it "argon," after the Greek word *argos,* which means "inert," "inactive," or "lazy." Both were credited with the discovery of argon, which was identified by the then-new technology called **spectroscopy.** The spectroscope can identify each element, when heated, by the color and lines it forms on a light spectrum. Each element's spectrum is unique. Lord Rayleigh located a position on the periodic table where the atomic weight of argon matched the weight predicted for that vacant spot, and thus in 1894 he and Ramsay concluded that they had discovered a new element. This was the discovery of the first of the noble gases.

Common Uses

Argon is used when an inert atmosphere is required. Individually, or as mixture with other inert gases, it is used to fill electric light bulbs, fluorescent tubes, lasers, and so forth. By replacing oxygen in incandescent light bulbs, it prevents oxygen from corroding the bulb's filament. It is also used as a nonoxidizing gas for welding and to decarbonize steel and as an inert atmosphere in which to grow semiconductor crystals.

Examples of Compounds

As mentioned, it has recently been discovered that argon can only form fragile compounds with a few other elements at extremely low temperatures. For example, under extreme cold, argon can combine with fluorine and hydrogen to form the fragile molecule HArF.

Hazards

Argon is nontoxic, but as an asphyxiant gas, it can smother by replacing oxygen in the lungs

KRYPTON

SYNBOL: Kr   **PERIOD:** 4   **GROUP:** 18 (VIIIA)   **ATOMIC NO:** 36
ATOMIC MASS: 83.798 amu   **VALENCE:** 0 and 2   **OXIDATION STATE:** 0   **NATURAL STATE:** Gas
ORIGIN OF NAME: The name "krypton" is derived from the Greek word *kryptos,* meaning "hidden."

ISOTOPES: There are a total of 37 isotopes of krypton. Six of these are stable: Kr-78, Kr-80, Kr-82, Kr-83, Kr-84, and Kr-86. The isotope Kr-78 has such a long half-life ($0.9 \times 10^{+20}$ years) that it is considered stable even though it contributes only 0.35% to the natural krypton in the Earth's atmosphere. All the others are radioactive, man-made by-products of nuclear power plants and radioactive isotopes with half-lives ranging from 107 nano-seconds to $2.29 \times 10^{+15}$ years.

ELECTRON CONFIGURATION

Energy Levels/Shells/Electrons	Orbitals/Electrons
1-K = 2	s2
2-L = 8	s2, p6
3-M = 18	s2, p6, d10
4-N = 8	s2, p6

Properties

Krypton is a rather dense, tasteless, colorless, odorless gas. Its critical temperature is between that of oxygen and carbon dioxide. It is extracted during fractional distillation of liquid oxygen at a temperature of about –63.8°C. At one time it was thought that krypton, as well as the other noble gases, were completely inert. However, in 1967 scientists were able to combine fluorine with krypton at low temperatures to form the compound krypton difluoride (KrF_2). In this case krypton has a valence of 2.

Krypton's melting point is –156.6°C, its boiling point is –152.30°C, and its density is 0.003733g/cm^3.

Characteristics

Krypton is the fourth element in group 18 (VIIIA), which is also known as group 0 because the elements is this group were thought to have a zero oxidation point. Krypton has many of the chemical properties and characteristics of some of the other noble gases.

The fragile compounds formed by noble gases at low temperatures, such as KrF_2, are called clathrates.

Abundance and Source

Krypton is the 81st most abundant element on Earth and ranks seventh in abundance of the gases that make up Earth's atmosphere. It ranks just above methane (CH_4) in abundance in the atmosphere. Krypton is expensive to produce and thus has limited use. The gas is captured commercially by fractional distillation of liquid air. Krypton shows up as an impurity in the residue. Along with some other gases, it is removed by filtering through activated charcoal and titanium.

There are traces of krypton in some minerals and meteorites. Krypton is found beyond Earth in space.

History

In 1898 Sir William Ramsay (1852–1916), working with his graduate assistant, Morris W. Travers (1872–1961), realized that there must be some other elements between the atomic weights of helium and argon (4 to 40) that would fit in group 18 (VIIIA), or what was then known as group 0. Thus, they attempted to identify the gases that would fill in this group. They improved their liquid-air instruments in an effort to capture the residues after liquid helium and argon were removed. It did not take them long to identify a new noble gas—krypton—and a few weeks later, they discovered neon and xenon.

Common Uses

Krypton is expensive to produce, which limits its use as an inert gas. It is used in a mixture with argon to fill incandescent light bulbs, fluorescent lamps, lasers, and high-speed photography lamps. Radioactive Kr-85 is used as a source of radiation to measure the thickness of industrial materials. It is also used to test for "leakage" of scientific instruments.

Since 1960 the wavelength of the spectral lines of the krypton-86 isotope has been used as the standard for the length of the meter. One meter is now defined as 1,650,762.73 wavelengths of the reddish-orange spectral line of the Kr-86 isotope.

Examples of Compounds

There are no common compounds of krypton, but a few exotic compounds have been formed at low temperatures. At temperatures of –220°C, krypton can be forced to combine with fluorine to form two different molecules: KrF_2 and KrF_4. Both are unstable and decompose at temperatures higher than –220°C. They have been produced only in gram quantities.

Hazards

Being an inert gas, krypton is nontoxic. However, the man-made radioisotopes of krypton can cause radiation poisoning.

XENON

SYMBOL: Xe   **PERIOD:** 5   **GROUP:** 18 (VIIIA)   **ATOMIC NO:** 54
ATOMIC MASS: 131.293 amu   **VALENCE:** 2, 4, 6, and 8   **OXIDATION STATE:** 0
NATURAL STATE: Gas
ORIGIN OF NAME: The word "xenon" is derived from the Greek word *xenon,* meaning "stranger."
ISOTOPES: There are 46 isotopes of xenon. Nine of these are stable. Two of the stable isotopes are radioactive, but with half-lives long enough to be considered stable. They are Xe-124 ($1.1\times10^{+17}$years) and Xe-136 ($3.6\times10^{+20}$ years). The 47 man-made artificial radioactive isotopes have half-lives ranging from 150 nanoseconds to 11.9 days.

ELECTRON CONFIGURATION

Energy Levels/Shells/Electrons	Orbitals/Electrons
1-K = 2	s2
2-L = 8	s2, p6
3-M = 18	s2, p6, d10
4-N = 18	s2, p6, d10
5-O = 8	s2, p6

Properties

Xenon has a relatively high atomic weight and is about 4.5 times heavier than air. It is colorless, tasteless, and odorless. Its critical temperature is comparatively high at 16.6°C, which is far above oxygen (–188°C). This means that xenon will boil away from commercial fractional distillation of liquid oxygen.

Xenon's melting point is –111.79°C, its boiling point is –108.12°C, and its density is 0.005887g/cm^3.

Characteristics

Xenon is noncombustible, and even though it is considered inert, it will combine with a few elements (i.e., oxygen, fluorine, and platinum). Xenon is the only member of group 18 that exhibits all of the even valence states of +2, +4, +6, and +8. It has similar oxidation states even though most periodic tables list a single oxidation state of zero.

Abundance and Source

Xenon is found in trace amounts in the atmosphere. It makes up just 0.086 ppm by volume of air. Xenon is the rarest of the noble gases. For every thousand-million atoms of air, there are only 87 atoms of xenon. Even so, it is recovered in commercial amounts by boiling off the xenon from fractional distillation of liquid air. Small amounts of xenon have been found in some minerals and meteorites, but not in amounts great enough to exploit.

History

Sir William Ramsay (1852–1916) and Morris William Travers (1872–1961) discovered three new elements in just three months in 1898. They were krypton (May), neon (June), and xenon (July). The most difficult to identify was xenon because Ramsay and Travers needed to produce 10,000 pounds of liquid krypton in their refrigeration equipment in order to obtain just one pound of xenon. This was possible because of xenon's high critical temperature and because xenon's density is greater than oxygen's.

Common Uses

When excited electrically, xenon (sometimes mixed with krypton) produces a brilliant white flash of light that makes it useful as the gas in strobe lights. The flash used in photography can repeatedly be used to provide a well-balanced light for illumination. The xenon in flash tubes is not consumed and can be flashed over and over again.

Xenon lamps are also used as an antiseptic to kill bacteria, to power lasers, and as tracers. Because of its high atomic mass, xenon ions are preferred as fuel for ion engines to power spacecraft in deep space.

Examples of Compounds

In the 1960s, scientists first produced compounds of xenon and some other noble gases at the Argonne National Laboratory located near Chicago. Xenon and krypton are the only noble gases that readily form compounds with oxygen and fluorine. For instance, when xenon combines with fluorine, it can form a series of compounds, such as *xenon difluoride* (XeF_2), *xenon tetrafluoride* (XeF_4), and *xenon hexafluoride* (XeF_6). These and other compounds of xenon are formed within metal containers at high temperatures and pressures. They are not stable.

Xenon tetraoxide (XeO_4) exhibits xenon with a +8 oxidation state. It is a very unstable and explosive gas. The ion of xenon has also been compounded with platinum to form $XePtF_6$.

Hazards

As a noble gas that is mostly inert, xenon is nontoxic and noncombustible. Some of its compounds are toxic and potentially explosive, but there is little chance of coming into contact with them on a day-to-day basis.

RADON

SYMBOL: Rn **PERIOD:** 6 **GROUP:** 18 (VIIIA) **ATOMIC NO:** 86
ATOMIC MASS: 222 amu **VALENCE:** +2, +4, and +6 **OXIDATION STATE:** 0
  NATURAL STATE: Gas
ORIGIN OF NAME: Originally named "niton" after the Latin word for "shining," it was given the name "radon" in 1923 because it is the radioactive decay gas of the element radium.
ISOTOPES: There are 37 isotopes of radon. All are radioactive. None are stable. They range in mass numbers from Rn-196 to Rn-228. Their half-lives range from a few microseconds to 3.8235 days for Rn-222, which is the most common. It is a gas that is the result of alpha decay of radium, thorium, or uranium ores and underground rocks.

ELECTRON CONFIGURATION

Energy Levels/Shells/Electrons	Orbitals/Electrons
1-K = 2	s2
2-L = 8	s2, p6
3-M = 18	s2, p6, d10
4-N = 32	s2, p6, d10, f14
5-O = 18	s2, p6, d10
6-P = 8	s2, p6

Properties

Radon gas fits the criteria to be classed as a noble element located in group 18(VIIIA) or group 0. It is the only noble "inert" gas that is naturally radioactive. It is the heaviest of the gases in group 18.

Radon gas is easily converted to a liquid and will become solid at the relatively high temperature of –71°C. As a solid, it glows with a yellow light. Its melting point is –71°C, its boiling point is –62°C, and its density is 0.00973g/cm^3.

Characteristics

Radon is the heaviest of the noble gases and is the only one that is radioactive. It is the decay product of radium, thorium, and uranium ores and rocks found underground. As it decays, it emits alpha particles (helium nuclei) and is then **transmuted** to polonium and finally lead. The Earth's atmosphere is just 0.0000000000000000001% radon, but because radon is 7.5 times heavier than air, it can collect in basements and low places in buildings and homes.

Abundance and Source

Radon's source is a step in the transmutation of several elements: uranium → thorium → radium → radon → polonium → lead. (There are a number of intermediate decay products and steps involved in this process.) Radon-222 forms and collects just a few inches below the surface of the ground and is often found in trapped pockets of air. It escapes through porous soils and crevices.

History

While studying radium, Friedrich Ernst Dorn (1848–1916) found that it gave off a radioactive gas that, when studied in more detail, proved to be the sixth noble gas. Dorn was given credit for its discovery in 1900. He called it "radon," a variation of the word "radium." Sir William Ramsay and R. W. Whytlaw-Gray, who also investigated the properties of radon, called it "niton" from the Latin word *nitens,* which means "shining." Several other scientists who worked with radon named it "thoron" because of the transmutation of radon-220 from the decay of thorium. However, since 1923, the gas has been known as radon because it is the radioactive decay gas of the element radium. The name is derived from the Latin word *radius,* which means "ray."

Common Uses

Radon's main use is as a short-lived source of radioactivity for medical purposes. It is collected from the decay of radium as a gas and sealed in small glass capsules that are then inserted at the site of the cancer. It is also used to trace leaks in gas and liquid pipelines and to measure their rate of flow. The rate at which radon gas escapes from the Earth is one measurement that helps scientists predict earthquakes.

Examples of Compounds

Because radon is inert and radioactive, there are not many useful compounds. The only one confirmed so far is *radon fluoride* (RnF).

Hazards

The major hazard from radon stems from its radiation of alpha particles, even though alpha particles (helium nuclei) can be stopped by a sheet of cardboard. When Rn-222 is inhaled, it decays into lead-210, which is also radioactive, and because it not easily exhaled, it remains in the lungs for long periods, causing lung cancer. It is estimated that about 10% of all lung cancers are cause by radon.

Homes with cracked concrete slabs or dirt basements are at risk for radon contamination. Radon is generated just under the surface of the Earth and can seep through the floors and walls. If the ventilation is poor in the basement, the gas can accumulate to a dangerous level. Inexpensive kits that measure the levels of radon in homes and businesses are available commercially.

Tobacco plants accumulate radon from the soil. Uranium from the phosphate fertilizer used on the plants is also another source of radiation. Small amounts of lead-210 are spread on the tobacco leaves. Thus, smokers are exposed to levels of radiation that is about 1,000 times higher than the radiation exposure of workers in nuclear power plants.

Some authorities attribute the increase in lung cancer in nonsmokers to radon gas poisoning. Several studies found no positive correlation between the rate of cancer and the number of homes that had been exposed to radon for large groups of individuals in several countries. A seeming incongruity is the fact that radon does have some usefulness as a radiation source for some cancer treatments.

Lanthanide Series (Rare-Earth Elements): Period 6

Introduction

The lanthanide series is composed of metallic elements with similar physical properties, chemical characteristics, and unique structures. These elements are found in period 6, starting at group 3 of the periodic table. The lanthanide series may also be thought of as an extension of the transition elements, but the lanthanide elements are presented in a separate row of period 6 at the bottom of the periodic table.

The elements in the lanthanide series are also called rare-earth elements; they are not scarce or rare, but at one time they were thought to be rare because they were very difficult to find and extract from their ores, difficult to separate from each other, and difficult to identify. Chemical elements that have similar physical and chemical properties tend to occur together in the same ores and minerals.

The rare-earth metals are sometimes divided into two groups. The first group is called the yttrium group. It consists of Y, La, Ce, Pr, Nd, Pm, and Sm. The second group is call the dysprosium group and consists of Eu, Gd, Tb, Dy, Ho, Er, Tm, Yb, and Lu.

The lanthanide series can begin either with lanthanum ($_{57}$La) or with cerium ($_{58}$Ce) and continue through lutetium ($_{71}$Lu).

The element yttrium ($_{39}$Y), located just above lanthanum in group 3, is sometimes included in this series because its physical properties and chemical characteristics are similar to those of other elements in the series.

In this book, the series begins with lanthanum, which is located in group 3 of period 6, following barium ($_{56}$Ba). Lanthanum is not always considered as part of this series that bears its name, but it closely resembles the other elements in the series. Thus, this is a logical place to start the series given that we already include yttrium in group 3.

As mentioned, all of the elements in the lanthanide series possess similar physical properties and chemical characteristics. One of the major properties of these elements is that their valence electrons are not in their outer shells. In all 15 of the lanthanides, the outer shell is the sixth, or P, shell, which contains two electrons. For most of the 15 lanthanides, the fifth, or O, shell contains eight electrons (with three exceptions). It is the fourth, or N, shell (third

PERIODIC TABLE OF THE ELEMENTS

GROUPS

PERIODS

TRANSITION ELEMENTS

GROUPS	IA	IIA	IIIB	IVB	VB	VIB	VIIB		VIII		IB	IIB	IIIA	IVA	VA	VIA	VIIA	VIIIA
	1	2	3	4	5	6	7	8	9	10	11	12	13	14	15	16	17	18
1	1 H 1.0079																	2 He 4.00260
2	3 Li 6.941	4 Be 9.01218											5 B 10.81	6 C 12.011	7 N 14.0067	8 O 15.9994	9 F 18.9984	10 Ne 20.179
3	11 Na 22.9898	12 Mg 24.305											13 Al 26.9815	14 Si 28.0855	15 P 30.9738	16 S 32.066(6)	17 Cl 35.453	18 Ar 39.948
4	19 K 39.0983	20 Ca 40.08	21 Sc 44.9559	22 Ti 47.88	23 V 50.9415	24 Cr 51.996	25 Mn 54.9380	26 Fe 55.847	27 Co 58.9332	28 Ni 58.69	29 Cu 63.546	30 Zn 65.39	31 Ga 69.72	32 Ge 72.59	33 As 74.9216	34 Se 78.96	35 Br 79.904	36 Kr 83.80
5	37 Rb 85.4678	38 Sr 87.62	39 Y 88.9059	40 Zr 91.224	41 Nb 92.9064	42 Mo 95.94	43 Tc (98)	44 Ru 101.07	45 Rh 102.906	46 Pd 106.42	47 Ag 107.868	48 Cd 112.41	49 In 114.82	50 Sn 118.71	51 Sb 121.75	52 Te 127.60	53 I 126.905	54 Xe 131.29
6	55 Cs 132.905	56 Ba 137.33	★	72 Hf 178.49	73 Ta 180.948	74 W 183.85	75 Re 186.207	76 Os 190.2	77 Ir 192.22	78 Pt 195.08	79 Au 196.967	80 Hg 200.59	81 Tl 204.383	82 Pb 207.2	83 Bi 208.980	84 Po (209)	85 At (210)	86 Rn (222)
7	87 Fr (223)	88 Ra 226.025	▲	104 Unq (261)	105 Unp (262)	106 Unh (263)	107 Uns (264)	108 Uno (265)	109 Une (266)	110 Uun (267)	111 Uuu (272)	112 Uub	113 Uut	114 Uuq	115 Uup	116 Uuh	117 Uus	118 Uuo

6 ★ Lanthanide Series (RARE EARTH)

57 La 138.906	58 Ce 140.12	59 Pr 140.908	60 Nd 144.24	61 Pm (145)	62 Sm 150.36	63 Eu 151.96	64 Gd 157.25	65 Tb 158.925	66 Dy 162.50	67 Ho 164.930	68 Er 167.26	69 Tm 168.934	70 Yb 173.04	71 Lu 174.967

7 ▲ Actinide Series (RARE EARTH)

89 Ac 227.028	90 Th 232.038	91 Pa 231.036	92 U 238.029	93 Np 237.048	94 Pu (244)	95 Am (243)	96 Cm (247)	97 Bk (247)	98 Cf (251)	99 Es (252)	100 Fm (257)	101 Md (258)	102 No (259)	103 Lr (260)

from the outside) that becomes the defining shell in which electrons are added as each element increases in **atomic number.**

The elements in the lanthanide series are soft, silver-like metals that mostly form trivalent compounds. As metals, they are reactive, they tarnish when exposed to the atmosphere, and they react with most nonmetals when heated. Europium ($_{68}$Er) is the most reactive. The sizes of the elements do not increase much when moving from left to right, but the hardness of the elements does increase.

The ores from which rare-earth elements are extracted are monazite, bastnasite, and oxides of yttrium and related fluorocarbonate minerals. These ores are found in South Africa, Australia, South America, India, and in the United States in California, Florida, and the Carolinas. Several of the rare-earth elements are also produced as **fission** by-products during the decay of the radioactive elements uranium and plutonium. The elements of the lanthanide series that have an even atomic number are much more abundant than are those of the series that have an odd atomic number.

A commercial mixture of several of the rare-earth elements is called didymium (Di). It is neither an element nor a compound, but is used to name the mixture of oxides and salts of most of the rare-earth elements that are extracted from the ore monazite. Another unique substance, called **misch metal,** is an alloy of iron and several rare-earth elements (La, Ce, and Pr). This mixture is **pyrophoric,** which means it sparks when scratched. This is why it is used for cigarette-lighter flints.

Several rare-earth elements are used as **control rods** in nuclear reactors because of their ability to absorb neutrons.

LANTHANUM

SYMBOL: La   **PERIOD:** 6   **SERIES NAME:** Lanthanide   **ATOMIC NO:** 57
ATOMIC MASS: 138.9055 amu   **VALENCE:** 3   **OXIDATION STATE:** +3   **NATURAL STATE:** Solid
ORIGIN OF NAME: From the Greek word *lanthanein,* meaning "to be hidden."
ISOTOPES: There are 49 isotopes of lanthanum. One, La-139, is stable and makes up 99.910% of the known amount found on Earth. Another isotope has such a long half-life that is considered stable: with a half-life of $1.05\times10^{+11}$ years, La-138 makes up just 0.090% of the known abundance on Earth. All the other isotopes are radioactive and have half-lives ranging from 150 nanoseconds to several thousand years.

ELECTRON CONFIGURATION

Energy Levels/Shells/Electrons	Orbitals/Electrons
1-K = 2	s2
2-L = 8	s2, p6
3-M = 18	s2, p6, d10
4-N = 18	s2, p6, d10
5-O = 9	s2, p6, d1
6-P = 2	s2

Properties

Lanthanum is a soft silvery-white metal that, when cut with a knife, forms an oxide with the air (tarnishes) on the exposed area. It is the most reactive of the elements in the series. It reacts slightly with cold water but rapidly with hot water, producing hydrogen gas (H_2) and lanthanum oxide (La_2O_3). It directly interacts with several other elements, including nitrogen, boron, the halogens, carbon, sulfur, and phosphorus.

Its melting point is 918°C, its boiling point is 3,464°C, and its density is 6.15 g/cm³.

Characteristics

Lanthanide, as a pure metal, is difficult to separate from its ores, and it is often mixed with other elements of the series. It is mostly obtained through an ion-exchange process from the sands of the mineral monazite, which can contain as much as 25% lanthanum as well as the oxides of several other elements of the series. The metal is malleable and ductile and can be formed into many shapes. Lanthanum is considered the most basic (alkaline) of the rare-earth elements.

Abundance and Source

The main ore in which lanthanum is found is monazite sands, and it is also found in the mineral bastnasite. Monazite sands contain all of the rare-earth elements as well as some elements that are not rare-earths. Its ores are found in South Africa, Australia, Brazil, and India and in California, Florida, and the Carolinas in the United States

The prices of lanthanide elements are somewhat reasonable and are less than gold per kilogram. (Gold is about $1,800 per kg.) Cesium (Ce), which is relatively common, is often alloyed with La, Nd, and Pr and iron to form misch metal. This alloy has several uses based on its unique ability to spark when scratched. The most common use is as flints for cigarette lighters.

Lanthanum is the fourth most abundant of the rare-earths found on the Earth. Its abundance is 18 ppm of the Earth's crust, making it the 29th most abundant element on Earth. Its abundance is about equal to the abundance of zinc, lead, and nickel, so it is not really rare. Because the chemical and physical properties of the elements of the lanthanide series are so similar, they are quite difficult to separate. Therefore, some of them are often used together as an alloy or in compounds.

History

Cerium was the first rare-earth element discovered, and its discovery came in 1803 by Jöns Jakob Berzelius in Vienna. Johann Gadolin (1760–1852) also studied some minerals that were different from others known at that time. Because they were different from the common "earth elements" but were all very similar to each other, he named them "rare-earth elements." However, he was unable to separate or identify them. In the 1800s only two rare-earths were known. At that time, they were known as yttria and ceria. Carl Gustav Mosander (1797–1858) and several other scientists attempted to separate the impurities in these two elements. In 1839 Mosander treated cerium nitrate with dilute nitric acid, which yielded a new rare-earth oxide he called "lanthanum." Mosander is credited with its discovery. This caused a change in the periodic table because the separation produced two new elements. Mosander's method for separating rare-earths from a common mineral or from each other led other chemists to use the same process, and by the early nineteenth century, all of the rare-earths were identified. Over the years many chemists explored and sought to identify rare-earths.

Common Uses

Carl Auer Baron van Welsbach (1858–1929) of Austria developed misch metal as a method of igniting a gas flame. In 1903 he patented an alloy of 70% Ce and 30% Fe that gave off sparks when scratched by steel. Baron van Welsbach is also the inventor of the gas mantle. Today, China manufactures most of the misch metal used in the world. The alloys that China uses consist of Ce, La, and Nd. They use whatever mixture of these elements are found in their ores, and thus there is no need to refine them. Lanthanum is used to make electrodes for high-intensity, carbon-arc lights that are used in motion picture studios and searchlights. It also used in the refining of high-grade europium metal and the creation of glass with a high refractive index as well as for quality lenses in cameras and scientific instruments. It is also used in the manufacture of strong permanent magnets.

Lanthanum is used for electronic instruments, as a rocket fuel, as a **reducing agent,** and in automobile **catalytic converters.**

Examples of Compounds

Lanthanum arsenide ($La^{1+} + As^{1-} \rightarrow LaAs$) is used as a binary semiconductor. It is very toxic.

Lanthanum fluoride ($La^{3+} + 3\ F^{1-} \rightarrow LaF_3$) is a white powder used to coat the inside of phosphorus lamps and lasers.

Lanthanum oxide ($2La^{3+} + 3O^{2-} \rightarrow La_2O_3$) is used to make high-quality glass, ceramics, carbon-arc electrodes, and fluorescent lamps.

Lanthanum carbide ($La^{8+} + 2C^{4-} \rightarrow LaC_2$) has an unusual oxidation state of 8+.

Hazards

In powder form, lanthanum will ignite spontaneously. If ingested, it can cause liver damage and prevent blood from clotting. Many of its compounds are toxic.

CERIUM

SYMBOL: Ce   **PERIOD:** 6   **SERIES NAME:** Lanthanide   **NUMBER:** 58
ATOMIC MASS: 140.116 amu **VALENCE:** 3 and 4   **OXIDATION STATE:** +3 and +4
  NATURAL STATE: Solid
ORIGIN OF NAME: Named for the asteroid *Ceres,* which was discovered two years before the element.
ISOTOPES: There are 44 isotopes of cerium, four of which are considered stable. Ce-140 accounts for most of the cerium (88.450%) found in the Earth's crust, and Ce-138 makes up just 0.251% of the element in the crust. There are two isotopes with half-lives long enough to be considered stable: Ce-136 (0.185%), with a half-life of $0.7 \times 10^{+14}$ years, and Ce-142 (11.14%), with a half-life of $5 \times 10^{+16}$ years. All the other isotopes are radioactive with half-lives ranging from 150 nanoseconds to 137.641 days. All are made artificially.

ELECTRON CONFIGURATION

Energy Levels/Shells/Electrons	Orbitals/Electrons
1-K = 2	s2
2-L = 8	s2, p6
3-M = 18	s2, p6, d10
4-N = 19	s2, p6, d10, p1
5-O = 9	s2, p6, d1
6-P = 2	s2

Properties

Cerium is a grayish/iron-colored, very reactive metallic element that is attacked by both acids and alkalies. Pure cerium will ignite if scratched with a knife, but it can be combined safely with many other elements and materials. It is relatively soft and both malleable and ductile.

Its melting point is 798°C, its boiling point is 3,443°C, and its density is 6.770g/cm³.

Characteristics

As a pure metal, cerium is unstable and will decompose rapidly in moist air. It also decomposes in hot water to form hydrogen. Its oxide compounds and halides are stable and have a number of uses.

Cerium is separated from other rare-earth elements by an ion-exchange process in which it reacts with fluoride. This compound is then reduced with calcium metal ($3Ca + 2CeF_3 \rightarrow 2Ce + 3CaF_3$). Cerium can also be produced by the electrolysis of molten cerium salts. The metal ion collects at the cathode, and the chlorine or fluorine gases of the salt compound at the anode.

Abundance and Source

Cerium is the 25th most abundant element on Earth. It is also the most abundant rare-earth metal in the lanthanide series. Its major ores are monazite and bastnasite. Cerium is found in the Earth's crust in 46 ppm, which is about 0.0046% of the Earth's crust. Cerium is mixed with other elements in its ores, making it difficult to find, isolate, and identify. Its existence was unknown until about 1803

Monazite sands contain most of the rare-earths. The sands of the beaches of Florida and parts of California contain monazite. Monazite is also found in South Africa, India, and Brazil. Bastnasite is found in southern California and New Mexico.

History

Similar to the discovery of many other elements, cerium was detected simultaneously by several different scientists. In 1803 the Swedish chemist Jöns Jakob Berzelius (1770–1848) and the German chemists Wilhelm Hisinger (1766–1852) and Martin Klaproth (1743–1817) identified cerium as an impurity of bastnasite.

However, it was not until 1875 that W. F. Hillebrand and T. H. Norton purified the metal cerium.

Common Uses

The compound cerium oxide (either Ce_2O_3 or CeO_2) is used to coat the inside of ovens because it was discovered that food cannot stick to oven walls that are coated with cerium oxide. Cerium compounds are used as electrodes in high-intensity lamps and film projectors used by the motion picture industry. Cerium is also used in the manufacturing and polishing of high-refraction lenses for cameras and telescopes and in the manufacture of incandescent lantern mantles. It additionally acts as a chemical reagent, a misch metal, and a chemical catalyst. Cerium halides are an important component of the textile and photographic industries, as an additive to other metals, and in automobile catalytic converters. Cerium is also used as an alloy to make special steel for jet engines, solid-state instruments, and rocket propellants.

Examples of Compounds

There are two oxides of cerium with oxidation states of 3 and 4. Cerium can also react with fluorine and chlorine.

Cerium (III) Oxide: $2Ce^{3+} + 3O^{2-} \rightarrow 2Ce_2O_3$.

Cerium (IV) oxide: $Ce^{4+} + 2O^{2-} \rightarrow CeO_2$.

Following is an example of a Ce^{8+} ion compound:

Cerium carbide (CeC_2): $Ce^{8+} + 2C^{4-} \rightarrow CeC_2$.

There is one radioactive isotope of cerium that is used in medicine. It is Ce-141, with a half-life of 32,641 days.

Hazards

Most of the compounds of cerium are toxic if ingested or if the fumes are inhaled. Cerium will ignite when heated.

PRASEODYMIUM

SYMBOL: Pr   **PERIOD:** 6   **SERIES NAME:** Lanthanide   **ATOMIC NO:** 59
ATOMIC MASS: 140.9075 amu   **VALENCE:** 3   **OXIDATION STATE:** +3   **NATURAL STATE:** Solid
ORIGIN OF NAME: The name is derived from two Greek words, *prasios* and didymos, which together mean "green twins."
ISOTOPES: There are 45 isotopes of praseodymium. All are artificially produced and radioactive with half-lives ranging from several hundred nanoseconds to 23.6 days. Only one is stable (Pa-141), and it makes up 100% of the praseodymium found in the Earth's crust.

ELECTRON CONFIGURATION

Energy Levels/Shells/Electrons	Orbitals/Electrons
1-K = 2	s2
2-L = 8	s2, p6
3-M = 18	s2, p6, d10
4-N = 21	s2, p6, d10, f3
5-O = 8	s2, p6
6-P = 2	s2

Properties

Praseodymium is a silvery-white, soft metal that is easily formed into various shapes. When the pure metal is exposed to the air, a green oxide coating forms on its surface. To prevent oxidation, praseodymium is usually kept in oil in a covered container.

Its melting point is 931°C, its boiling point is 3,520°C, and its density is $6.77 g/cm^3$.

Characteristics

As a metal, Pr is **hygroscopic** (adsorbs water) and tarnishes in the atmosphere. It will react with water to liberate hydrogen. It is soluble in acids and forms greenish salts, along with some other rare-earths. It is used to fabricate the electrodes for high-intensity lights.

Abundance and Source

Praseodymium is the 41st most abundant element on Earth and is found in the ores of monazite, cerite, bastnasite, and allanite along with other rare-earths. Praseodymium is also the stable isotope resulting from the process of **fission** of some other heavy elements, such as uranium.

Praseodymium is mainly found in monazite sands and bastnasite ores. The monazite sands contain all of the rare-earths and are found in river sand in India and Brazil as well as in Florida beach sand. A large deposit of bastnasite exists in California.

Praseodymium is separated from its ore and other rare-earths by a process called ion exchange, which exchanges one type of ion for another.

History

At first praseodymium was called didymium, which is Greek for "twin," because it was always found with another rare-earth element. Using spectroscopic analysis, the two different color bands, one green and one yellow, indicated that there were two elements in didymium, but no one could identify the new elements.

In 1885 Carl Auer Baron von Welsbach (1858–1929) separated the oxides of two similar elements from didymium. He named one praseodymium from the Greek word *prasios,* which means "green" or the "green twin," and he named the other element "neodymium," which is derived from "new" and "dymium" and means "new twin."

Common Uses

Praseodymium's major use is as an alloying agent along with magnesium to produce high-strength steel that is used in airplane engines and automobiles parts. Notwithstanding its greenish color, another important use of praseodymium is as a yellow pigment to color glass and ceramics. Along with several other rare-earths, it is also used to form the electrodes for high-intensity arc lamps.

It is used to manufacture safety goggles that filter out strong yellow light (used in welding, for example). Misch metal uses about 5% Pr in the manufacture of cigarette lighter flints.

Examples of Compounds

Since the main oxidation state of praseodymium is +3, most of its stable compounds are built by the Pr^{3+} ion. A major example follows:

Praseodymium (III) oxide: $2Pr^{3+} + 3O^{2-} \rightarrow Pr_2O_3$. This compound is a yellow powder used to color glass and ceramics.

Under certain conditions, the oxidation state of +4 exists for Pr, as in the following example:

Praseodymium (IV) dioxide: $Pr^{4+} + 2O^{2-} \rightarrow PrO_2$.

Hazards

If praseodymium gets wet or is submerged in water, the hydrogen released may explode. It must be kept dry and protected from the atmosphere.

NEODYMIUM

SYMBOL: Nd   **PERIOD:** 6 **SERIES NAME:** Lanthanide   **ATOMIC NO:** 60
ATOMIC MASS: 144.24 amu   **VALENCE:** 3   **OXIDATION STATE:** +3
   NATURAL STATE: Solid
ORIGIN OF NAME: Derived from the two Greek words *neos* and didymos. When combined, they mean "new twin."
ISOTOPES: There are 47 isotopes of neodymium, seven of which are considered stable. Together the stable isotopes make up the total abundance in the Earth's crust. Two of these are radioactive but have such long half-lives that they are considered stable because they still exist on Earth. They are Nd-144 (half-life of $2.29 \times 10^{+15}$ years) and Nd-150 (half-life of $6.8 \times 10^{+18}$ years). All the other isotopes are synthetic and have half-lives ranging from 300 nanoseconds to 3.37 days.

ELECTRON CONFIGURATION

Energy Levels/Shells/Electrons	Orbitals/Electrons
1-K = 2	s2
2-L = 8	s2, p6
3-M = 18	s2, p6, d10
4-N = 22	s2, p6, d10, f4
5-O = 8	s2, p6
6-P = 2	s2

Properties

Neodymium is the third most abundant rare-earth element in the Earth's crust (24 ppm). It is reactive with moist air and tarnishes in dry air, forming a coating of Nd_3O_3, an oxide with a blue tinge that flakes away, leaving bare metal that then will continue to oxidize.

Its melting point is 1,021°C, its boiling point is 3,074°C, and its density is 7,01 g/cm^3.

Characteristics

As an element, neodymium is a soft silver-yellow metal. It is malleable and ductile. It can be cut with a knife, machined, and formed into rods, sheets, powder, or ingots. Neodymium can form trivalent compounds (salts) that exhibit reddish or violet-like colors.

Neodymium reacts with water to form $Nd(OHO)_3$ and hydrogen (H_2), which can explode if exposed to a flame or spark. It is shipped and stored in containers of mineral oil.

Abundance and Source

Although neodymium is the 28th most abundant element on Earth, it is third in abundance of all the rare-earths. It is found in monazite, bastnasite, and allanite ores, where it is removed by heating with sulfuric acid (H_2SO_4). Its main ore is monazite sand, which is a mixture of Ce, La, Th, Nd, Y, and small amounts of other rare-earths. Some monazite sands are composed of over 50% rare-earths by weight. Like most rare-earths, neodymium can be separated from other rare-earths by the **ion-exchange process.**

History

In 1885 Carl Auer Baron van Welsbach separated a common rare-earth called didymium into two distinct rare-earths. One he called "green didymia" (praseodymium) and the other he named "new didymia" (neodymium). The green color of "green didymia" (praseodymium) is caused by contamination of iron.

Common Uses

Misch metal is composed of about 18% neodymium, from which cigarette-lighter flints are made. Because neodymium absorbs the yellow "sodium" line in the visible light **spectrum,** it can be added to glass to produce violet-, red-, or gray-colored glass. Neodymium glass is used to calibrate spectrometers and other optical devices in astronomical and laboratory observation instruments. It is also used in the production of artificial rubies used in lasers. Its salts are used as pigments for ceramic enamels and glazes.

Neodymium is magnetic and is used in many of the most powerful magnets in the world. Some types of steel contain up to 18% neodymium as an alloy. It is also used as a color for TV tubes and as a tint for eyeglasses.

Examples of Compounds

Examples of neodymiun with a +3 ion and a neodymiun ion with an oxidation state of +8 follow:

Neodymiun (III) chloride: $Nd^{3+} + 3Cl^{1-} \rightarrow NdCl_3$.

Neodymium (VIII) neodymium carbide: $Nd^{8+} + 2C^{4-} \rightarrow NdC_2$.

Neodymium oxide (Nd_2O_3) is a light-blue powder used to color glass and as a pigment for ceramics. It is also used to make color TV tubes.

Neodymium compounds have different colors (e.g., blue, olive green, rose, and black).

Hazards

Many of the compounds (salts) of neodymium are skin irritants and toxic if inhaled or ingested. Some are explosive (e.g., neodymium nitrate [$Nd(NO_3)_3$]).

PROMETHIUM

SYMBOL: Pm   PERIOD: 6   SERIES NAME: Lanthanide   ATOMIC NO: 61
ATOMIC MASS: 145 amu   VALENCE: 2 and 3   OXIDATION STATE: +3   NATURAL
   STATE: Solid
ORIGIN OF NAME: Named for the Greek mythological god Prometheus, who stole fire
   from Olympus and gave it to human beings.
ISOTOPES: There are a total of 64 isotopes of promethium with half-lives ranging from two
   milliseconds to over 17 years. There is no stable isotope, but Pm-147 with a half-life
   of 2.64 years is considered the most stable. No promethium is found naturally in the
   Earth's crust. All of it is produced artificially from the leftover residue in nuclear reactors.

ELECTRON CONFIGURATION

Energy Levels/Shells/Electrons	Orbitals/Electrons
1-K = 2	s2
2-L = 8	s2, p6
3-M = 18	s2, p6, d10
4-N = 23	s2, p6, d10, f5
5-O = 8	s2, p6
6-P = 2	s2

Properties

Promethium is a silvery-white, radioactive metal that is recovered as a by-product of uranium fission. Promethium-147 is the only isotope generally available for study. The spectral lines of promethium can be observed in the light from a distant star in the constellation Andromeda. Even so, it is not found naturally on Earth, and scientists consider it to be an artificial element. Its melting point is 1,042°C, its boiling point is estimated at 3,000°C, and its density is 7.3 g/cm³.

Characteristics

Promethium was the missing element in the lanthanide rare-earth series in the periodic table. Since it does not exist on Earth, it was not recovered until nuclear reactors were common. Even so, scientists found it difficult to isolate it from other rare-earths.

When neodymiun-146 is bombarded with and captures neutrons, it becomes Nd-147 with a half-life of 11 days. Through beta decay, Nd-147 then becomes Pm-147 with a half-life of 2.64 years. Other complicated neutron and beta decay reactions from these radioactive elements are possible.

Abundance and Source

Promethium is not found in nature. Therefore, it is by far the least abundant on Earth: none exists on the Earth. All of it is man-made in nuclear reactors. It is found only in the transmuted decay by-products ("ashes") from the fission of radioactive uranium.

Promethium ($_{61}$Pm) was predicted to fill a space between the rare-earths neodymium ($_{60}$Nd) and samarium ($_{62}$Sm) in the periodic table in 1902. Although a few scientists claimed to have produced it, separating promethium from other rare-earths proved to be difficult, and thus identifying it was elusive. Only small amounts are produced and exist.

History

The identification of promethium-147 was verified by using the mass **spectrometer,** which analyzes light rays. Several radioisotopes of promethium were identified by a **spectrograph,** which is something like a photograph depicting the lines represented by different wavelengths of light rays. Although promethium was discovered in 1944 by Jacob A. Marinsky, Lawrence E. Glendenin, and Charles D. Coryell (1912–1995), they did not claim its discovery until 1946. It was Coryell's wife, Grace Mary Coryell, who suggested the name "promethium" after the Titan Prometheus, who in Greek mythology stole fire from the gods to give to mankind. Charles Coryell was the scientist who led the team working with the by-products of the decay of uranium. Such a small amount was produced in nuclear reactors that it took more than 10 years before there was evidence that element 61 actually existed, and even longer to confirm its existence. Very small amounts are produced today.

Common Uses

Promethium produces beta rays (high-energy electrons). These beta rays are used to produce nuclear-powered batteries to provide electricity for spacecraft, as well as long-term usage for up to five years in regions without electricity. It also could be used as a source of portable X-rays, as a gauge to measure the thickness of various materials, and to produce special lasers that can communicate with submarines.

Promethium-147 is used in the manufacture of luminescent paint for watch dials, as well as being a source of beta rays.

Examples of Compounds

The main oxidation state for promethium is +3. At least 30 compounds of promethium have been identified, but none are commercially available. Three examples of typical rare-earth metal compounds are as follows:

Promethium (III) nitride: $Pm^{3+} + N^{3-} \rightarrow PmN$.

Promethium (III) trichloride: $Pm^{3+} + 3Cl^{1-} \rightarrow PmCl_3$.

Promethium (III) sesquioxide: $2Pm^{3+} + 3O^{2-} \rightarrow Pm_2O_3$.

Hazards

Promethium is an extremely strong radiation hazard. Because it is so rare, few people will come in contact with it, but special precautions must be used when working with its isotopes.

SAMARIUM

SYMBOL: Sm **PERIOD:** 6 **SERIES NAME:** Lanthanide **ATOMIC NO:** 62
ATOMIC MASS: 150.36 amu **VALENCE:** 2 and 3 **OXIDATION STATE:** +2 and +3 **NATURAL STATE:** Solid
ORIGIN OF NAME: It is named after the mineral samarskite.
ISOTOPES: There are 41 known isotopes of samarium. Seven of these are considered stable. Sm-144 makes up just 3.07% of the natural occurring samarium, Sm-150 makes up 7.38% of natural samarium found on Earth, Sm-152 constitutes 26.75%, and Sm-154 accounts for 22.75%. All the remaining isotopes are radioactive and have very long half-lives; therefore, they are considered "stable." All three contribute to the natural occurrence of samarium: Sm-147 = 14.99%, Sm-148 = 11.24%, and Sm-149 = 13.82%.

Samarium is one of the few elements with several stable isotopes that occur naturally on Earth.

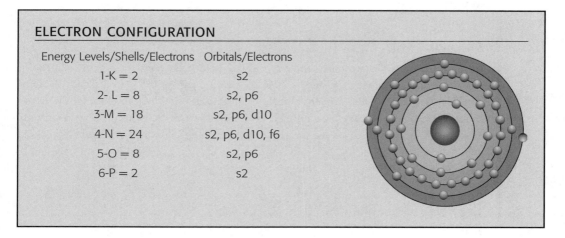

ELECTRON CONFIGURATION

Energy Levels/Shells/Electrons	Orbitals/Electrons
1-K = 2	s2
2- L = 8	s2, p6
3-M = 18	s2, p6, d10
4-N = 24	s2, p6, d10, f6
5-O = 8	s2, p6
6-P = 2	s2

Properties

Samarium is a hard, brittle, silver-white metal. When freshly cut, it does not tarnish significantly under normal room temperature conditions. Four of its isotopes are radioactive and emit alpha particles (helium nuclei). They are Sm-146, Sm-147, Sm-148, and Sm-149.

Its melting point is 1,074°C, its boiling point is 1,794°C, and its density is 7.52g.cm³

Characteristics

Samarium is somewhat resistant to oxidation in air but will form a yellow **oxide** over time. It ignites at the rather low temperature of 150°C. It is an excellent reducing agent, releases hydrogen when immersed in water, and has the capacity to absorb neutrons in nuclear reactors.

Abundance and Source

Samarium is the 39th most abundant element in the Earth's crust and the fifth in abundance (6.5 ppm) of all the rare-earths. In 1879 samarium was first identified in the mineral samarskite [(Y, Ce U, Fe)$_3$ (Nb, Ta, Ti$_5$)O$_{16}$]. Today, it is mostly produced by the ion-exchange process from monazite sand. Monazite sand contains almost all the rare-earths, 2.8% of which is samarium. It is also found in the minerals gadolinite, cerite, and samarskite in South Africa, South America, Australia, and the southeastern United States. It can be recovered as a by-product of the fission process in nuclear reactors.

History

Using a spectrometer in 1853, Jean Charles-Gallisard de Marignac (1817–1894) suspected that dydimia was a mixture of yet-to-be-discovered elements. However, it was not until 1879 that Paul-Emile Locoq de Boisbaudran (1838–1912), using a difficult chemical fractionation process, discovered samarium in a sample of samarskite, calling it "samarium" after the mineral, which was named for a Russian mine official, Colonel von Samarski. Samarskite ore is found where didymia is found. Didymia ("twins") was the original name given to a combination of the two rare-earths (praseodymium and neodymium) before they were separated and identified.

Common Uses

Samarium is easy to magnetize, but very difficult to demagnetize. This makes it ideal for the manufacture of permanent magnets (SmCo$_5$) that are part of the hard disks for computers. Samarium is also used as a neutron absorber in nuclear reactors, as well as for lasers and metallurgical research. It makes up about 1% of the metals in misch metal, an alloy in cigarette lighter flints. It is also one of several rare-earths used in floodlights and carbon-arc lights used by the motion picture industry. Samarium is used as a catalyst in several industries, including the **dehydrogenation** of ethanol alcohol.

Examples of Compounds

Following are two examples of the +2 and +3 oxidation states of samarium:
Samarium (II) chloride (dichloride): $Sm^{2+} + 2Cl^{1-} \rightarrow SmCl_2$.
Samarium (III) sesquioxide: $Sm^{3+} + 3O^{2-} \rightarrow Sm_2O_3$.
There is also the possibility of the oxidation state of +8 for samarium:
Samarium carbide: $Sm^8 + 2C^{4-} \rightarrow SmC_2$.
Samarium (II) compounds have a tendency to be reddish-brown, whereas samarium (III) compounds are yellow.

Hazards

The salts of samarium are toxic if ingested. These salts react with water, liberating hydrogen, which may explode.

EUROPIUM

SYMBOL: Eu   **PERIOD:** 6   **SERIES NAME:** Lanthanide   **ATOMIC NO:** 63
ATOMIC MASS: 151.964 amu   **VALENCE:** 2 and 3   **OXIDATION STATE:** +2 and
   +3   **NATURAL STATE:** Solid
ORIGIN OF NAME: Named for the continent of Europe.
ISOTOPES: There are a total of 45 isotopes of europium. Two are considered stable and
   account for 100% of the europium found on Earth: Eu-151 (47.81%) and Eu-153
   (52.19%). All the other 53 isotopes are radioactive and artificially produced, primarily
   through electron capture.

ELECTRON CONFIGURATION

Energy Levels/Shells/Electrons	Orbitals/Electrons
1-K = 2	s2
2-L = 8	s2, p6
3-M = 18	s2, p6, d10
4-N = 25	s2, p6, d10, f7
5-O = 8	s2, p6
6-P = 2	s2

Properties

Europium is one of the most rare of the rare-earths. Its abundance on Earth is only 1.1 ppm. It is a soft, shiny, steel-gray metal that is quite ductile and malleable, which means it can be worked and formed into many shapes. It looks like and feels like the element lead (Pb), but is somewhat heavier. It is the most chemically active of all the rare-earths.

Europium's melting point is 822°C, its boiling point is 1,529°C, and its density is 5.243 g/cm^3.

Characteristics

Pure europium will slowly oxidize in air at room temperature and will produce hydrogen when placed in water. It will ignite spontaneously if the air temperature is over 150°C. In some ways europium resembles the elements calcium, strontium, and barium found in group 2 of the periodic table.

Abundance and Source

Europium is the 13th most abundant of all the rare-earths and the 55th most abundant element on Earth. More europium exists on Earth than all the gold and silver deposits. Like many other rare-earths, europium is found in deposits of monazite, bastnasite, cerite, and allanite ores located in the river sands of India and Brazil and in the beach sand of Florida. It has proven difficult to separate europium from other rare-earths. Today, the ion-exchange process is used to extract europium from the other rare-earths found in monazite sand (ore).

History

In 1896 Eugene-Anatole Demarcay (1852–1904), a French chemist, was working with a sample of samarium when he realized that it was contaminated by an unknown element. He was able to separate the two (samarium and europium) in 1901 by a long and tedious process. He is given credit for the discovery of europium and was the one to give the new element its name.

Common Uses

There are only a few commercial uses for europium. Europium oxide, (Eu_2O_3), a compound of europium, is added to infra-sensitive **phosphors** to enhance the red colors on TV and computer-monitor picture tubes. It is also added to fluorescent light tubes to increase their efficiency, as well as to some materials to make lasers. Since it is a good neutron absorber, it is part of nuclear reactor **control rods.** Europium is an additive to the glue used on postage stamps, thus making it possible for the electronic sorting machines in U.S. postal offices to "read" the stamps.

Examples of Compounds

Two examples of compounds using the oxidation states of 2 and 3 follow:
Europium (II) oxide: $2Eu + 3H_2O \rightarrow Eu_2O_3 + 3H_2$.
Europium (III) chloride: $2Eu + 3Cl \rightarrow EuCl_3$.

Hazards

Europium is very reactive and, in powder form, may burst into flames spontaneously at room temperature. Most of the salts of europium are toxic when inhaled or ingested.

GADOLINIUM

SYMBOL: Gd   **PERIOD:** 6   **SERIES NAME:** Lanthanide   **ATOMIC NO:** 64
ATOMIC MASS: 157.25 amu   **VALENCE:** 3   **OXIDATION STATE:** +3   **NATURAL STATE:** Solid
ORIGIN OF NAME: Named for the mineral gadolinite, which was named for the French chemist Johann Gadolin.
ISOTOPES: There are 39 isotopes of gadolinium. Seven of these are stable. They are: Gd-54, which makes up 2.18% of all the gadolinium found in the Earth's crust; Gd-55, supplying 14.80%; Gd-156, making up 20.47%; Gd-157, constituting 15.56%; and Gd-158, accounting for 24.85%. In addition, there are two isotopes of gadolinium that are radioactive and with such long half-lives that they still exist in the Earth's crust. They are regarded as stable isotopes along with the other seven. They are Gd-152 ($1.08 \times 10^{+14}$ years), which exists in just 0.20% in abundance, and Gd-160 ($1.3 \times 10^{+21}$ years), found in 21.86% abundance.

ELECTRON CONFIGURATION

Energy Levels/Shells/Electrons	Orbitals/Electrons
1-K = 2	s2
2-L = 8	s2, p6
3-M = 18	s2, p6, d10
4-N = 25	s2, p6, d10, f7
5-O = 9	s2, p6, d1
6-P = 2	s2

Properties

Gadolinium is silvery-white, soft, malleable, and ductile with a metallic luster. It is the second of what is referred to as the dysprosium, subgroup in the middle of the lanthanide series of rare-earths. It tarnishes in air, forming the oxide (Gd_2O_3) on the surface, which flakes off the surface, exposing a fresh metal that in turn oxidizes.

Its melting point is 1,313°C, its boiling point is 3,273°C, and its density is 7.90g/cm^3.

Characteristics

Gadolinium, unlike most of the rare earths in the dysprosium subgroup, reacts slowly with water, releasing hydrogen. It is strongly magnetic at low temperatures. Two of its stable isotopes (Gd-155 and Gd-157) have the greatest ability of all natural elements to absorb thermal neutrons to control the fission chain reaction in nuclear reactors. However, few of these isotopes are found in the ores of gadolinium.

Abundance and Source

Gadolinium is the 40th most abundant element on Earth and the sixth most abundant of the rare-earths found in the Earth's crust (6.4 ppm). Like many other rare-earths, gadolinium is found in monazite river sand in India and Brazil and the beach sand of Florida as well as in bastnasite ores in southern California. Similar to other rare-earths, gadolinium is recovered from its minerals by the ion-exchange process. It is also produced by nuclear fission in atomic reactors designed to produce electricity.

History

In 1878 Jean-Charles Gallissard de Marignac discovered one of the "missing" rare-earths, which he named "ytterbium." Two years later, Marignac discovered another, which was later named "gadolinium" for Johann Gadolin, the "father of the rare-earths." Gadolin was given this honor because he discovered the first rare-earth and, in general, described the similar chemical and physical properties of all the rare-earths.

Other chemists also worked to separate gadolinium from the mineral dydimia. Paul-Emile Locoq de Boisbaudran, following clues provided by Marignac, isolated element 62 (samarium) from dydimia in 1879. In 1886, using the same source, he was able to isolate element 64, named in Johann's honor. Both men are generally credited with its discovery, which was confirmed in 1886.

Common Uses

Gadolinium's main use is based on its ability to absorb neutrons, thus making it ideal as a neutron-shielding and neutron-absorbing metal. It is also used as an alloying agent for steel and other metals to make the metals more workable and to be able to withstand low temperatures.

Gadolinium is used in the manufacture of electronics and can be combined with yttrium to make garnets used in microwaves. Gadolinium is used as a catalyst to speed up chemical reactions, and to activate phosphor compounds in TV screens and cast **filaments** in electrical devices. It is also used in high-temperature furnaces. Gadolinium is **paramagnetic** at normal room temperatures (weaker than **ferromagnetic**) and becomes strongly ferromagnetic at very cold temperatures.

Examples of Compounds

One way to produce gadolinium is by heating anhydrous gadolinium chloride with calcium (e.g., $2GdCl_3 + 3Ca \rightarrow 2Gd + 3CaCl_3$). The commercial process of ion exchange is used to produce most gadolinium.

Following are two examples of the +3 oxidation state of gadolinium:

Gadolinium (III) chloride: $Gd^{3+} + 3Cl^{1-} \rightarrow GdCl_3$.

Gadolinium (III) oxide: $2Gd^{3+} + 3O^{2-} \rightarrow Gd_2O_3$.

Hazards

The halogens of gadolinium are very toxic, and gadolinium nitrate is explosive. As with most rare-earths, care should be taken not to inhale fumes or ingest particles of gadolinium.

TERBIUM

SYMBOL: Tb   **PERIOD:** 6   **SERIES NAME:** Lanthanide   **ATOMIC NO:** 65
ATOMIC MASS: 158.925 amu   **VALENCE:** 3 and 4   **OXIDATION STATE:** +3   **NATURAL STATE:** Solid
ORIGIN OF NAME: Named for a village in Sweden.
ISOTOPES: There are a total of 52 isotopes of terbium, and only one of these is stable (Tb-159). Terbium-59 makes up 100% of the element found in the Earth's crust

ELECTRON CONFIGURATION

Energy Levels/Shells/Electrons	Orbitals/Electrons
1-k = 2	s2
2-L = 8	s2, p6
3-M = 18	s2, p6, d10
4-N = 27	s2, p6, d10, f9
5-O = 8	s2, p6
6-P = 2	s2, p6

Properties

There are two allotropic (crystal forms) of terbium, both of which are dependent on its temperature. The alpha ((α) form exists at room temperatures and up to temperatures of 1,298°C, and the beta (β) form exists beyond these temperatures. Although terbium is a silvery metal that resembles aluminum and feels like lead, it is much heavier than either of these two elements. It is placed in the yttrium subgroup (lanthanide series) of the rare-earths. It is also resistant to corrosion.

Its melting point is 1,356.9°C, its boiling point is 3,230°C, and its density is 8.23g/cm^3.

Characteristics

Terbium is not found in great quantities on Earth. In fact, minerals where terbium is found contain about 0.03% terbium. Not much of the stable isotope is found as a free metal; rather most of it is mixed with other rare-earths or are in compound forms.

Abundance and Source

Of all the 17 rare-earths in the lanthanide series, terbium is number 14 in abundance. Terbium can be separated from the minerals xenotime (YPO_4) and euxenite, a mixture of the following: (Y, Ca, Er, La, Ce, Y, Th)(Nb, Ta, Ti_2O_6). It is obtained in commercial amount from monazite sand by the ion-exchange process. Monazite may contain as much as 50% rare-earth elements, and about 0.03% of this is terbium.

History

A stone quarry near the town of Ytterby in Sweden produces a large number of rare-earth elements. Carl Gustaf Mosander (1797–1858) discovered several rare-earths, including the rare-earth mineral gadolinite in this quarry in 1843. He was able to separate gadolinite into three separate, but closely related, rare-earth minerals that he named *yttria* (which was colorless), *erbia* (yellow color), and *terbia* (rose-colored). From these minerals, Mosander identified two new rare-earth elements, terbium and erbium. The terbia that was found was really a compound of terbium: terbium oxide (Tb_2O_3)

Common Uses

There are few uses for terbium. However, terbium can be used as an activator for green phosphor in TV tubes, and some of its compounds are used to produce laser lights. It is also used to "dope" (coat) some forms of solid-state instruments, as a stabilizer in fuel cells so that they can operate at high temperatures, and as a metal for control rods in nuclear reactors.

Examples of Compounds

Terbium has several oxidation states, but +3 is the most common, as in the following examples:

Terbium (III) fluoride: $Tb^{3+} + 3Fl^{1-} \rightarrow TbFl_3$.
Terbium (III) oxide: $2Tb^{3+} + 3O^{2-} \rightarrow Tb_2O_3$.

An interesting method for obtaining a pure sample of terbium is to place one of the terbium halides (fluorine or chlorine) in a crucible and heat it in a helium atmosphere. The two elements will separate as a result of different densities. When the sample cools, the terbium can be separated from the halide.

The oxide terbium peroxide has a unique formula and exhibits a very rare oxidation state for terbium that is not a whole number. It is one of the rare cases in which the valence is not a whole integer: $4Te^{3.5} + 7O^{2-} \rightarrow Tb_4O_3$

Hazards

The halogens (group VIIA) of terbium are strong irritants. Most of the compounds are toxic and some are explosive. A vacuum or inert atmosphere must be maintained when working with the metal because of its strong oxidation properties.

DYSPROSIUM

SYMBOL: Dy **PERIOD:** 6 **SERIES NAME:** Lanthanide **ATOMIC NO:** 66
ATOMIC MASS: 162.50 amu **VALENCE:** 3 **OXIDATION STATE:** +3
NATURAL STATE: Solid
ORIGIN OF NAME: The word dysprosium was derived from the Greek word *dysprositos,* which means "difficult to approach."
ISOTOPES: There are a total of 39 isotopes of dysprosium, seven of which are stable. The atomic mass of the stable isotopes ranges from 156 to 164 amu (atomic mass units or atomic weight). The unstable isotopes of dysprosium have half-lives ranging from 150 milliseconds to $3.0\times10^{+6}$ years. All of the unstable isotopes are radioactive and are produced artificially.

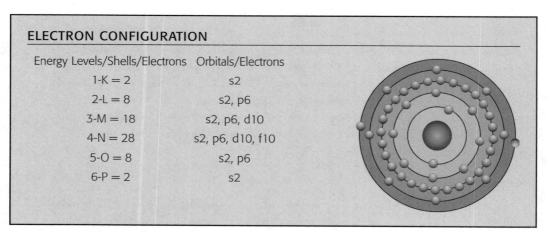

ELECTRON CONFIGURATION

Energy Levels/Shells/Electrons	Orbitals/Electrons
1-K = 2	s2
2-L = 8	s2, p6
3-M = 18	s2, p6, d10
4-N = 28	s2, p6, d10, f10
5-O = 8	s2, p6
6-P = 2	s2

Properties

Dysprosium is a dense (specific gravity = 8.540) metal. It is soft, and when cut with a knife, it appears as a silvery metal that oxidizes slowly at room temperatures. The white oxide (Dy_2O_3) that forms on the outside of the metal sloughs off, exposing a fresh surface of the metal for more oxidation. The oxide of dysprosium is also called dysprosia.

Its melting point is 1,412°C, its boiling point is 2,567°C, and its density is 8.540g/cm³.

Characteristics

Dysprosium, with characteristics similar to most of the other rare-earths, was difficult to discover. Although dysprosium does not react rapidly with moist air at low temperatures, it does react with water and the halogens at high temperatures. It also reacts in solutions of weak acids. At low temperatures, dysprosium is strongly magnetic.

Abundance and Source

Dysprosium is the 43rd most abundant element on Earth and ranks ninth in abundance of the rare-earths found in the Earth's crust. It is a metallic element that is usually found as an oxide (disprosia). Like most rare-earths, it is found in the minerals monazite and allanite, which are extracted from river sands of India, Africa, South America, and Australia and the beaches of Florida. It is also found in the mineral bastnasite in California.

History

Dysprosium was first discovered in 1886 by the chemist, Paul-Emile Lecoq de Boisbaudran (1838–1912) as he analyzed a sample of the newly discovered erbium oxide (element 68). Boisbaudran was able to separate erbium oxide from a small sample of a new oxide of a metal. He identified this new element as element 66 on the periodic table and called it "dysprosium."

Common Uses

There are not many uses for dysprosium. Scientists continue to experiment with it as a possible alloy metal (it has a high melting point) to be mixed with steel to make control rods that absorb neutrons in nuclear reactors. There are only a few commercial uses for dysprosium, such as a laser material and as a fluorescence activator for the phosphors used to produce the colors in the older TV and computer cathode ray tubes (CRTs). When combined with steel or nickel as an alloy, it makes strong magnets.

Examples of Compounds

Dysprosium has an oxidation state of +3, which forms the Dy^{3+} metallic ion that is limited to a small group of compounds. A general example that demonstrates how the ion of dysprosium combines with halogen anions follows: $Dy^{3+} + 3Cl^{1-} \rightarrow DyCl_3$.

Dysprosium^{3+} *+ oxygen*$^{2-}$ *(disprosia):* $2Dy^{3+} + O^{2-} \rightarrow Dy_2O_3$. This is a white compound that is more magnetic than iron oxide. It turns yellow when dissolved in acid.

Hazards

Dysprosium nitrate $[Dy_2(NO_3)_3]$ is a strong oxidizing agent and will ignite when in contact with organic material. Most dysprosium salts are toxic if ingested or inhaled.

HOLMIUM

SYMBOL: Ho **PERIOD:** 6 **SERIES NAME:** Lanthanide **ATOMIC NO:** 67
ATOMIC MASS: 164.903 amu **VALENCE:** 3 **OXIDATION STATE:** +3 **NATURAL STATE:** Solid
ORIGIN OF NAME: Derived from the Latin word for the ancient city named Holmia (present-day Stockholm, located in Sweden).
ISOTOPES: There are a total of 57 isotopes of holmium. Only one of these, Ho-165, is stable, and it is the only isotope found in the Earth's crust. All the other 56 isotopes have half-lives of a few milliseconds to $1.20 \times 10^{+3}$ years, the half-life of Ho-166.

ELECTRON CONFIGURATION

Energy Levels/Shells/Electrons	Orbitals/Electrons
1-K = 2	s2
2-L = 8	s2, p6
3-M = 18	s2, p6, d10
4-N = 29	s2, p6, s10, f11
5-O = 8	s2, p6
6-P = 2	s2

Properties

Holmium is a crystal-like, solid rare-earth with a metallic luster. It is one of the more scarce elements of the lanthanide series. It is soft, like lead, and can be hammered and pounded into thin sheets.

Its melting point is 1,474°C, its boiling point is 2,700°C, and its density is 8.79g/cm^3.

Characteristics

Although stable at room temperatures, holmium will corrode at higher temperatures and humidity. Its oxide coating is a yellowish film that reacts slowly with water and dissolves in weak acids. Holmium has one of the highest magnetic properties of any substance, but it has little commercial use.

Abundance and Source

Holmium is the 12th most abundant of the rare-earths found in the Earth's crust. Although it is the 50th most abundant element on Earth, it is one of the least abundant lanthanide metals. It is found in gadolinite and the monazite sands of South Africa and Australia and in the beach sands of Florida and the Carolinas in the United States. Monazite sand contains about a 50% mixture of the rare-earths, but only 0.05% by weight is holmium. Today, small quantities of holmium are produced by the ion-exchange process.

History

In the 1800s chemists searched for new elements by fractionating the oxides of rare-earths. Carl Gustaf Mosander's experiments indicated that "pure" ceria ores were actually contaminated with oxides of lanthanum, a new element. Mosander also fractionated the oxides of yttria into two new elements, erbium and terbium. In 1878 J. Louis Soret (1827–1890) and Marc Delafontaine (1837–1911), through spectroscopic analysis, found evidence of the element holmium, but it was contaminated by the rare-earth dysprosia. Since they could not isolate it and were unable to separate holmium as a pure rare-earth, they did not receive credit for its discovery.

In 1879 holmium was discovered, independently, by Per Theodor Cleve (1840–1905), who managed to isolate holmium from the other rare-earths. Cleve received credit for the discovery of holmium and named it for the Latin word *holmia,* which means "Stockholm," a city in his native country, Sweden.

Common Uses

Holmium has just a few commercial uses, but it could be developed to produce items requiring strong permanent magnets. It is used for filaments in vacuum tubes and in electrochemistry. It is also used to help identify the atomic weights of elements by spectroscopy, which identifies the unique lines produced by each element when viewed through a spectroscope. It also has limited use as neutron absorber in nuclear reactors and as coloring agent for glass.

Examples of Compounds

Holmium has an oxidation state of +3 that, on a limited basis, can form a few compounds with the halogens and oxygen. They are of no commercial uses, and most holmium is used for research purposes.

Holmium chloride: $Ho^{3+} + Cl^{1-} \rightarrow HoCl_3$.

Generic compounds of halogens are expressed as follows, with "X" standing for any of the halide ions: $Ho^{3+} + X^{1-} \rightarrow HoX_3$.

There is only one oxide of holmium, which sometimes is called "holmia": the oxide ($2Ho^{3+} + 3O^{2-} \rightarrow Ho_2O_3$) is a pale yellow solid used as a special catalyst to speed up chemical reactions and as a **refractory material** to line laboratory and industrial ovens.

Hazards

Because holmium is produced in such very small amounts, it is not a potential hazard to the general public. However, professional chemists take the same precautions as they do with other rare-earths.

ERBIUM

SYMBOL: Er   **PERIOD:** 6   **SERIES NAME:** Lanthanide   **ATOMIC NO:** 68
ATOMIC MASS: 167.259 amu   **VALENCE:** 3   **OXIDATION STATE:** +3
   NATURAL STATE: Solid
ORIGIN OF NAME: Named for the quarry in Ytterby, Sweden, where ores and minerals of many elements are found.
ISOTOPES: There are 39 isotopes of erbium, six of which are stable: Er-162, Er-164, Er-166, Er-167, Er-168, and Er-170. These six isotopes make up the total atomic weight (mass) of erbium, and all the other isotopes are artificially made and short-lived. Their half-lives range from 200 nanoseconds to 49 hours.

ELECTRON CONFIGURATION

Energy Levels/Shells/Electrons	Orbitals/Electrons
1-K = 2	s2
2-K = 8	s2, p6
3-M = 18	s2, p6, d10
4-N = 30	s2, p6, d10, f12
5-O = 8	s2, p6
6-P = 2	s2

Properties

Erbium is a soft, malleable metal with a silvery metallic luster that only tarnishes (oxidizes) slightly in air. It is one of the rare-earths of the yttrium subgroup of the lanthanide series.

Its melting point is 1,529°C, its boiling point is 2,868°C, and its density is 9.07g/cm^3.

Characteristics

Although erbium is magnetic at very low temperatures, it is antiferromagnetic and becomes a superconductor at temperatures near absolute zero. It is insoluble in water but soluble in acids. Its salts range from pink to red. Erbium and some of the other rare-earth elements are considered to be "impurities" in the minerals in which they are found. Small quantities of erbium can also be separated from several other rare-earths.

Abundance and Source

Erbium ranks 17th in abundance among the rare-earths, and it is the 46th most abundant element found in the Earth's crust. It exists in only 2.5 ppm, meaning that about 2.5 pounds of erbium could be extracted from one million pounds of dirt in the Earth's crust. Higher concentrations are found in some areas, but in general, the oxides of erbium are rather scarce.

It is found in ores such as monazite, gadolinite, and bastnasite. It was first separated into three elements in 1843 (yttria, erbia, and terbia). Erbium is also produced as a by-product of nuclear fission of uranium.

History

Carl Gustaf Mosander, a Swedish chemist, successfully separated two rare-earths from a sample of lanthanum found in the mineral gadolinite. He then tried the same procedure with the rare-earth yttria. He was successful in separating this rare-earth into three separate rare-earths with similar names: yttia, erbia, and terbia. For the next 50 years scientists confused these three elements because of their similar names and very similar chemical and physical properties. Erbia and terbia were switched around, and for some time the two rare-earths were mixed up. The confusion was settled ostensibly in 1877 when the chemistry profession had the final say in the matter. However, they also got it wrong. What we know today as erbium was originally terbium, and terbium was erbium.

Common Uses

Erbium has limited commercial use, but it is used as an alloy metal for vanadium to make it easier to work and to form spring steel. The oxide of erbium is pink, which is used to color glass and to make lasers that will operate at normal room temperatures. It has limited use as control rods in nuclear fission reactors.

Examples of Compounds

Erbium has a +3 oxidation state that easily combines with the halogens and oxygen. The "X" here can be used to represent the halide ion combining with the metallic ion of erbium, as follows:

$Er^{3+} + 3X^{1-} \rightarrow ErX_3$.

There is only one oxide of erbium, which is called *erbium oxide* or *erbia* ($2Er^{3+} + 3O^{2-} \rightarrow Er_2O_3$).

Hazards

Erbium nitrate [$Er(NO_3)_3$] may explode when "shocked" or at high temperatures. As with other rare-earths, erbium and its compounds should be handled with care because they can be toxic.

THULIUM

SYMBOL: Tm **PERIOD:** 6 **SERIES NAME:** Lanthanide **ATOMIC NO:** 69
ATOMIC MASS: 168.9342 amu **VALENCE:** 3 **OXIDATION STATE:** +3 **NATURAL STATE:** Solid
ORIGIN OF NAME: Named for Thule, the Greek word for Scandinavia, the most northerly habitable land in ancient mythology.
ISOTOPES: There are a total of 46 isotopes of thulium. One of these, Tm-169 is the only stable isotope of thulium and accounts for the total atomic mass of the element. All the other isotopes are artificially produced and radioactive and have half-lives ranging from a few microseconds to two years.

ELECTRON CONFIGURATION

Energy Levels/Shells/Electrons	Orbitals/Electrons
1-K = 2	s2
2-L = 8	s2, p6
3-M = 18	s2, p6, d10
4-N = 31	s2, p6, d10, f13
5-O = 8	s2, p6
6-P = 2	s2

Properties

Thulium is a naturally occurring rare metal that exists is very small amounts mixed with other rare-earths. It is a bright silvery metal that is malleable and ductile and can be cut easily with a knife. Its melting point is so high that it is difficult to force it into a melted state. Its vapor pressure is also high, and thus, much of the molten thulium evaporates into the atmosphere. Its melting point is 1,545°C, its boiling point is 2,950°C, and its density is 9.32g/cm^3.

Characteristics

Thulium is near the end of the lanthanide series, where the metals tend to be heavier than the ones located near the beginning of the series. It is so scarce that it requires the processing of about 500 tons of earth to extract four kilograms of thulium. The only element that is scarcer is promethium, which is not found naturally on Earth.

Abundance and Source

Thulium is the 61st most abundant element in the Earth's crust and is found along with other rare-earths in monazite sand, which is about 50% rare-earths by weight. Only about 0.007% of this is thulium. It is also found in bastnasite ore. It ranks 16th out of the 17 rare-earths in abundance. Thulium is usually found as an oxide along with other rare-earths. Like most rare-earths, thulium can be separated from its ore by the ion-exchange process, where its positive ion reacts with elements with negative ions like fluorine, chlorine, or oxygen to form binary compounds (e.g., Tm_2O_2). It can also be recovered as a by-product of the nuclear fission reaction in nuclear reactors.

History

In 1879 Per Theodor Cleve, a Swedish chemist, discovered thulium by looking for impurities in the oxides of other rare-earths. This is the same method first used by Carl Gustaf Mosander to discover lanthanum, erbium, and terbium. After Cleve processed the oxide of erbium (Er_2O_3), two new substances remained, one green, the other brown. Cleve named the brown substance "holmia," and the green substance he named "thulia" after the Greek word *Thule,* which in legend represented the most northern region that was inhabited by humans. It is now known as Scandinavia.

Common Uses

Thulium is relatively scarce and expensive, which limits its commercial uses. Thulium-170, which is a radioactive isotope of thulium produced by fission in nuclear reactors, can be used as small, portable X-ray sources. It also has limited use as an alloy metal with other metals and has experimentally been used in lasers. (Note: Of all the isotopes of thulium, only thulium-169 is stable and nonradioactive.)

Examples of Compounds

Similar to other rare-earths, thulium has a single oxidation state of +3. A general formula for the positive ion of thulium and the elements found in group 7 (fluorides) with a negative ion is expressed as follows:

$Tm^{3+} + 3X^{1-} \rightarrow TmX_3$, where X represents one of the fluorides.

The only oxide of thulium, which sometimes is called *thulia,* is produced as follows:

$2Tm^{3+} + 3O^{2-} \rightarrow Tm_2O_3$.

Hazards

The dust and powder of thulium are explosive and toxic if inhaled or ingested. As with all radioactive elements, thulium can cause radiation poisoning.

YTTERBIUM

SYMBOL: Yb   **PERIOD:** 6   **SERIES NAME:** Lanthanide   **ATOMIC NO:** 70
ATOMIC MASS: 173.04 amu   **VALENCE:** 2 and 3   **OXIDATION STATE:** +2 and +3   **NATURAL STATE:** Solid
ORIGIN OF NAME: Ytterbium is named for the Ytterby quarry located in Sweden.

ISOTOPES: There are a total of 37 isotopes of ytterbium. Seven of these are stable, and they make up all of the natural ytterbium found on Earth. One of these isotopes (Yb-176) has such a long half-life ($1.6 \times 10^{+17}$ years) that it contributes 12.76% of the natural ytterbium existing on Earth, and thus it is considered stable. All the other 30 isotopes are artificially radioactive and produced by nuclear fission in nuclear reactors with half-lives ranging from a fraction of a second to 32 days.

ELECTRON CONFIGURATION

Energy Levels/Shells/Electrons	Orbitals/Electrons
1-K = 2	s2
2-L = 8	s2, p6
3-M = 18	s2, p6, d10
4-N = 32	s2, p6, d10, f14
5-O = 8	s2, p6
6-P = 2	s2

Properties

Ytterbium is a silvery, soft, malleable, and ductile metal with a lustrous metallic shine. It is slightly reactive in air or water at room temperatures. Ytterbium is located next to last of the rare-earths in the lanthanide series. It slowly oxidizes as it reacts with oxygen in the atmosphere, forming a somewhat duller coating. Ytterbium was the first rare-earth to be discovered by Carl Gustof Mosander in 1843. More of it exists in the Earth's crust than once was believed.

It was often confused with other rare-earths and was known by two other names, aldebaranium and cassiopeium. Ytterbium's melting point is 819°C, its boiling point is 1,196°C, and its density is 6.9654g/cm^3.

Characteristics

In the past there was some confusion about the rare-earths because they are not really earths at all, but rather binary compounds of oxides of metals. Compounding the confusion was the fact that they were always found combined with several other rare-earths.

The salts of ytterbium are paramagnetic, which exhibit weaker magnetic fields than do iron magnets.

Abundance and Source

Ytterbium is the 45th most abundant element, and it ranks 10th in abundance (2.7 ppm) among the 17 rare-earths found in the Earth's crust.

It is found in ores along with other rare-earths that were first found in the Ytterby quarry of Sweden. These ores are xenotime, euxenite, gadolinite, and monazite. Monazite river sand is the main source of ytterbium, which is found in India and Brazil and the beaches of Florida. Ytterbium is also found as a decay product of the fission reaction in nuclear reactors.

History

In 1843 Carl Gustaf Mosander separated gadolinite into three distinct materials, to which he gave the names yttria, erbia, and terbia. There was some confusion as to the composition of erbia. In 1878 Jean Charles Galissard de Marignac separated erbia into two rare-earths. One he called "ytterbia," and the other maintained the name "erbia." Marignac is credited with the discovery of ytterbium. Some years later, in 1907, Georges Urbain (1872–1938) believed that ytterbium was not a single element, so he experimentally separated it into what he called "neoytterbium" (new ytterbium) and a new element he called "lutecium," whose name was later changed to "lutetium."

Common Uses

There is not much commercial use for ytterbium. Radioactive ytterbium can be used for a small portable X-ray source and as an alloy to make special types of strong steel. The oxides of ytterbium are used to make lasers and some synthetic gemstones.

Examples of Compounds

Since ytterbium has both a +2 and +3 oxidation state, it can form two different compounds with the halogens. See the following examples:

Ytterbium (II) chloride: $Yb^{2+} + 2Cl^{1-} \rightarrow YbCl_2$.
Ytterbium (III) chloride: $Yb^{3+} + 3Cl^{1-} \rightarrow YbCl_3$.

Ytterbium oxide (Yb_2O_3) is used to make special alloys, ceramics, and glass. It can be used for carbon arc-lamp electrodes that produce a very bright light.

Hazards

Ytterbium dust and powder can explode and may be toxic if inhaled. The compound. ytterbium arsenate is a poison.

LUTETIUM

SYMBOL: Lu   **PERIOD:** 6   **SERIES NAME:** Lanthanide   **ATOMIC NO:** 71
ATOMIC MASS: 174.967 amu   **VALENCE:** 3   **OXIDATION STATE:** +3   **NATURAL STATE:** Solid
ORIGIN OF NAME: Lutetium's name is derived from the ancient Latin name for Paris, France: Lutecia.
ISOTOPES: There are a total of 59 isotopes of Lutetium. Only two of these are stable: Lu-175, which makes up 97.41% of all the natural abundance found on Earth. The other is a long-lived radioisotope (Lu-176) with such a long half-life ($4.00 \times 10^{+10}$ years) that it is considered stable: Lu-176 contributes 2.59% to the natural abundance of lutetium.

ELECTRON CONFIGURATION

Energy Levels/Shells/Electrons	Orbitals/Electrons
1-K = 2	s2
2-L = 8	s2, p6
3-M = 18	s2, p6, d10
4-N = 32	s2, p6, d10, f14
5-O = 9	s2, p6, d1
6-P = 2	s2

Properties

In the last (17th) position in the lanthanide series, lutetium is the heaviest and largest molecule of all the rare-earths as well as the hardest and most corrosion-resistant. It has a silvery-white color and is somewhat stable under normal atmospheric conditions.

Its melting point is 1,663°C, its boiling point is 3,402°C, and its density is 9.84g/cm³.

Characteristics

Lutetium has had a number of different names over the years. At one time or another, it was called neoytterbium, lutecium, lutetia, lutetium, and cassiopium. Some scientists in Germany still refer to it as cassiopium.

Lutetium reacts slowly with water and is soluble in weak acids. Its crystals exhibit strong magnetic properties, which are important to the study of magnetism.

Abundance and Source

Lutetium is the 60th most abundant element on Earth, and it ranks 15th in the abundance of the rare-earths. It is one of the rarest of the lanthanide series. It is found in monazite sand (India, Australia, Brazil, South Africa, and Florida), which contains small amounts of all the rare-earths. Lutetium is found in the concentration of about 0.0001% in monazite. It is difficult to separate it from other rare-earths by the ion-exchange process. In the pure metallic form, lutetium is difficult to prepare, which makes is very expensive.

History

At about the turn of the nineteenth century, much progress was made in separating and identifying individual rare-earths. This had always been difficult since they all have very similar chemical and physical properties. During the years 1907 and 1908, both Carl Auer von Welsbach and Georges Urbain independently demonstrated that ytterbium was a mixture of two formerly unknown elements. Welsbach gave the new elements the names "aldebaramium" and "cassiopium," and Urbain gave them the names of "neoytterbium" and "lutecium." Because the men worked separately, there was some confusion as to who actually discovered lutetium. Thus, both are generally credited as its discoverers. The prefix "neo," which means "new," was eliminated and the spelling was changed to "lutetium" (atomic number 71), as it is known today by most of the world.

An interesting bit of history is that the American chemist Charles James, of the University of New Hampshire, and his students also discovered lutetium in 1907. They processed many tons of ore, and by using the crystallization process, produced a small sample of lutetium. James's work was recognized in 1999 by the ACS (American Chemical Society). This is the only example of a rare-earth being discovered in the United States.

Common Uses

Because lutetium is difficult to prepare on a large scale, its practical uses are limited. Some of its radioisotopes are used as catalysts in the cracking (refining) process of crude oil, which produces lighter fractions such as diesel fuel and gasoline. It can also be used as a catalyst to speed up the reaction in some hydrogenation processes wherein hydrogen is added to vegetable oils to make more solid products. Some of its isotopes have been used to determine the age of meteorites.

Examples of Compounds

The result of the ion-exchange process in which the +3 ion of lutetium combines with the −1 ion of chlorine to form the binary compound 3LiCl is written as follows:

$Lu^{3+} + 3Cl^{1-} \rightarrow 3LuCl$.

Lutetium oxide (Lu_2O_3), the oxide found in monazite ore, is a white solid. It is **hygroscopic** and also absorbs carbon dioxide, making it useful to remove CO_2 in closed atmospheres.

Lutetium nitrate [$Lu(NO_3)_3$] is a fire and explosion hazard when heated.

Hazards

Lutetium fluoride is a skin irritant, and its fumes are toxic if inhaled. The dust and powder of the oxides of some rare-earths, including lutetium, are toxic if inhaled or ingested.

Actinide Series (Period 7) and Transuranic Elements

Introduction

The actinide elements are radioactive metals that begin with the element actinium ($_{89}$Ac) and continue through the element lawrencium ($_{103}$Lr), and there is a subseries of the actinide series known as the transuranic elements. They are the elements with atomic numbers larger than $_{92}$U. It is important to remember that the transuranic elements are part of the actinide series. In other words, the actinide series includes the transuranic elements, which are considered a continuation of period 7 (at group 3, VIIIB), following radium ($_{88}$Ra) on the periodic table. However, because this actinide series and transuranic subseries contain so many elements—15 including actinium—they do not fit this space following $_{88}$Ra in the periodic table. Therefore, the actinide series is placed on the lower part of the table below the lanthanide series of rare-earths. In some references the actinide series is called the second rare-earth series since they are homologous to the lanthanide rare-earth elements just above them in the periodic table.

Two of the 15 elements in the actinide series are found in nature [thorium ($_{90}$Th) and uranium ($_{92}$U)]. Actinium ($_{89}$Ac), protactinium ($_{91}$Pa), and neptunium ($_{93}$Np) are also found in nature, but in very small quantities and only as by-products of the radioactive decay of some of the isotopes of thorium and uranium. All the other radioactive elements in the actinide series are synthetically produced and exist in very small amounts for short periods of time, making them difficult to study. They all resemble the first one in the series (actinium), as well as each other in their chemical and physical characteristics. As mentioned, they are also somewhat homologous (similar) to the elements just above them in the lanthanide series. An unusual characteristic of these elements is that as they react with other elements, they add electrons to the 5f electron orbital of the O shell. See the individual "Electron Configuration" for each atom. Note that these elements have two electrons in their Q shell, and most, but not all, have nine electrons in their P shell. As a proton is added to each of the atoms of the actinide series, their atomic number increases progressively from left to right. Although they may have several different valences as they form compounds, their main oxidation state is +3.

As mentioned, a subset of the actinide series is referred to as the transuranic elements, which are the heavy elements with atomic numbers greater than uranium ($_{92}$U). This actinide

PERIODIC TABLE OF THE ELEMENTS

TRANSITION ELEMENTS

GROUPS/PERIODS	1 IA	2 IIA	3 IIIB	4 IVB	5 VB	6 VIB	7 VIIB	8	9 VIII	10	11 IB	12 IIB	13 IIIA	14 IVA	15 VA	16 VIA	17 VIIA	18 VIIIA
1	1 H 1.0079																	2 He 4.00260
2	3 Li 6.941	4 Be 9.01218											5 B 10.81	6 C 12.011	7 N 14.0067	8 O 15.9994	9 F 18.9984	10 Ne 20.179
3	11 Na 22.9898	12 Mg 24.305											13 Al 26.9815	14 Si 28.0855	15 P 30.9738	16 S 32.066(6)	17 Cl 35.453	18 Ar 39.948
4	19 K 39.0983	20 Ca 40.08	21 Sc 44.9559	22 Ti 47.88	23 V 50.9415	24 Cr 51.996	25 Mn 54.9380	26 Fe 55.847	27 Co 58.9332	28 Ni 58.69	29 Cu 63.546	30 Zn 65.39	31 Ga 69.72	32 Ge 72.59	33 As 74.9216	34 Se 78.96	35 Br 79.904	36 Kr 83.80
5	37 Rb 85.4678	38 Sr 87.62	39 Y 88.9059	40 Zr 91.224	41 Nb 92.9064	42 Mo 95.94	43 Tc (98)	44 Ru 101.07	45 Rh 102.906	46 Pd 106.42	47 Ag 107.868	48 Cd 112.41	49 In 114.82	50 Sn 118.71	51 Sb 121.75	52 Te 127.60	53 I 126.905	54 Xe 131.29
6	55 Cs 132.905	56 Ba 137.33	★	72 Hf 178.49	73 Ta 180.948	74 W 183.85	75 Re 186.207	76 Os 190.2	77 Ir 192.22	78 Pt 195.08	79 Au 196.967	80 Hg 200.59	81 Tl 204.383	82 Pb 207.2	83 Bi 208.980	84 Po (209)	85 At (210)	86 Rn (222)
7	87 Fr (223)	88 Ra 226.025	▲	104 Unq (261)	105 Unp (262)	106 Unh (263)	107 Uns (264)	108 Uno (265)	109 Une (266)	110 Uun (267)	111 Uuu (272)	112 Uub	113 Uut	114 Uuq	115 Uup	116 Uuh	117 Uus	118 Uuo

★ 6 Lanthanide Series (RARE EARTH)

57 La 138.906	58 Ce 140.12	59 Pr 140.908	60 Nd 144.24	61 Pm (145)	62 Sm 150.36	63 Eu 151.96	64 Gd 157.25	65 Tb 158.925	66 Dy 162.50	67 Ho 164.930	68 Er 167.26	69 Tm 168.934	70 Yb 173.04	71 Lu 174.967

▲ 7 Actinide Series (RARE EARTH)

89 Ac 227.028	90 Th 232.038	91 Pa 231.036	92 U 238.029	93 Np 237.048	94 Pu (244)	95 Am (243)	96 Cm (247)	97 Bk (247)	98 Cf (251)	99 Es (252)	100 Fm (257)	101 Md (258)	102 No (259)	103 Lr (260)

subgroup includes neptunium ($_{93}$Np) through lawrencium ($_{103}$Lr). These elements were discovered as synthetic radioactive isotopes at the Lawrence National Laboratory at the University of California at Berkeley and the Joint Institute for Nuclear Research in Dubna, Russia. Both institutions are credited with the discovery of some of the heavier radioactive elements. Neptunium and plutonium were the first and second transuranic elements produced artificially. They are the products of radioactive decay of uranium in nuclear reactors. All the other transuranic elements are also synthesized (man-made) by bombarding (slamming together) particles or nuclei of atoms of different elements that have two different atomic numbers and lighter atomic weights. These reactions require very high energies and are accomplished in cyclotrons, nuclear reactors, or high-energy **particle accelerators,** which create a few atoms at a time of new, heavier, and unstable elements. The elements in the transuranic subgroup, up to and including fermium ($_{100}$Fm), are produced by the capture of neutrons.

ACTINIUM

SYMBOL: Ac   **PERIOD:** 7   **SERIES NAME:** Actinide   **ATOMIC NO:** 89
ATOMIC MASS: 227.028 amu   **VALENCE:** 3   **OXIDATION STATE:** +3   **NATURAL STATE:** Radioactive Solid
ORIGIN OF NAME: The name "actinium" is derived from the Greek word(s) *aktis* or *aktinos,* meaning "beam" or "ray."
ISOTOPES: There are a total of 35 isotopes of actinium, none of which are stable. All are radioactive, and none exist in the Earth's crust in any large amounts, although a few can be extracted from large quantities of pitchblende and other minerals. All are extremely scarce. Those produced artificially in nuclear reactors, cyclotrons, or linear accelerators have relatively short half-lives, ranging from 69 nanoseconds to 21 years.

ELECTRON CONFIGURATION

Energy Levels/Shells/Electrons	Orbitals/Electrons
1-K = 2	s2
2-L = 8	s2, p6
3-M = 18	s2, p6, d10
4-N = 32	s2, p6, d10, f14
5-O = 18	s2, p6, d10
6-P = 9	s2, p6, d1
7-Q = 2	s2

Properties

Actinium is an extremely radioactive, silvery-white, heavy metal that glows in the dark with an eerie bluish light. It decays rapidly which makes it difficult to study, given that it changes into thorium and francium through electron capture and alpha decay. Its melting point is 1,051°C, its boiling point is 3,198°C, and its density is 10.07g/cm^3.

Characteristics

Actinium is the last (bottom) member of group 3 (IIIB) of elements in the periodic table and the first of the actinide series of metallic elements that share similar chemical and physical characteristics. Actinium is also closely related in its characteristics to the element lanthanum, which is located just above it in group 3. The elements in this series range from atomic number 89 (actinium) through 103 (lawrencium). Actinium's most stable isotope is actinium-227, with a half-life of about 22 years. It decays into Fr-223 by alpha decay and Th-227 through beta decay, and both of these isotopes are decay products from uranium-235.

Abundance and Source

Actinium is a rare element that is found in very small amounts in uranium ore (pitchblende), making it difficult and expensive to extract even a small quantity. It is less expensive and easier to produce small amounts by bombarding the element radium with neutrons in a nuclear reactor. Actinium has few commercial uses.

History

Actinium is another example of an element being discovered independently by two different men. After the Curies had separated radium ($_{88}$Ra) from uranium ($_{92}$U) ore in 1899, their friend Andre-Louis Debierne (1874–1949) discovered another radioactive element (actinium) mixed with uranium in the ore pitchblende. A few years later, in 1902, Friedrich Otto Giesel also discovered element 89, an element that resembled the rare-earths but was much too heavy to be a rare-earth metal. Although Giesel named element 89 "emanium," his records were predated by Debierne's, which gave Debierne credit for the discovery as well as naming rights. He chose the name "actinium," meaning "ray."

Common Uses

There are no significant uses for actinium because of its scarcity and the expense of producing it. The only practical use for small amounts of actinium is as a **tracer** in medicine and industry. It is too difficult to produce in substantial quantities to make it useful. Actinium can be used as a source of neutrons to bombard other elements to produce isotopes of those elements, but other neutron sources are less expensive.

Examples of Compounds

Most of the radioactive actinium isotopes that are produced in nuclear reactors are in milligram quantities. There are not many common compounds.

The metallic ion of actinium has an oxidation state of +3. Two examples of Ac^{3+} compounds follow:

Actinium (III) trifluoride: $Ac^{3+} + 3F^{1-} \rightarrow AcF_3$.
Actinium sequioxide: $2Ac^{3+} + 3O^{2-} \rightarrow Ac_2O_3$.

Hazards

Most of the radioactive isotopes of actinium pose an extreme radiation hazard. They are bone-seeking radioactive poisons.

THORIUM

SYMBOL: Th   **PERIOD:** 7   **SERIES NAME:** Actinide   **ATOMIC NO:** 90
ATOMIC MASS: 232.0381 amu   **VALENCE:** 4   **OXIDATION STATE:** +4   **NATURAL STATE:**
  Solid
ORIGIN OF NAME: Thorium was named for Thor, the Scandinavian (Norse) god of "thunder."
ISOTOPES: There are 30 radioisotopes of thorium. One isotope in particular, thorium-232, although a weak source of radiation, has such a long half-life ($1.405 \times 10^{+10}$ years, or about 14 billion years) that it still exists in nature and is considered stable.

ELECTRON CONFIGURATION

Energy Levels/Shells/Electrons	Orbitals/Electrons
1-K = 2	s2
2-L = 8	s2, p6
3-M = 18	s2, p6, d10
4-N = 32	s2, p6, d10, f14
5-O = 18	s2, p6, d10
6-P = 10	s2, p6, d2
7-Q = 2	s2

Properties

Thorium is a radioactive, silvery-white metal when freshly cut. It takes a month or more for it to tarnish in air, at which point it forms a coating of black oxide. Although it is heavy, it is also a soft and malleable actinide metal. The metal has a rather low melting point, but its oxide has a very high melting point of about 3,300°C. Thorium reacts slowly with water but reacts more vigorously with hydrochloric acid (HCl).

Thorium's melting point is 1,750°C, its boiling point is 4,788°C, and its density is 11.79g/cm³.

Characteristics

Thorium is chemically similar to hafnium ($_{72}$Hf) and zirconium ($_{40}$Zr), located just above it in group 4 (IVB). Thorium-232 is found in nature in rather large quantities and goes through a complicated decay process called the thorium decay series. This series involves both alpha and beta emissions, as follows: Th-232 →Ra-228→Ac-228→Th-228→Ra-224→Rn-220→ Po-216→Po-212→Pb-212→Bi-212→Ti-208→Pb-208. Thorium-232 can also be converted into thorium-233 or uranium-233 by bombarding it with neutrons. This results in Th-232 adding a neutron to its nucleus, thus increasing its atomic weight. It then decays into uranium-233. This makes it potentially useful as an experimental new type of fissionable material for use in nuclear reactors designed to produce electricity.

Abundance and Source

Thorium is the 37th most abundant element found on Earth, and it makes up about 0.0007% of the Earth's crust. It is mostly found in the ores of thorite, thorianite (the oxide of thorium), and monazite sand. It is about as abundant as lead in the Earth's crust. As a potential fuel for nuclear reactors, thorium has more energy potential than the entire Earth's supply of uranium, coal, and gas combined.

History

Jöns Jakob Berzelius (1779–1848) found in 1815 what he considered a "new earth," which he thought was similar to the oxide of another metallic element. He named his new element after the Scandinavian god of thunder, Thor. He was mistaken, as his new element was later proven to be Yttrium phosphate.

About four years later in 1819, the Reverend Hans Morten Thrane Esmark (1801–1882), an amateur mineralogist, found a black mineral in Norway and gave a sample of it to his father, a geology professor, for analysis. Unable to identify it, Professor Jens Esmark sent the sample for chemical analysis to Berzelius, who found that it contained 60% of a new type of earth oxide not recognized before. It was identified as the mineral thorite ($ThSiO_4$). Berzelius reported his discovery in an 1829 publication and retained the name "thorium," in honor of Thor, the Norse god of war. Berzelius is thus credited with thorium's discovery.

Common Uses

Thorium has several commercial uses. For example, thorium oxide (ThO_2) has several uses, including in the Welsbach lantern mantle that glows with a bright flame when heated by a gas burner. Because of the oxide's high melting point, it is used to make high-temperature crucibles, as well as glass with a high index of refraction in optical instruments. It is also used as a catalyst in the production of sulfuric acid (H_2SO_4), in the cracking procedures in the petroleum industry, and in the conversion of ammonia (NH_3) into nitric acid (HNO_3). Thorium is used as a "jacket" around the core of nuclear reactors, where it becomes fissionable uranium-233 that is then used for the nuclear reaction to produce energy. Additionally, it is used in photoelectric cells and X-ray tubes and as a coating on the tungsten used to make filaments for light bulbs. It has great potential to supplant all other nonrenewable energy sources (i.e., coal, gas, and atomic energy). Thorium-232 can be converted into uranium-233, a fissionable fuel available in much greater quantities than other forms of fissionable materials used in nuclear reactors.

Examples of Compounds

Thorium's main oxidation is +4. Therefore, its metallic ion is Th^{4+} as follows:

Thorium chloride: $Th^{4+} + 4Cl^{1-} \rightarrow ThCl_4$.

Thorium dioxide: $Th^{4+} + 2O^{2-} \rightarrow ThO_2$.

Thorium nitride: $3Th^{4+} + 4N^{3-} \rightarrow Th_3N_4$.

Thorium forms many compounds, including the following example compound:

Thorium carbide: $Th^{4+} + 2C^{4-} \rightarrow ThC_2$. This compound is used as a nuclear fuel.

Hazards

As thorium undergoes natural radioactive decay, a number of products, including gases, are emitted. These decay products are extremely dangerous radioactive poisons if inhaled or ingested.

PROTACTINIUM

SYMBOL: Pa **PERIOD:** 7 **SERIES NAME:** Actinum **ATOMIC NO:** 91
ATOMIC MASS: 231.0358 amu **VALENCE:** 4 and 5 **OXIDATION STATE:** +4 and
+5 **NATURAL STATE:** Solid
ORIGIN OF NAME: A combination of the Greek word *protos,* meaning first, combined with
the element actinium, which together means "before actinium."
ISOTOPES: There are a total of 30 isotopes of protactinium. All are radioactive, and none
are stable. Their decay modes are either alpha or beta decay or electron capture. Their
half-lives range from 53 nanoseconds to $3.276 \times 10^{+4}$ years.

ELECTRON CONFIGURATION

Energy Levels/Shells/Electrons	Orbitals/Electrons
1-K = 2	s2
2-L = 8	s2, p6
3-M = 18	s2, p6, d10
4-N = 32	s2, p6, d10, f14
5-O = 20	s2, p6, d10, f2
6-P = 9	s2, p6, d1
7-Q = 2	s2

Properties

Protactinium is a relatively heavy, silvery-white metal that, when freshly cut, slowly oxidizes in air. All the isotopes of protactinium and its compounds are extremely radioactive and poisonous. Proctatinium-231, the isotope with the longest half-life, is one of the scarcest and most expensive elements known. It is found in very small quantities as a decay product of uranium mixed with pitchblende, the ore of uranium. Protactinium's odd atomic number ($_{91}$Pa) supports the observation that elements having odd atomic numbers are scarcer than those with even atomic numbers.

Its melting point is just under 1,600°C, its boiling point is about 4,200°C, and its density is 15.37g/cm^3.

Characteristics

Because the proportion of protactinium to its ores is of the magnitude of one part in ten million, it takes many truckloads of ore to extract a small quantity of the metal. About 30 years ago, approximately 125 grams of protactinium was extracted from over 60 tons of ore at a cost of over $500,000. These 125 grams represent the total amount of protactenium that exists in the entire world today.

Abundance and Source

As mentioned, protactinium is one of the rarest elements in existence. Although protactinium was isolated, studied, and identified in 1934, little is known about its chemical and physical properties since only a small amount of the metal was produced. Its major source is the fission by-product of uranium found in the ore pitchblende, and only about 350 milligrams can be extracted from each ton of high-grade uranium ore. Protactinium can also be produced by the submission of samples of throrium-230 ($_{90}$Th) to radiation in nuclear reactors or particle accelerators, where one proton and one or more neutrons are added to each thorium atom, thus changing element 90 to element 91.

History

It was first identified and named *brevium,* meaning "brief," by Kasimir Fajans and O. H. Gohring in 1913 because of its extremely short half-life. In 1918 Otto Hahn (1879–1968) and Lise Meitner (1878–1968) independently discovered a new radioactive element that decayed from uranium into $_{89}$Ac (actinium). Other researchers named it "uranium X2." It was not until 1918 that researchers were able to identify independently more of the element's properties and declare it as the new element 91 that was then named "protactinium." This is another case in which several researchers may have discovered the same element. Some references continue to give credit for protactinium's discovery to Frederich Soddy (1877–1956) and John A. Cranston (dates unknown), as well as to Hahn and Meitner.

Common Uses

Protactinium is very rare, and not enough of it is available for commercial use. It is used only in laboratory research.

Examples of Compounds

Examples of compounds of protactinium's oxidation states of +4 and +5 follow:

Protactinium (IV) oxide: $Pa^{4+} + 2O^{2-} \rightarrow PaO_2$.
Protactinium (V) oxide: $2Pa^{5+} + 5O^{2-} \rightarrow Pa_2O_5$.

Hazards

All the isotopes of protactinium are highly radioactive poisons and therefore very dangerous.

URANIUM

SYMBOL: U   PERIOD: 7   SERIES NAME: Actinide   ATOMIC NO: 92
ATOMIC MASS: 238.0291 amu   VALENCE: 3, 4, 5, and 6   OXIDATION STATES: +3, +4, +5, and +6   NATURAL STATE: Solid
ORIGIN OF NAME: Named for the planet Uranus.
ISOTOPES: There are total of 26 isotopes of uranium. Three of these are considered stable because they have such long half-lives and have not all decayed into other elements

and thus still exist in the Earth's crust. The three are uranium-234, with a half-life of $2.455 \times 10^{+5}$ years, which makes up 0.0054% of the uranium found on Earth; uranium-235, with a half-life of $703.8 \times 10^{+6}$ years, which accounts for 0.724% of the Earth's uranium; and uranium-238m with a half-life of $4.468 \times 10^{+9}$ years, which makes up most of the Earth's supply of uranium at 99.2742% of the uranium found naturally.

ELECTRON CONFIGURATION

Energy Levels/Shells/Electrons	Orbitals/Electrons
1-K = 2	s2
2-L = 8	s2, p6
3-M = 18	s2, p6, d10
4-N = 32	s2, p6, d10, f14
5-O = 21	s2, p6, d10, f3
6-P = 9	s2, p6, d1
7-Q = 2	s2

Properties

Uranium is the fourth metal in the actinide series. It looks much like other actinide metallic elements with a silvery luster. It is comparatively heavy, yet malleable and ductile. It reacts with air to form an oxide of uranium. It is one of the few naturally radioactive elements that is fissionable, meaning that as it absorbs more neutrons, it "splits" into a series of other lighter elements (lower atomic weights) through a process of alpha decay and beta emission that is known as the uranium decay series, as follows: U-238→ Th-234→Pa-234→U-234→ Th-230→Ra-226→Rn-222→Po-218→Pb-214 & At-218→Bi-214 & Rn-218→Po-214→ Ti-210→Pb-210→Bi-210 & Ti-206→Pb-206 (stable isotope of lead, $_{82}$Pb).

Uranium's melting point is 1,135°C, its boiling point is about 4,100°C, and its density is about 19g/cm^3, which means it is about 19 times heavier than water.

Characteristics

Uranium reacts with most nonmetallic elements to form a variety of compounds, all of which are radioactive. It reacts with hot water and dissolves in acids, but not in alkalis (bases). Uranium is unique in that it can form solid solutions with other metals, such as molybdenum, titanium, zirconium, and niobium.

Because the isotope uranium-235 is fissionable, meaning that it produces free neutrons that cause other atoms to split, it generates enough free neutrons to make it unstable. When the unstable U-235 reaches a **critical mass** of a few pounds, it produces a self-sustaining fission **chain reaction** that results in a rapid explosion with tremendous energy and becomes a nuclear (atomic) bomb. The first nuclear bombs were made of uranium and plutonium. Today, both of these "fuels" are used in reactors to produce electrical power. **Moderators** (control rods) in nuclear power reactors absorb some of the neutrons, which prevents the mass from becoming critical and thus exploding. Although some countries have overcome their fear of nuclear power and generate a large portion of their electricity through nuclear reactors, the United States, after developing nuclear power plants 40 to 50 years ago, has stopped the continued expansion of nuclear power plants. Despite the experience of the Three-Mile Island event that spread no more radiation than what people living at high altitudes receive, nuclear power plants are safer than coal-fired electrical generation plants (there are fewer accidents) and they are far less damaging to the quality of air. Plans are being currently developed in the United States for the construction of nuclear power plants that utilize improved technologies to meet the ever-increasing energy demands of U.S. citizens, while improving the quality of our air and water.

Abundance and Source

Uranium is the 44th most abundant element on Earth. It is found mainly in the ore pitchblende, but can also be extracted from ores such as uraninite (UO_2), carnotite [$K2(UO2)_2VO_4$], autunite [$Ca(UO_2)_2(PO_4)_2$], phosphate rock [$Ca_3(PO_4)_2$], and monazite sand. These ores are found in Africa, France, Australia, and Canada, as well as in Colorado and New Mexico in the United States. Today, most uranium is sold both to governments and on the black market as "yellow cake" (triuranium octoxide U_3O_8). This form can be converted to uranium dioxide (UO_2), which is a fissionable compound of uranium mostly used in nuclear electrical power plants. Only 0.7204% of uranium is the isotope U-235, which is fissionable and can be used in nuclear power plants. Although U-235 is capable of producing enough free neutrons to sustain a nuclear chain reaction, it is very difficult to obtain enough U-235 for this purpose. To produce an adequate supply for the first atomic (nuclear) bombs, a large gaseous diffusion plant was constructed that separated small amounts of U-235 from nonfissionable isotopes and their ores by using the differences in their atomic weights. The plant used porous membranes that, through diffusion, allow the lighter U-235 atoms to pass through the pores while the heavier U-238 does not. Thus, the U-235 is separated and concentrated from the heavier U-238. The common uranium isotope U-238 can be converted to plutonium-239 in "breeder" nuclear reactors. Pu-239 is fissionable and is often used in the production of nuclear bombs as well as in nuclear power plants. Another form of uranium (U-233) that is not found in nature can be artificially produced by bombarding thorium-232 with neutrons to produce thorium-233, which has a half-life of 22 minutes and decays into protactinium-233 with a half-life of 27 days. Pa-233 then, through beta decay, transmutes into uranium-233. Just one pound of U-233 in nuclear reactors produces energy equal to 1,500 tons of coal.

History

At the end of the eighteenth century, scientists thought that pitchblende was a mixture of iron and zinc compounds. In 1789 Martin Heinrich Klaproth (1743–1817) discovered a new metallic element in a sample of pitchblende, which he named "uranus" after the recently discovered planet. Although what he actually discovered was the compound uranous oxide (UO_2), it was adequate to establish him as the discoverer of uranium. For almost a century, scientists believed that the compound uranous oxide (UO_2) was the elemental metal uranium. In 1841 Eugene-Melchoir Peligot (1811–1890) finally isolated the metal uranium from its compound. Even so, no one knew that both the compounds and metal of uranium were radio-active until 1896, when Henri Becquerel (1852–1908) mistakenly placed a piece of potassium uranyl sulfate on a photographic plate that was wrapped in black paper. When the plate was developed, it was fogged. This proved to be a newly discovered source of radiation in nature other than sunlight or fluorescence. Becquerel also invented an electrometer instrument that could be used to detect the radiation.

In 1898 Marie Sklodowska Curie (1867–1934), while experimenting with thorium and uranium, coined the word "radioactivity" to describe this newly discovered type of radiation. She went on to discover polonium and radium. Madam Curie and her husband Pierre Curie (1859–1906), who discovered the piezoelectric effect, which is used to measure the level of radiation, and Henri Becquerel jointly received the 1903 Nobel Prize in Physics for their work on radioactivity.

Of some interest is that after uranium ($_{92}$U) was named after the planet Uranus, neptunium ($_{93}$Np), which was discovered next, was named after Neptune, the next planet in our solar system. And finally, plutonium ($_{94}$Pu) the next transuranic element discovered, was named after Pluto, the last planet discovered so far in our solar system.

Common Uses

The most common use of uranium is to convert the rare isotope U-235, which is naturally fissionable, into plutonium through neutron capture. Plutonium, through controlled fission, is used in nuclear reactors to produce energy, heat, and electricity. Breeder reactors convert the more abundant, but nonfissionable, uranium-238 into the more useful and fissionable plutonium-239, which can be used for the generation of electricity in nuclear power plants or to make nuclear weapons.

Although uranium forms compound with many nonmetallic elements, there is not much use for uranium outside the nuclear energy industry. Depleted uranium has had most of the U-235 removed from it through decay processes. It finds uses as armor-piercing antitank shells, ballast for missile reentry systems, glazes for ceramics, and shielding against radiation. An increasing concern is the possibility that terrorists can construct "dirty bombs" that use conventional explosives to spread "spent" radioactive materials that are still radioactive enough to inflict harm to people exposed to the bomb's blast and those downwind of the explosion, as well at to the environment.

Uranium-238 has a half-life of 4.468 billion years over which time it decays into stable lead-206. This process can be used to date ancient rocks by comparing the ratio of the isotope lead-206, the last isotope in the uranium decay series, to the level of uranium-238 in the sample of rock to determine its age. This system has been used to date the oldest rocks on Earth as being about 4.5 billion years old, which is about the time of the formation of our planet.

Examples of Compounds

The most stable oxidation states of uranium are +4 and +6. For instance uranium can combine with chlorine using both the U^{4+} and U^{6+} ions, as follows:

Uranium (IV) chloride: $U^{4+} + 4Cl^{1-} \rightarrow UCl_4$.

Uranium (VI) chloride: $U^{6+} + 6Cl^{1-} \rightarrow UCl_6$.

Uranium also combines with oxygen in various ratios. For instance, *uranium dioxide* (UO_2) is a brownish-black powder that was once thought to be pure uranium. *Uranium trioxide* (UO_3), a heavy orangish-powder, was once referred to as *uranyl oxide*.

Hazards

All compounds as well as metallic uranium are radioactive—some more so than others. The main hazard from radioactive isotopes is radiation poisoning. Of course, another potential hazard is using fissionable isotopes of uranium and plutonium for other than peaceful purposes, but such purposes involve political decisions, not science.

NEPTUNIUM

SYMBOL: Np   **PERIOD:** 7   **SERIES NAME:** Actinide   **ATOMIC NO:** 93

ATOMIC MASS: 237.0482 amu   **VALENCE:** 3, 4, 5, and 6   OXIDATION STATES: +3, +4, +5, and +6   **NATURAL STATE:** Solid

ORIGIN OF NAME: Named for the planet Neptune.

ISOTOPES: There are a total of 23 isotopes of neptunium. None are stable. All are radioactive with half-lives ranging from two microseconds to $2.144 \times 10^{+6}$ years for the isotope Np-237, which spontaneously fissions through alpha decay.

ELECTRON CONFIGURATION

Energy Levels/Shells/Electrons	Orbitals/Electrons
1-K = 2	s2
2-L = 8	s2, p6
3-M = 18	s2, p6, d10
4-N = 32	s2, p6, d10, f14
5-O = 22	s2, p6, d10, f4
6-P = 9	s2, p6, d1
7-Q = 2	s2

Properties

The chemistry of neptunium ($_{93}$Np) is somewhat similar to that of uranium ($_{92}$U) and plutonium ($_{94}$Pu), which immediately precede and follow it in the actinide series on the periodic table. The discovery of neptunium provided a solution to a puzzle as to the missing decay products of the thorium decay series, in which all the elements have mass numbers evenly divisible by four; the elements in the uranium series have mass numbers divisible by four with a remainder of two. The actinium series elements have mass numbers divisible by four with a remainder of three. It was not until the neptunium series was discovered that a decay series with a mass number divisible by four and a remainder of one was found. The neptunium decay series proceeds as follows, starting with the isotope plutonium-241: Pu-241→Am-241→Np-237→Pa-233→U-233→Th-229→Ra-225→Ac-225→Fr-221→At-217→Bi-213→Ti-209→Pb-209→Bi-209.

Neptunium is a silvery-white radioactive, heavy metal. Its melting point is 644°C, its boiling point is 3,902°C, and its density is 20.25g/cm³.

Characteristics

Neptunium is the first of the subseries of the actinide series known as the **transuranic elements**—those heavy, synthetic (man-made) radioactive elements that have an atomic number greater than uranium in the actinide series of the periodic table. An interesting fact is that neptunium was artificially synthesized before small traces of it were discovered in nature. More is produced by scientists every year than exists in nature.

Neptunium has an affinity for combining with nonmetals (as do all transuranic elements) such as oxygen, the halogens, sulfur, and carbon.

Abundance and Source

At one time, neptunium's entire existence was synthesized by man. Sometime later, in the mid-twentieth century, it was discovered that a very small amount is naturally produced in uranium ore through the actions of neutrons produced by the decay of uranium in the ore pitchblende. Even so, a great deal more neptunium is artificially produced every year than ever did or does exist in nature. Neptunium is recovered as a by-product of the commercial production of plutonium in nuclear reactors. It can also be synthesized by bombarding uranium-238 with neutrons, resulting in the production of neptunium-239, an isotope of neptunium with a half-life of 2.3565 days.

History

In 1940 Edwin Mattison McMillan (1907–1991) and a former graduate student, Philip Hauge Abelson (1913–1995), jointly experimented at the physics laboratory of the University of California, Berkeley, by bombarding uranium oxide with high-speed neutrons in a cyclotron. Their experiment indicated that a new element exhibited oxidation states of +4 and +6 and that it exhibited chemical and physical properties similar to uranium. They prepared their samples of neptunium-239 by bombarding uranium-238 with high-energy neutrons, which resulted in uranium-239 that, in turn, decayed by beta radiation, resulting in neptunium-239. The reaction follows: U-238 + (neutrons, gamma rays) $\rightarrow$ U-239 (decays by beta radiation) $\rightarrow$ Np-239. Their discovery was confirmed and their data was published in 1940. Because $_{93}$Np's atomic number followed $_{92}$U, McMillan decided to name the new element "neptunium" for the planet Neptune, which was next in line following Uranus in the solar system. McMillan's and Abelson's work was interrupted during World War II and was later continued by Arthur C. Wahl and Joseph W. Kennedy, who were able to determine the physical reactions that resulted in neptunium.

Common Uses

The most important radioactive isotope of neptunium is Neptunium-237, with a half-life of $2.144 \times 10^{+6}$ years, or about 2.1 million years, and decays into protactinium-233 through alpha decay. Neptunium's most important use is in nuclear research and for instruments designed to detect neutrons.

Examples of Compounds

Some examples of halogen compounds of neptunium that are formed by its ions with oxidation states of +3, +4, +5, and +6 follow:

Neptunium (III) fluoride: $Np^{3+} + 3F^{1-} \rightarrow NpF_3$.
Neptunium (IV) chloride: $Np^{4+} + 4Cl^{1-} \rightarrow NpCl_4$.
Neptunium (V) iodide: $Np^{5+} + 5I^{1-} \rightarrow NpI_5$.
Neptunium (VI) fluoride: $Np^{6+} + 6F^{1-} \rightarrow NpF_6$.
Neptunium also forms compounds with oxygen. An example follows:

Neptunium (IV) didioxide: $Np^{4+} + 2O^{2-} \rightarrow NpO_2$. Neptunium dioxide is a powder used to form metal targets that are to be radiated by plutonium.

Hazards

All isotopes of neptunium are highly radioactive and are hazardous and thus need to be carefully used in controlled laboratory settings. These isotopes as well as neptunium's compounds are radioactive poisons.

PLUTONIUM

SYMBOL: Pu   **PERIOD:** 7   **SERIES NAME:** Actinide   **ATOMIC NO:** 94
ATOMIC MASS: 239.11 amu   **VALENCE:** 3, 4, 5, and 6   **OXIDATION STATE:** +3, +4, +5, and +6   **NATURAL STATE:** Solid
ORIGIN OF NAME: Named for the planet Pluto.
ISOTOPES: There are a total of 24 isotopes of plutonium. All of them are unstable and radioactive. Their half-lives range from 28 nanoseconds to $8.00 \times 10^{+7}$ years.

ELECTRON CONFIGURATION

Energy Levels/Shells/Electrons	Orbitals/Electrons
1-K = 2	s2
2-L = 8	s2, p6
3-M = 18	s2, p6, d10
4-N = 32	s2, p6, d10, f14
5-O = 24	s2, p6, d10, f6
6-P = 8	s2, p6
7-Q = 2	s2

Properties

All isotopes of plutonium are radioactive. The two isotopes that have found the most uses are Pu-238 and Pu-239. Pu-238 is produced by bombarding U-238 with **deuterons** in a cyclotron, creating neptunium-238 and two free neutrons. Np-238 has a half-life of about two days, and through beta decay it transmutates into plutonium-238. There are six allotropic metallic crystal forms of plutonium. They all have differing chemical and physical properties. The alpha (α) allotrope is the only one that exists at normal room temperatures and pressures. The alpha allotrope of metallic plutonium is a silvery color that becomes yellowish as it oxidizes in air. All the other allotropic forms exist at high temperatures.

The most stable isotope of plutonium is Pu-244, with a half-life of $8.00 \times 10^{+7}$ years (about 82,000,000 years). Being radioactive, Pu-244 can decay in two different ways. One way involves alpha decay, resulting in the formation of the isotope uranium-240, and the other is through spontaneous fission.

The melting point of plutonium is 640°C, its boiling point is 3,232°C, and its density is over 19 times that of the same volume of water ($19.84 g/cm^3$).

Characteristics

A single kilogram of radioactive metallic plutonium-238 produces as much as 22 million kilowatt-hours of heat energy. Larger amounts of Pu-238 produce more heat. However, Pu-238 is not fissionable, and thus it cannot sustain a chain reaction. However, plutonium-239 is fissionable, and a 10-pound ball can reach a critical mass sufficient to sustain a fission chain reaction, resulting in an explosion, releasing the equivalent of over 20,000 tons of TNT. This 10-pound ball of Pu-239 is only about one-third the size of fissionable uranium-235 required to reach a critical mass. This makes plutonium-239 the preferred fissionable material for nuclear weapons and some nuclear reactors that produce electricity.

Plutonium has some peculiar qualities. In its molten state it will corrode any container in which it is stored. It has an unusual ability to combine with almost all the other elements listed on the periodic table, and it can change its density by as much as 25% according to its environment. Small pieces of it can spontaneously ignite at temperatures as low as 150°C. A modern atomic bomb has three main parts: the primary, the secondary, and a radiation container. The primary is formed as a small ball of plutonium called a "pit," which is used for the "core" of an atomic (nuclear) bomb. This pit is surrounded by the secondary component, which is a chemical explosive material, which is then surrounded by a shell of uranium. When the chemical explosive is triggered, it implodes, compressing the plutonium core, which results in an increase in the pit's density. This forces the nuclei of the atoms closer together, producing "free" neutrons. This compression causes the plutonium nuclei to fission in a chain reaction that results in the release of tremendous energy. This reaction also causes the surrounding uranium to fission, which releases more nuclear energy—enough to wipe out New York City or Washington, D.C.

The first nuclear bomb that was tested was called "Trinity" and weighed 10,000 pounds. Today's bombs weigh about 250 pounds and are about the size of an average suitcase.

Abundance and Source

Plutonium exists in trace amounts in nature. Most of it isotopes are radioactive and man-made or produced by the natural decay of uranium. Plutonium-239 is produced in nuclear reactors by bombarding uranium-238 with deuterons (nuclei of deuterium, or heavy hydrogen). The transmutation process is as follows: ^{238}U + deuterons → 2 nuclei + ^{239}Np + β → decays to → ^{238}Pu + β-.

There is more than an adequate supply of plutonium-239 in the world because it is a "waste" product of the generation of electricity in nuclear power plants. One of the objections to developing more nuclear reactors is the dilemma of either eliminating or storing all the excess plutonium. In addition, there is always the risk of terrorists' obtaining a supply of Pu-239 to make nuclear weapons.

History

The discovery and early use of plutonium is a unique story. In 1941 Glenn T. Seaborg (1912–1999) and his colleagues, Arthur Wahl (1917–2006) and Joseph Kennedy (1915–1956) , found trace amounts (about 1 gram) of plutonium ($_{94}$Pu) in neptunium ($_{93}$Np) at the physics labs of the University of California at Berkeley. They used the university's cyclotron to produce plutonium by bombarding uranium oxide with deuterons (nuclei of heavy hydrogen), which first produced neptunium that decayed into plutonium as it gained a proton that increased its atomic number to 94. They recorded their discovery but did not publish it until 1946 because of World War II. Once plutonium was determined to be fissionable and once it was seen that a critical mass of the element could maintain a sustained chain reaction, it was "classified." A large factory was constructed to produce kilogram quantities of this new element that required several kilograms less than the fissionable but scarce uranium-235 to build atomic (nuclear) bombs. This mass production of the newly discovered plutonium and its use in early atomic weapons became a reality years before most scientists and the public knew it even existed. Today, plutonium is a by-product of the operations of nuclear power reactors, and there is an excess of it to the point that how to store this extremely radioactive waste product has become problematic. Scientists have yet to discover a way to make plutonium in a stable form, and they are still studying the strange properties and characteristics of this artificially produced element.

Common Uses

The most common use of plutonium is as a fuel in nuclear reactors to produce electricity or as a source for the critical mass required to sustain a fission chain reaction to produce nuclear weapons. Plutonium also is used to convert nonfissionable uranium-238 into the isotope capable of sustaining a controlled nuclear chain reaction in nuclear power plants. It takes only 10 pounds of plutonium-239 to reach a critical mass and cause a nuclear explosion, as compared with about 33 pounds of fissionable, but scarce, uranium-235.

Plutonium-238 and plutonium-239 are two isotopes that can be used outside of the nuclear weapons industry. Plutonum-238 is currently used in small thermoelectric generators to provide electricity for space probes that are sent far beyond the region where the sun could be used to generate electric power. Two early instruments sent beyond our solar system are the *Galileo* and *Cassini* probes. Plutonium-239's critical mass undergoes a fissionable chain reaction, making it ideal for use as fuel for some types of nuclear reactors as well as to produce nuclear weapons. In the future it may be possible to use all the waste plutonium produced in the world to power small thermal electrical power plants that could be installed in each house to provide inexpensive and continuous household electrical power. This probably will not happen until the public overcomes its fear of nuclear energy.

Examples of Compounds

Plutonium can form compounds with many nonmetals such as oxygen, the halogens, and nitrogen. It can also be used as an alloy with other metals. A few examples of plutonium compounds exhibiting the oxidation states of +3 and +4 follow:

Plutonium (III) fluoride: $Pu^{3+} + 3F^{1-} \rightarrow PuF_3$. Also called plutonium trifluoride.

Plutonium (IV) oxide: $Pu^{4+} + 2O^{2-} \rightarrow PuO_2$. Also known as plutonium dioxide.

Hazards

Plutonium is by far one of the most toxic radioactive poisons known. The metal, its alloys, and its compounds must be handled in a shielded and enclosed "glove box" that contains an inert argon atmosphere. It is a carcinogen that can cause radiation poisoning leading to death.

AMERICIUM

SYMBOL: Am   **PERIOD:** 7   **SERIES NAME:** Actinide   **ATOMIC NO:** 95
ATOMIC MASS: 243 amu   **VALENCE:** 3, 4, 5, and 6   **OXIDATION STATE:** +3, +4, +5 and +6   **NATURAL STATE:** Solid
ORIGIN OF NAME: Named after the continent America because Europium was named after the European continent.
ISOTOPES: There are 24 isotopes of americium. All are radioactive with half-lives ranging from 72 microseconds to over 7,000 years. Five of americium's isotopes are fissionable with spontaneous alpha decay.

ELECTRON CONFIGURATION

Energy Levels/Shells/Electrons	Orbitals/Electrons
1-K = 2	s2
2-L = 8	s2, p6
3-M = 18	s2, p6, d10
4-N = 32	s2, p6, d10, f14
5-O = 25	s2, p6, d10, f7
6-P = 8	s2, p6
7-Q = 2	s2

Properties

All the isotopes of americium belonging to the transuranic subseries of the actinide series are radioactive and are artificially produced. Americium has similar chemical and physical characteristics and is **homologous** to europium, located just above it in the rare-earth (lanthanide) series on the periodic table. It is a bright-white malleable heavy metal that is somewhat similar to lead. Americium's melting point is 1,176°C, its boiling point is 2,607°C, and its density is 13.68g/cm³.

Characteristics

Although americium's main valence (oxidation state) is +3, it is **tetravalent.** It can form compounds with its ions of +4, +5, and +6, particularly when oxidized. Its most stable isotope is americium-243, with a half-life of 7,379 years, which, over time through alpha decay, transmutates into neptunium-239.

Abundance and Source

Americium does not exist in nature. All of its isotopes are man-made and radioactive. Americium-241 is produced by bombarding plutonium-239 with high-energy neutrons, resulting in the isotope plutonium-240 that again is bombarded with neutrons and results in the formation of plutonium-241, which in turn finally decays into americium-241 by the process of beta decay. Both americium-241 and americium-243 are produced within nuclear reactors. The reaction is as follows: ^{239}Pu + (neutron and λ gamma rays) $\rightarrow$ ^{240}Pu + (neutron and λ gamma rays) $\rightarrow$ $^{241}Pu \rightarrow$ ^{241}Am + beta minus (β-); followed by $^{241}Am \rightarrow _{93}Np\text{-}237$ + $^{4}He_2$ (helium nuclei).

History

At the end of World War II, the American physicists Glenn T. Seaborg (1912–1999), Ralph A. James, Leon O. Morgan, and Albert Ghiorso discovered americium (and curium) in the Metallurgical Laboratory at the University of Chicago, now known as the Argonne National Laboratory. The element curium ($_{96}Cm$) was first discovered as they bombarded plutonium with high-speed neutrons. This event was followed by the detection of the element americium ($_{95}Am$). These two elements were discovered in reverse order in 1944, but they both fell into place for the transuranic subseries elements of the actinide series. Although finding new elements is one thing, positively identifying their chemical and physical properties to the satisfaction of other scientists as "new" and fitting the periodic table is another. It required some effort and time for the team to identify all the properties and characteristics of curium and americium, which led a wag to suggest the names "pandemonium" and "delirium" for the two new elements because of the difficulties the team had in the confirmation of their identities.

Common Uses

Americium-141, with a half-life of 432 years, which produces both alpha (helium nuclei) and gamma rays (somewhat like X-rays), has found some commercial uses. One use is as the detector used in household smoke alarms The americium-141 isotope emits alpha particles that ionize air by removing electrons from the molecules of air and improving its electrical conductivity. Smoke reduces this conductivity, thus triggering the alarm. Another use is americium-241's high-energy gamma rays in **radiography,** in which portable X-ray instruments can easily be transported for emergency use. As a gamma ray source, americium-241 is used as a diagnostic aid to check the quality of welds in metals.

Examples of Compounds

Only a few compounds of americium exist. The most important is *americium oxide* (AmO_2), whose main use is in the preparation of other compounds. Americium can also form compounds with the halogens, similar to other transuranic elements—for example, *americium iodide* (AmI_3), *americium fluoride,* and *americium chloride* (AmF_3 and $AmCl_3$).

Hazards

All the isotopes and compounds of americium are deadly sources of radiation and cause radiation poisoning and death. Precautions must be taken when working with it. The small amount of americium-241 found in smoke detectors in household smoke alarms is harmless unless the isotope is removed and swallowed.

CURIUM

SYMBOL: Cm   **PERIOD:** 7   **SERIES NAME:** Actinides   **ATOMIC NO:** 96
ATOMIC MASS: 247 amu   **VALENCE:** 3 and 4   **OXIDATION STATE:** +3 and +4   **NATU-RAL STATE:** Solid
ORIGIN OF NAME: Named after Pierre and Marie Curie.
ISOTOPES: There are 23 isotopes of curium. All of them are man-made and radioactive. The most stable is curium-247, with a half-life of $1.56 \times 10^{+7}$ years (156,600,000 years), which through alpha decay transmutates into plutonium-243.

ELECTRON CONFIGURATION

Energy Levels/Shells/Electrons	Orbitals/Electrons
1-K = 2	s2
2-L = 8	s2, p6
3-M = 18	s2, p6, d10
4-N = 32	s2, p6, d10, f14
5-O = 25	s2, p6, d10, f7
6-P = 9	s2, p6, d1
7-Q = 2	s2

Properties

After the discovery of plutonium and before elements 95 and 96 were discovered, their existence and properties were predicted. Additionally, chemical and physical properties were predicted to be homologous (similar) to europium ($_{63}$Eu) and gadolinium ($_{64}$Gd), located in the rare-earth lanthanide series just above americium ($_{95}$Am) and curium ($_{96}$Cm) on the periodic table. Once discovered, it was determined that curium is a silvery-white, heavy metal that is chemically more reactive than americium with properties similar to uranium and plutonium. Its melting point is 1,345°C, its boiling point is ⁻1,300°C, and its density is 13.51g/cm³.

Characteristics

Curium is a synthetic (not natural) transuranic element of the actinide series. It was determined that curium's major valence and oxidation state was +3, similar to other elements of this series. The most stable isotope of curium is curium-247, with a half-life of $1.56 \times 10^{+7}$ years.

Abundance and Source

There is no natural curium on Earth. All of its isotopes are man-made and artificially produced through nuclear reactions with other elements. The curium isotope Cm-242 was first produced by bombarding plutonium-239 with helium nuclei (alpha particles), which contributed neutrons that changed $_{94}$Pu to $_{96}$Cm.

History

Glenn T. Seaborg, Ralph A. James, Leon Morgan, and Albert Ghiorso first produced curium in 1944 after returning to their laboratory located at the University of California in Berkeley from the Argonne National Laboratory in Chicago. They used the laboratory's cyclotron to accelerate alpha particles to bombard the isotope plutonium-239. This resulted in the formation of curium-242, which has a half-life of 163 days, and which in turn, through alpha decay (or spontaneous fission) is converted to plutonium-238. The reaction follows: ^{239}Pu + neutrons → ^{242}Cm—alpha decay (or fission) → ^{238}Pu.

Common Uses

There are no major commercial uses for curium because of the extremely small amount produced. In the future, the most important use of curium may be to provide the power for small, compact **thermoelectric** sources of electricity, by generating heat through the nuclear decay of radioisotope curium-241. These small, efficient power sources can be used in individual homes or remote regions to provide electricity to areas that cannot secure it from other sources. It could also be used as a source of electricity in spacecraft. However, today curium's main use is for basic scientific laboratory research.

Examples of Compounds

Considering the small amount of curium produced, a surprising number of compounds have been formed and identified. For example, there are two common oxides, *curium dioxide* (CmO_2) and *curium trioxide* (Cm_2O_3). Curium also combines with other nonmetals such as the halogens as follows: *curium iodide* (CmI_3), *curium bromide* ($CmBr_3$), and *curium tetrafluoride* (CmF_4).

Hazards

Curium metal and its compounds are radioactive bone-seeking poisons that attack the skeletal system of humans and animals. Care must be used in handling them.

BERKELIUM

SYMBOL: Bk   **PERIOD:** 7   **SERIES NAME:** Actinide   **ATOMIC NO:** 97
ATOMIC MASS: 247 amu   **VALENCE:** 3 and 4   **OXIDATION STATE:** +3 and +4
   NATURAL STATE: Solid
ORIGIN OF NAME: Named after the nuclear laboratory and the town in which the laboratory is located: Berkeley, California.
ISOTOPES: There are a total of 23 isotopes of berkelium, none of which are found in nature or are stable. Their half-lives range from 600 nanoseconds for Bk-242 to 1,389 years for Bk-247, which is also the most stable isotope that by alpha decay transmutates into americium-243.

ELECTRON CONFIGURATION

Energy Levels/Shells/Electrons	Orbitals/Electrons
1-K = 2	s2
2-L = 8	s2, p6
3-M = 18	s2, p6, d10
4-N = 32	s2, p6, d10, f14
5-O = 27	s2, p6, d10, f9
6-P = 8	s2, p6
7-Q = 2	s2

Properties

Berkelium is a metallic element located in group 11 (IB) of the transuranic subseries of the actinide series. Berkelium is located just below the rare-earth metal terbium in the lanthanide series of the periodic table. Therefore, it has many chemical and physical properties similar to terbium ($_{65}$Tb). Its isotopes are very reactive and are not found in nature. Only small amounts have been artificially produced in particle accelerators and by alpha and beta decay.

Its melting point is ˉ1050°C, its boiling point is unknown, and its density is 14g/cm^3.

Characteristics

Only milligram amounts of ^{249}Bk have been produced, and most of its other isotopes have short half-lives. Therefore, not all of its properties and characteristics and fully known.

Abundance and Source

The pure metal of berkelium does not exist in nature and has never been directly artificially produced, although the first isotope of berkelium produced was berkelium-243. It was artificially formed by bombarding americium-241 with the nuclei of helium (alpha particles), as follows: ^{241}Am+(alpha particle = 2 protons + 2 neutron)→ ^{243}Bk. (Note: Two protons as well as two neutrons are found in the nucleus of helium, and thus the two protons changed the atomic number of americium [$_{95}$Am] to berkelium [$_{97}$Bk].) Today a different process is used to produce berkelium in small amounts, as follows: ^{244}Cm+(5n = neutrons & λ = gamma rays) → (becomes) ^{249}Cm → ^{249}Bk + β- = (beta-minus decay).

History

Glenn T. Seaborg and his colleagues, S. G. Thompson and A. Ghiorso continued to use the cyclotron in their laboratory located at the University of California, Berkeley, to create new transuranic elements. Between 1949 and 1950, they produced their fourth artificially radioactive element of the actinide series by using the cyclotron to bombard americium-241 with alpha particles (helium nuclei), resulting in the transmutation of the Am-241 to Bk-243 by adding not only two neutrons but also two protons to americium.

Common Uses

Because such small amounts of berkelium have been produced, not many uses for it have been found. One use is as a source for producing the element californium by bombarding isotopes of berkelium with high-energy neutrons in nuclear reactors. Berkelium is also used in some laboratory research.

Examples of Compounds

Some berkelium compounds have been produced in fractions of a gram. For instance, a sample of berkelium chloride ($BkCl_3$) was artificially produced. It weighed only three bil-lionths of a gram (0.00000003 grams). Since that time, several other compounds have been synthesized. Although difficult to produce, they are *berkelium oxychloride* (ByOCl), *berkelium fluoride* (BkF_3), and *berkelium trioxide* (BkO_3).

Hazards

Like the radioactive isotopes of berkelium, its compounds are also extremely dangerous radioactive poisons. Because of the extremely small amounts of berkelium isotopes and com-pounds that exist and are produced, it is unlikely that many people will be exposed to them.

CALIFORNIUM

SYMBOL: Cf   **PERIOD:** 7   **SERIES NAME:** Actinides   **ATOMIC NO:** 98
ATOMIC MASS: 252 amu   **VALENCE:** 3 and 4   **OXIDATION STATE:** +3 and +4
    NATURAL STATE: Solid
ORIGIN OF NAME: Named for both the state of California and the University of California.
ISOTOPES: There are a total of 21 isotopes of californium. None are found in nature and all are artificially produced and radioactive. Their half-lives range from 45 nanoseconds for californium-246 to 898 years for californium-251, which is its most stable isotope and which decays into curium-247 either though spontaneous fission or by alpha decay.

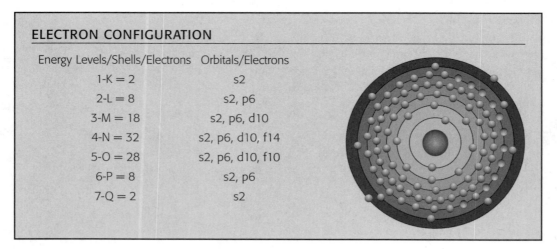

ELECTRON CONFIGURATION

Energy Levels/Shells/Electrons	Orbitals/Electrons
1-K = 2	s2
2-L = 8	s2, p6
3-M = 18	s2, p6, d10
4-N = 32	s2, p6, d10, f14
5-O = 28	s2, p6, d10, f10
6-P = 8	s2, p6
7-Q = 2	s2

Properties

Californium is a synthetic radioactive transuranic element of the actinide series. The pure metal form is not found in nature and has not been artificially produced in particle accelerators. However, a few compounds consisting of californium and nonmetals have been formed by nuclear reactions. The most important isotope of californium is Cf-252, which fissions spontaneously while emitting free neutrons. This makes it of some use as a portable neutron source since there are few elements that produce neutrons all by themselves. Most transuranic elements must be placed in a nuclear reactor, must go through a series of decay processes, or must be mixed with other elements in order to give off neutrons. Cf-252 has a half-life of 2.65 years, and just one microgram (0.000001 grams) of the element produces over 170 million neutrons per minute.

Californium's melting point is ⁻900°C, its boiling point is unknown, and its density is also unknown.

Characteristics

Californium is a transuranic element of the actinide series that is homologous with dysprosium ($_{66}$Dy), just above it in the rare-earth lanthanide series. Cf-245 was the first isotope of californium that was artificially produced. It has a half-life of just 44 minutes. Isotopes of californium are made by subjecting berkelium to high-energy neutrons within nuclear reactors, as follows: ^{249}Bk + (neutrons and λ gamma rays) $\rightarrow$ ^{250}Bk $\rightarrow$ ^{250}Cf + β- (beta particle emission)

Abundance and Source

Neither californium nor its compounds are found in nature. All of its isotopes are produced artificially in extremely small amounts, and all of them are extremely radioactive. All of its isotopes are produced by the transmutation from other elements such as berkelium and americium. Following is the nuclear reaction that transmutates californium-250 into californium-252: ^{250}Cf + (neutron and λ gamma rays) $\rightarrow$ ^{251}Cf + (neutron and λ gamma rays) $\rightarrow$ ^{252}Cf.

History

In 1950 the team of nuclear scientists at the University of California at Berkeley led by Stanley Thompson, which included Kenneth Street, Jr., Albert Ghiorso, and Glenn T. Seaborg, artificially produced californium, which was the sixth transuranic element formed in their large cyclotron, by bombarding curium-242 with alpha particles (He⁺⁺). They had a dilemma with following their former precedence of naming their newly discovered elements with a name related to the elements' analogues located in the rare-earth series above them in the lanthanide series. Therefore, since a term related to "dysprosium," meaning "hard to find" did not seem to be appropriate for the new element in the actinide series, they named it after their laboratory and the state in which it was discovered.

Common Uses

Californium's uses are limited, which is why the U.S. Nuclear Regulatory Commission, which controls the output and use of radioisotopes, has made californium-252 available for commercial use at the cost of only $10 per millionth of a gram. This small quantity is adequate for many sources of free neutrons to be used commercially. For example, free neutrons can be used in devices to measure moisture in products, including the Earth's crust, to find water or supplies of underground oil. Cf-252's ability to produce neutrons has also found uses in medicine. Cf-252's natural spontaneous fission makes it an ideal and accurate counter for electronic systems.

Examples of Compounds

Only a few compounds of californium have been prepared. All are extremely radioactive and have not found many common uses. Californium will combine with several nonmetals as follows: *californium oxide* (CfO_3), *californium trichloride* ($CfCl_3$), and *californium oxychloride* ($CfOCl$).

Hazards

Californium's greatest danger is as a biological bone-seeking radioactive element, which can be both a radiation hazard and a useful treatment for bone cancer. If mishandled, all of californium's isotopes and compounds can be a potential radiation poison.

EINSTEINIUM

SYMBOL: Es   **PERIOD:** 7   **SERIES NAME:** Actinide   **ATOMIC NO:** 99
ATOMIC MASS: 252 amu   **VALENCE:** 2 and 3   **OXIDATION STATE:** +2 and +3
   NATURAL STATE: Solid
ORIGIN OF NAME: Named after and in honor of the famous physical scientist Albert Einstein.
ISOTOPES: There are total of 20 isotopes of einsteinium. Einsteinium is not found in nature. All the isotopes are radioactive and are produced artificially. Their half-lives range from eight seconds to 472 days. None have exceptionally long half-lives.

ELECTRON CONFIGURATION

Energy Levels/Shells/Electrons	Orbitals/Electrons
1-K = 2	s2
2-L = 8	s2, p6
3-M = 18	s2, p6, d10
4-N = 32	s2, p6, d10, f14
5-O = 29	s2, p6, d10, f11
6-P = 8	s2, p6
7-Q = 2	s2

Properties

Einsteinium belongs to group 13 (IIIA) of the heavy transuranic subseries of elements found in the actinide series. It was discovered after World War II, sometime in 1952, as a trace element in the residue from the massive explosion of the hydrogen bomb on Eniwetok

Atoll in the Marshall Islands, located in the West Central Pacific Ocean. Although the atoll was obliterated, literally wiped off the face of the Earth, several heavy elements, both known and unknown at that time, were detected in the aftermath of the explosion by a team of scientists led by Albert Ghiorso of the Berkeley laboratory. Einsteinium was one of these trace elements that was detected. Its existence, as well as several other discovered elements, was not announced until 1955, due to secrecy related to this new type of thermonuclear bomb. The melting and boiling points as well as the density of einsteinium are not known because of the extremely small amounts that have been produced.

Characteristics

Einsteinium's most stable isotope, einsteinium-252, with a half-life of 472 days, decays into berkelium-248 through alpha decay, and then into californium-252 through beta capture. It can also change into fermium-252 through beta decay.

Einsteinium has homologous chemical and physical properties of the rare-earth holmium ($_{67}$Ho), located just above it in the lanthanide series in the periodic table.

Abundance and Source

Einsteinium does not exist in nature and is not found in the Earth's crust. It is produced in small amounts by artificial nuclear transmutations of other radioactive elements rather than by additional explosions of thermonuclear weapons. The formation of einsteinium from decay processes of other radioactive elements starts with plutonium and proceeds in five steps as follows:

1. ^{239}Pu $\rightarrow$ 2 neutrons + gamma rays $\rightarrow$ ^{241}Pu $\rightarrow$ ^{241}Am + β-.
2. ^{241}Am $\rightarrow$ 1 neutron + gamma $\rightarrow$ ^{242}Am $\rightarrow$ ^{242}Cm + β-.
3. ^{242}Cm $\rightarrow$ 7 neutrons + gamma $\rightarrow$ ^{249}Cm $\rightarrow$ ^{249}Bk + β-.
4. ^{249}Bk $\rightarrow$ 1 neutron + gamma $\rightarrow$ ^{250}Bk $\rightarrow$ ^{250}Cf + β-.
5. ^{250}Cf $\rightarrow$ 3 neutrons + gamma $\rightarrow$ ^{253}Cf $\rightarrow$ ^{253}Es + β-.

History

As mentioned, Einsteinium was first discovered in 1952 in residue material from the first thermonuclear (hydrogen) bomb. The team that discovered this element included the Americans Gregory Choppin, Stanley Thompson, and Albert Ghiorso, as well as British physicist Bernard Harvey. This was a **fusion** (combining) nuclear reaction of heavy hydrogen (deuterium), wherein its nuclei were driven together by the great force of an atomic bomb inside the hydrogen bomb. This process is the opposite of **fission** nuclear reaction (splitting of heavy nuclei) as in the so-called atomic bombs. The explosion of this fusion bomb was the first time some of the heavier elements beyond uranium, including einsteinium and fermium, were formed by the following nuclear reaction: ^{238}U $\rightarrow$ 15 neutrons + gamma rays $\rightarrow$ ^{253}U $\rightarrow$ 7β- $\rightarrow$ ^{253}Es. Einsteinium, of course, was named after Albert Einstein to honor him for developing the concept that energy and matter are essentially the same, represented by his famous formula $E = mc^2$, which, in theory, made nuclear energy possible.

Common Uses

Einsteinium does not really have any common uses except as related to research in nuclear and chemical laboratories.

Examples of Compounds

Einsteinium, as an actinide metal, has several compounds similar to other transuranic elements that are formed with some of the nonmetals, as follows: *einsteinium dioxide* (EsO_2), *einsteinium trioxide* (Es_2O_3), *einsteinium trichloride* ($EsCl_3$), *einsteinium dibromide* ($EsBr_2$), and *einsteinium triiodide* (EsI_3).

Hazards

The radioisotopes of einsteinium are highly unstable and radioactive. The small amount of the element and its compounds produced are not likely to be available in most laboratories. Thus, they do not pose any general hazard except in the case of scientists working with nuclear materials who must take precautions in handling exotic elements.

FERMIUM

SYMBOL: Fm   **PERIOD:** 7   **SERIES NAME:** Actinides   **ATOMIC NO:** 100
ATOMIC MASS: 257 amu   **VALENCE:** 3   **OXIDATION STATE:** +3   **NATURAL STATE:** Solid
ORIGIN OF NAME: Named after and to honor the scientist Enrico Fermi.
ISOTOPES: There are a total of 21 isotopes of fermium. Their half-lives range from fermium-258's 370 microseconds to fermium-257's 100.5 days, which is the longest of all its isotopes. None of fermium's isotopes exist in nature. All are artificially produced and are radioactive.

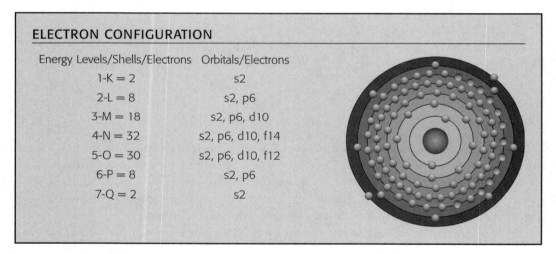

ELECTRON CONFIGURATION

Energy Levels/Shells/Electrons	Orbitals/Electrons
1-K = 2	s2
2-L = 8	s2, p6
3-M = 18	s2, p6, d10
4-N = 32	s2, p6, d10, f14
5-O = 30	s2, p6, d10, f12
6-P = 8	s2, p6
7-Q = 2	s2

Properties

Fermium is the eighth transuranic element in the actinide series of group 14 (IVA) of the periodic table. Similar to einsteinium, fermium was produced and discovered in the debris resulting from the explosion of the nuclear hydrogen bomb in 1952. Its existence was kept secret because of security measures that were established during World War II, thus keeping Albert Ghiorso and his colleagues at the University of California at Berkeley from receiving credit for the discovery until 1955. Fermium's melting point is thought to be about 1500+degrees Celsius, but its boiling point and density are unknown since so little of it is produced and because of the short half-lives of its isotopes.

Characteristics

The chemical characteristics of fermium are not very well known, but they are similar to its **homologue** erbium, the rare-earth element located just above it in the lanthanide series.

The nuclear reaction in the hydrogen bomb that produced fermium was the result of the acquisition of 17 neutrons by uranium from the explosion resulting in uranium-255 and some gamma radiation. U-255 decays by β-electron emission to form fermium-255, as depicted in the equation as follows: $_{92}$U-238 + 17 neutrons and gamma radiation $\rightarrow$ $_{92}$U-255 $\rightarrow$ $_{100}$Fm-255 + the emission of 8 electrons.

Abundance and Source

Fermium does not exist in nature. All of it is artificially produced in cyclotrons, isotope particle accelerators, or nuclear reactors by a very complicated decay process involving six steps of nuclear bombardment followed by the decay of beta particles, as follows:

1. Plutonium-239 (plus 2 neutrons) to $_{94}$Pu-241 $\rightarrow$ americium-241+ β-.
2. Americium-241 (plus 1 neutron) to $_{95}$Am-242 $\rightarrow$ to curium-242 + β-.
3. Curium-242 (plus 7 neutrons) to $_{96}$Cm-249 $\rightarrow$ to berkelium-249 + β-.
4. Berkelium-249 (plus 7 neutrons) to $_{97}$Bk-250 $\rightarrow$ to californium-250 + β-.
5. Califorium-250 (plus 3 neutrons) to $_{98}$Ca-253 $\rightarrow$ to einsteinium-253 + β-.
6. Einstenium-253 (plus 1 neutron) to $_{99}$Es-254 $\rightarrow$ to fermium-254 + β-.

History

As mentioned, fermium was first detected in the debris of the hydrogen bomb. The thermonuclear fusion (hydrogen) bomb of 1952 is somewhat opposite of the fission (atomic) bomb. Whereas fission splits the nuclei of one element into nuclei of smaller elements, fusion combines the nuclei of heavy hydrogen (deuterium) into nuclei of the element helium and other elements. Both reactions produce tremendous amounts of energy and radiation, as well as many different subatomic particles. Today, fermium is synthesized by intense nuclear bombardment in the High Flux Isotope Reactor at Oak Ridge National Laboratory located in Oak Ridge, Tennessee, where plutonium (or uranium) undergoes a series of decays to end up as an isotope of fermium. (See previous section titled "Abundance and Source.")

Common Uses

Because such small amounts of fermium are produced and because the half-lives of its isotopes are so short, there are no commercial uses for it except for basic scientific research.

Examples of Compounds

As with most other transuranic elements of the actinide series, fermium has an oxidation state of +3, as well as possibly a +2 oxidation state. Thus, this ion can combine with nonmetals, such as oxygen and the halogens, as do many of the other elements in this series. Two examples follow:

Fermium oxide: $2Fm^{3+} + 3O^{2-} \rightarrow Fm_2O_3.$
Fermium fluoride: $Fm^{3+} + 3F^{1-} \rightarrow FmF_3.$

Hazards

All the radioisotopes of fermium are dangerous radiation hazards. There is little chance of coming in contact with one of fermium's isotopes given that they all have very short half-lives and do not exist for long periods. In addition, very small amounts are produced and mainly available for research purposes.

MENDELEVIUM

SYMBOL: Md   **PERIOD:** 7   **SERIES NAME:** Actinium   **ATOMIC NO:** 101
ATOMIC MASS: 258 amu   **VALENCE:** 3   **OXIDATION STATE:** +2 and +3
   NATURAL STATE: Solid
ORIGIN OF NAME: Named after and to honor the Russian chemist Dmitri Mendeleev who developed the periodic table of the chemical elements.
ISOTOPES: There are a total of 19 isotopes of mendelevium. All of them are extremely radioactive and have half-lives ranging from 900 microseconds (for Md-245) to 51.5 days (for Md-258). They are produced in very small amounts.

ELECTRON CONFIGURATION

Energy Levels/Shells/Electrons	Orbitals/Electrons
1-K = 2	s2
2-L = 8	s2, p6
3-M = 18	s2, p6, d10
4-N = 32	s2, p6, d10, f14
5-O = 31	s2, p6, d10, f13
6-P = 8	s2, p6
7-Q = 2	s2

Properties

Mendelevium's chemical and physical properties are not well known because such small amounts with short half-lives have been produced. Many of its isotopes are produced just one atom at a time, making it difficult to weigh and measure samples. Its melting point is thought to be about 1,827°C, but its boiling point and density are unknown.

Characteristics

Mendelevium's most stable isotope is Md-258, with a half-life of 51.5 days. It decays into einsteinium-254 through alpha (helium nuclei) decay, or it may decay through the process of spontaneous fission to form other isotopes.

Abundance and Source

Only trace amounts of mendelevium have been artificially produced—much of it just one atom at a time—and thus to date, only several million atoms have been artificially made.

Therefore, there is not enough to measure by standard techniques. Other methods, such as ion-exchange **chromatography** and spectroscopy, are employed to study its chemical and physical properties.

History

Albert Ghiorso and his team of chemists that included Glenn T. Seaborg, Stanley G. Thompson, Bernard G. Harvey, and Gregory R. Choppin bombarded atoms of einsteinium-253 with helium ions in the cyclotron at the University of California at Berkeley. This resulted in a few atoms of mendelevium-256, which is one of the isotopes of mendelevium plus a free neutron.

Before their experiment that produced mendelevium, the team had speculated that this element number 101 must be somewhat similar to the element thulium ($_{69}$Tm) located just above it in the lanthanide series. Because they did not have a name for this new element, they referred to it as "eka-thulium," with an atomic number of 101. It was formally named mendelevium in 1955 only after they were able to produce a few atoms of einsteinium by the nuclear process as follows: $_{99}$Es-253 + $_2$H-4 → $_{101}$Md-256 + $_0$n-1 (a neutron with a mass of 1 and no charge). (Note: The isotope of einsteinium-258 instead of Es-253 can be used in a similar reaction to produce the more stable isotope mendelevium-258.)

Common Uses

Due to the small production and dearth of knowledge about mendelevium, there are no uses for it beyond basic laboratory research.

Examples of Compounds

Although little is known about mendelevium, it is possible to form compounds with some nonmetals such as oxygen and the halogen with its +3 ion, as follows:
Mendelevium oxide: $2Md^{3+} + 3O^{2-} \rightarrow Md_2O_3$.
Mendelevium chloride: $Md^{3+} + 3Cl^{1-} \rightarrow MdCl_3$.

Hazards

Similar to all artificially produced radioisotopes that go through natural decay process or spontaneous fission, mendelevium is an extreme radiation hazard. There is so little of it in existence and produced annually that there is no risk of individual or public radiation poisoning.

NOBELIUM

SYMBOL: No **PERIOD:** 7 **SERIES NAME:** Actinides **ATOMIC NO:** 102
ATOMIC MASS: 259 amu **VALENCE:** 3 **OXIDATION STATE:** +2 and +3 **NATURAL STATE:** Solid
ORIGIN OF NAME: Named after the scientist Alfred Nobel, who invented dynamite and used his fortune to fund and award the Nobel Prizes.
ISOTOPES: There a total of 15 isotopes of nobelium, ranging from 0.25 milliseconds (No-250) to 58 minutes (No-59). None are found in nature; all are unstable and are artificially produced in cyclotrons.

ELECTRON CONFIGURATION

Energy Levels/Shells/Electrons	Orbitals/Electrons
1-K = 2	s2
2-L = 8	s2, p6
3-M = 18	s2, p6, d10
4-N = 32	s2, p6, d10, f14
5-O = 32	s2, p6, d10, f14
6-P = 8	s2, p6
7-Q = 2	s2

Properties

Nobelium is the next to last transuranic element of the actinide series. The transuranic elements are those of the actinide series that are heavier than uranium. Nobelium is also the heaviest element of the vertical group 16 (VIA).

Because it is only produced in minute quantities and its isotopes have such sort half-lives, not much is known about its properties. It melting point is known and is about 827°C, but its boiling point and density are unknown.

Characteristics

Even though nobelium's chemical and physical properties are unknown, it is reasonable to assume that they resemble $_{70}$Yb, which is located just above it in the lanthanide series.

Abundance and Source

Nobelium does not exist in nature. All of its isotopes are radioactive with relatively short half-lives. Some are unstable and spontaneously fission, and all of them are artificial and man-made. Small quantities of nobelium are produced in cyclotrons by bombarding curium-246 with carbon-12 and neutrons to produce nobelium-254. The reaction follows: $_{96}$Cm-246 + $_{6}$C-12 → $_{102}$No-254. Carbon's six neutron and six protons are accelerated to high-speeds in a cyclotron as they hit the curium atoms with great energy that produces an additional four neutrons, thus producing the net gain in mass number by eight neutrons and an increase in atomic number by six protons, resulting in $_{102}$No-254.

History

Three groups had roles in the discovery of nobelium. First, scientists at the Nobel Institute of Physics in Stockholm, Sweden, used a cyclotron to bombard $_{96}$Cu-244 with heavy carbon $_{6}$C-13 (which is natural carbon-12 with one extra neutron). They reported that they produced an isotope of element 102 that had a half-life of 10 minutes. In 1958 the team at Lawrence Laboratory at Berkeley, which included Albert Ghiorso, Glenn Seaborg, John Walton, and Torbjorn Sikkeland, tried to duplicate this experiment and verify the results of the Nobel Institute but with no success. Instead, they used the Berkeley cyclotron to bombard cerium-

246 with carbon-12 to produce nobelium-254 with a half-life of three seconds. (See the nuclear process in the previous section titled "Abundance and Source.") Because the Russian group of scientists in Dubna, Russia, were able to reproduce the results of the Berkeley group, but not the results of the Noble group, the IUPAC, which first awarded credit for the discovery of nobelium to the group in Sweden, had to recall it and then award the credit for the discovery of nobelium to the Berkeley group.

Common Uses

There are no uses for nobelium except for laboratory research.

Examples of Compounds

Since nobelium has an oxidation state of +3, its ions are capable of forming compounds with a few nonmetals, as follows:

Mendelevium fluoride: $Md^{3+} + 3F^{1-} \rightarrow MdF_3$.
Mendelevium oxide: $2Md^{3+} + 3O^{2-} \rightarrow Md_2O_3$.

Hazards

Although nobelium poses a radiation hazard, the chances of being exposed to it are nil since there is little of it and its isotopes' half-lives are only a few seconds and minutes.

LAWRENCIUM

SYMBOL: Lr **PERIOD:** 7 **SERIES NAME:** Actinides **ATOMIC NO:** 103
ATOMIC MASS: 262 amu **VALENCE:** 3 **OXIDATION STATE:** +3 **NATURAL STATE:** Solid
ORIGIN OF NAME: Named for and in honor of Ernest O. Lawrence, who invented the cyclotron.
ISOTOPES: There are a total of 14 isotopes of lawrencium. Lawrencium-252 has the shortest half-life of just 0.36 of a second, and lawrencium-262 has the longest half-life of four hours. None are found in nature. All the isotopes of lawrencium are artificially manufactured in particle accelerators or nuclear reactors.

ELECTRON CONFIGURATION

Energy Levels/Shells/Electrons	Orbitals/Electrons
1-K = 2	s2
2-L = 8	s2, p6
3-M = 18	s2, p6, d10
4-N = 32	s2, p6, d10, f14
5-O = 32	s2, p6, d10, f14
6-P = 9	s2, p6, d1
7-Q = 2	s2

Properties

The nuclear chemists at the Lawrence Berkeley Laboratory worked with extremely small samples of lawrencium with short half-lives, which made it difficult to determine the new element's chemical and physical properties. Most of its isotopes spontaneously fission as they give off alpha particles (helium nuclei). Lawrencium's melting point is about 1,627°C, but its boiling point and density are unknown.

Characteristics

Lawrencium is the last of the transuranic elements and the 15th in the actinide series (there are 15 elements in the lanthanide series as well, assuming you start counting the series at the elements lanthanide and actinium, respectively.) It is assumed that lawrencium has some chemical and physical characteristics similar to lutetium, located just above it in the lanthanide series. It is also located at the bottom of the group 17 (VIIA) elements, which makes it the heaviest of the halides.

Abundance and Source

Because lawrencium does not exist in nature, it had to be produced artificially. This was done in 1961 by the team of scientists at Berkeley, using an ion accelerator to bombard three different isotopes of the element californium with heavy ions of the elements ^{10}boron and ^{11}boron along with some neutrons that produced the isotope $_{103}$Lr-258. The resulting product weighed only about two millionths of a gram and had a half-life of only 4.1 seconds, fissioning spontaneously.

History

Albert Ghiorso's team at the Lawrence Berkeley Laboratory included Torbjorn Sikkeland, Almon E. Larsh, and Robert M. Latimer. They artificially manufactured lawrencium in 1961. They placed an incredibly small amount of californium as the target in a new instrument called a "linear accelerator," which speeds up particles by electromagnetic propulsion in a long straight tube to hit a target element instead of having the particles following a circular path of the cyclotron. Californium was the target that was bombarded in a series of steps with both ^{10}boron and ^{11}boron to produce a microgram of the isotope of lawrencium-258. The reaction follows: $_{98}$Cf-249 + $_{98}$Cf-252 + ^{10}B and ^{11}B + several neutrons → $_{103}$La-257 and $_{103}$La-258 + alpha radiation.

Later in 1965, Russian scientists, using an ion accelerator, produced the isotope lawrencium-256 by bombarding americium-243 with oxygen-18 ions.

(Note: the original symbol for lawrencium was Lw, but it was changed by the IUPAC naming committee to Lr.)

Common Uses

There are no common uses. So little of it exists that it has no use outside of basic scientific research.

Examples of Compounds

A few atoms of lawrencium oxide (Lr_3O_2) have been produced for study, but it has no practical utility.

Hazards

Like the other short-lived radioactive isotopes, lawrencium is a radiation hazard. Also, as with the others, the danger to individuals and the public is small since there is not much of it in existence. Also, the small amount that has been produced has a short half-life, so over short periods of time it ceases to exist.

PERIODIC TABLE OF THE ELEMENTS

TRANSITION ELEMENTS

Groups / Periods	1 IA	2 IIA	3 IIIB	4 IVB	5 VB	6 VIB	7 VIIB	8 VIII	9 VIII	10 VIII	11 IB	12 IIB	13 IIIA	14 IVA	15 VA	16 VIA	17 VIIA	18 VIIIA
1	1 H 1.0079																	2 He 4.00260
2	3 Li 6.941	4 Be 9.01218											5 B 10.81	6 C 12.011	7 N 14.0067	8 O 15.9994	9 F 18.9984	10 Ne 20.179
3	11 Na 22.9898	12 Mg 24.305											13 Al 26.9815	14 Si 28.0855	15 P 30.9738	16 S 32.066(6)	17 Cl 35.453	18 Ar 39.948
4	19 K 39.0983	20 Ca 40.08	21 Sc 44.9559	22 Ti 47.88	23 V 50.9415	24 Cr 51.996	25 Mn 54.9380	26 Fe 55.847	27 Co 58.9332	28 Ni 58.69	29 Cu 63.546	30 Zn 65.39	31 Ga 69.72	32 Ge 72.59	33 As 74.9216	34 Se 78.96	35 Br 79.904	36 Kr 83.80
5	37 Rb 85.4678	38 Sr 87.62	39 Y 88.9059	40 Zr 91.224	41 Nb 92.9064	42 Mo 95.94	43 Tc (98)	44 Ru 101.07	45 Rh 102.906	46 Pd 106.42	47 Ag 107.868	48 Cd 112.41	49 In 114.82	50 Sn 118.71	51 Sb 121.75	52 Te 127.60	53 I 126.905	54 Xe 131.29
6	55 Cs 132.905	56 Ba 137.33	★	72 Hf 178.49	73 Ta 180.948	74 W 183.85	75 Re 186.207	76 Os 190.2	77 Ir 192.22	78 Pt 195.08	79 Au 196.967	80 Hg 200.59	81 Tl 204.383	82 Pb 207.2	83 Bi 208.980	84 Po (209)	85 At (210)	86 Rn (222)
7	87 Fr (223)	88 Ra 226.025	▲	104 Unq (261)	105 Unp (262)	106 Unh (263)	107 Uns (264)	108 Uno (265)	109 Une (266)	110 Uun (267)	111 Uuu (272)	112 Uub	113 Uut	114 Uuq	115 Uup	116 Uuh	117 Uus	118 Uuo

★ Lanthanide Series (RARE EARTH)

57 La 138.906	58 Ce 140.12	59 Pr 140.908	60 Nd 144.24	61 Pm (145)	62 Sm 150.36	63 Eu 151.96	64 Gd 157.25	65 Tb 158.925	66 Dy 162.50	67 Ho 164.930	68 Er 167.26	69 Tm 168.934	70 Yb 173.04	71 Lu 174.967

▲ Actinide Series (RARE EARTH)

89 Ac 227.028	90 Th 232.038	91 Pa 231.036	92 U 238.029	93 Np 237.048	94 Pu (244)	95 Am (243)	96 Cm (247)	97 Bk (247)	98 Cf (251)	99 Es (252)	100 Fm (257)	101 Md (258)	102 No (259)	103 Lr (260)

©1996 R.E. KREBS

Transactinide Series:
Period 7 (Continuation of Actinide Series)

Introduction

The **transactinide** series of elements (Z-104 to Z-113) are those elements that follow the actinide series (Z-89 to Z-103) and proceed to the superactinides, some of which are yet to be discovered. (Note: Z is the symbol used to represent the atomic numbers [protons] of elements in the transactinide series, as well as of other elements.) All elements of the transactinide series are radioactive, heavy metals that are unstable, and they usually decay by spontaneous fission or alpha decay into smaller nuclei of elements with less mass.

Dr. Glenn T. Seaborg proposed the term "actinide" for the new heavy elements that were predicted to follow the lanthanide series (Z-57 to Z-71). Dr. Seaborg believed that the actinides would be difficult to discover, and he proposed they would be trivalent homologues to the elements in the lanthanide series in which the 4f orbitals would be filled. His team at the Lawrence Berkeley National Laboratory (LBNL), located at the University of California's Berkeley campus, separated Z-95 (americium) and Z-96 (curium) as trivalent homologues of two of the elements in the lanthanide series located just above them in the periodic table.

The transactinide series is a logical extension of the actinide series and follows the principles of the standard periodic table. The "yet-to-be-found" heavy chemical elements are to be named using the IUPAC's temporary naming system designed to provide provisional or transitional names until the elements are discovered, confirmed, and officially assigned a permanent name, usually after the discoverer or his or her country. (See the subsequent tables for details regarding this transitional naming system. See Appendix A for names of the discoverers of the elements.) More recently, a subseries has been added to this provisional naming system. The transitional or temporary naming system can be used to name unknown radioactive heavy metal elements up to the 900s. Most elements beyond Z-200 will likely never be discovered, even one atom at a time, because to their extremely short half-lives and difficulty in producing amounts adequate for identifying their properties.

Table 4.1: Symbols used in naming elements not yet named or discovered in the actinide, transactinide, and superactinide series of heavy elements.

Arabic Numerals	Latin Roots	1-Letter Symbol
0	Nil	N
1	Un	U
2	Bi	B
3	Tri	T
4	Quad	Q
5	Pent	P
6	Hex	H
7	Sept	S
8	Oct	O
9	Enn	E

Table 4.2: Examples of provisional names for some of the heavy elements not yet named or discovered using the symbols from Table 4.1.

Z-number (atomic number)	Examples of Provisional Names	3-Letter Symbol
101	Unnilunium	Unu
111	Unununium	Uuu
121	Unbiunium	Ubu
231	Bitriunium	Btu
254	Bipentquadium	Bpq
105	Unnilpentium	Unp
320	Tribinilium	Tbn
171	Unseptunium	Usu
814	Octunquadium	Ouq
970	Ennseptnilium	Esn

Note: All provisional names end with the Latin suffix "ium."

Dr. Darleane C. Hoffman of the Nuclear Science Division of the Lawrence Berkeley National Laboratory and Department of Chemistry at the University of California at Berkeley has written and presented several papers documenting her work and that of her team on the laboratory production of transactinide and actinide elements "one-atom-at-a-time." She explains the difficulty of determining the chemistry of heavy elements: "How long does an atom need to 'exist' before it's possible to do any meaningful chemistry on it? Is it possible to learn anything at all about the reactions of an element for which no more than a few dozen atoms have ever existed simultaneously?"* (Note: Usually the isotopes of a heavy element need to have half-lives of at least one second in order to determine some of the element's chemical and physical properties.) Dr. Hoffman and her colleagues and other laboratories worldwide produce one atom at a time in accelerators by bombarding radioactive target elements with high-energy intense beams of neutrons and heavy ions (atoms that have either lost or gained a charge and particle from other elements). All the heavy elements of the actinide series and the transactinide series decay either by the emission of alpha or beta particles or through the process of spontaneous fission (SF), whereby they "split" into atoms of lighter elements. Most of these heavy elements, particularly as they increase in Z-numbers (the number of protons in their nuclei), have very short half-lives of a few microseconds to just a few minutes. Because of these conditions, limited chemical studies have been performed on these short-lived radioactive elements, but new nuclear chemistry techniques have been developed and used to investigate the properties of these fleeting artificial elements. At one time, scientists postulated that the chemical and physical properties of some of the yet-to-be-discovered heavy elements could be predicted by simple extrapolation from the properties of their lighter homologues located in the groups above them in the periodic table. But this idea proved impractical because many of the properties of their homologues were not similar enough for predicting what the heavier elements' chemistry would be like.

After several actinides and transactinides were discovered, a dilemma arose as to nomenclature for naming newly discovered elements. Generally, elements were named after deceased persons who had made some significant contribution related to the field of chemistry or physics, or the element was named after the country in which the element was discovered.

The discovery of element 106 occurred at about the same time in two different countries. In 1974 a Soviet team from the Joint Institute for Nuclear Research at Dubna, Russia, produced an isotope with a mass number of 259 and a half-life of seven microseconds. Also in 1974 the American Albert Ghiorso and his team at the Lawrence Berkeley National Laboratory created an isotope with a mass number of 263 and a half-life of 0.9 seconds. The American team's research was the first to be confirmed, and thus they suggested the name "seaborgium" to honor the nuclear chemist Glenn T. Seaborg. An international committee decided in 1992 that the laboratories at Berkeley and Dubna share the credit for the discovery of six isotopes of seaborgium ranging from $_{106}$Sg-260 to $_{106}$Sg-266. To confuse the issue, in 1994 the IUPAC decided to name element 106 "rutherfordium" because they had adopted a rule that no element could be named after a living person—which, at the time, was the case for Dr. Seaborg. Critics countered that element 99 (einsteinium) had been named after the famous scientist while he was alive, and thus the precedent had already been set. In 1997 a compromise was reached relative to naming elements 104 to 108. Element 104 became "rutherfordium," and element 106 was named "seaborgium," and both of these names are used on today's modern periodic tables.

---

*"Chemistry of the Heavy Elements: One Atom at a Time," *The Journal of Chemical Education,* March, 1999, Vol. 76, No. 3, p. 331

RUTHERFORDIUM (UNNILQUADIUM)

SYMBOL: Rf (Unq)   **PERIOD:** 7   **SERIES NAME:** Transactinide   **ATOMIC NO:** 104
ATOMIC MASS: ~253 to 263 amu   **VALENCE:** 4   **OXIDATION STATE:** +4 · **NATURAL**
  STATE: Solid (metal)
ORIGIN OF NAME: Named by the transition naming system of the IUPAC (unnilquadium)
  and later named after the New Zealand scientist Ernest Rutherford (rutherfordium).
ISOTOPES: There are a total of 15 isotopes for rutherfordium, ranging from Rf-253 to Rf-
  264. Their half-lives range from 23 microseconds to 10 minutes. They are all artificially
  made, radioactive, and unstable. Their decay modes are a combination of alpha decay
  and spontaneous fission (SF).

ELECTRON CONFIGURATION

Energy Levels/Shells/Electrons	Orbitals/Electrons
1-K = 2	s2
2-L = 8	s2, p6
3-M = 18	s2, p6, d10
4-N = 32	s2, p6, d10, f14
5-O = 32	s2, p6, d10, f14
6-P = 10	s2, p6, d2
7-Q = 2	s2

Properties and Characteristics

The chemical and physical properties of Unq (or rutherfordium) are homologous with the element hafnium ($_{72}$Hf), located just above it in group 4 (IVB) in the periodic table. It was first claimed to be produced artificially by the Joint Institute for Nuclear Research (JINR) located in Dubna, Russia. The Russian scientists used a cyclotron that smashed a target of plutonium-242 with very heavy ions of neon-22, resulting in the following reaction: $_{94}$Pu-242 + $_{10}$Ne-22 → $_{104}$Unq-260 + 4 $_0$n-1 (alpha radiation). The Russians named Unq-260 "kurcha-tovium" (Ku-260) for the head of their center, Ivan Kurchatov. (See details in the next section, "History.")

History

The Lawrence Berkeley Laboratory and other groups were unable to confirm the spontane-ous-fission reaction of Ku-260, so the Dubna group's discovery was disputed. The Berkeley equipment was unable to accelerate neon ions to the speeds required to produce Ku-260, and thus they tried a different reaction in a new "automated rapid chemistry apparatus" that identified and confirmed new isotopes of heavy metals. The procedure involved bombarding the element californium-239 with a mixture of the isotopes carbon-12 and carbon-13 ions, as follows:

$_{98}$Cf-249 + $_6$C-12—4 neutrons → $_{104}$Unq-257.

$_{98}$Cf-249 + $_6$C-13—3 neutrons → $_{104}$Unq-259.

In addition, some atoms of Unq-258 were produced by the same reaction, and Unq-261 was produced by bombarding isotopes of curium-248 with oxygen-18, as follows: $_{96}$Cm-248 + $_8$O-18 → $_{104}$Unq-261 + 5 alpha particles.

The International Union of Pure and Applied Chemistry (IUPAC), located in Oxford, England, devised the transitional naming system for new heavy metal elements. (See the IUPAC tables for an overview of the naming system.) Although the Russians claimed to have discovered the element they called $_{104}$Ku-260 (Unq-260), their research could not be replicated, so IUPAC's naming committee awarded credit to the Berkeley group, who had produced not only a wider range of isotopes, but also larger quantities of atoms for most of the isotopes than the Russian group had. The Berkeley group changed the Russian name "kurchatovium" and the name "unnilquadium" to "rutherfordium" after the scientist Ernest Rutherford, which is the name now accepted internationally.

Common Uses

There are no uses for unnilquadium (rutherfordium), except for high-energy nuclear-particle research.

Examples of Compounds

There are no compounds of unnilquadium (rutherfordium), but some might be possible using the element's ion's +4 oxidation state with some of the nonmetals.

Hazards

The hazards of unnilquadium are the same as for any radioactive element. Since there are only a few atoms with short half-lives produced at a time in scientific laboratories, there is little danger to the public.

DUBNIUM (UNNILPENTIUM)

SYMBOL: Db (Unp)   **PERIOD:** 7   **SERIES NAME:** Transactinide   **ATOMIC NO:** 105
ATOMIC MASS: ~262 amu   **VALENCE:** Unknown   **OXIDATION STATE:** Unknown   **NATURAL STATE:** Solid (metal)
ORIGIN OF NAME: Unnilquadium follows the transitional naming system of IUPAC but originally was named "hahnium" by the Berkeley group in honor of Otto Hahn, who discovered nuclear fission. The American Chemical Society endorsed the name "hahnium" for element 105, but as the Berkeley group continued its work and more isotopes of $_{105}$Unp were formed, the IUPAC changed the name "hahnium" to "dubnium" after the city Dubna, Russia, where the first isotopes of unnilpentium were formed.
ISOTOPES: There are a total of 15 isotopes of unnilpentium (dubnium). Their half-lives range from 0.76 seconds (for Db-257) to 16 hours (for Db-268). All of them decay by spontaneous fission and alpha decay.

ELECTRON CONFIGURATION

Energy Levels/Shells/Electrons	Orbitals/Electrons
1-K = 2	s2
2-L = 8	s2, p6
3-M = 18	s2, p6, d10
4-N = 32	s2, p6, d10, f14
5-O = 32	s2, p6, d10, f14
6-P = 11	s2, p6, d3
7-Q = 2	s2

Properties and Characteristics

Dubnium's (Unp) most stable isotope, Db-268, is unstable with a half-life of 16 hours. It can change into lawrencium-254 by alpha decay or into rutherfordium-268 by electron capture. Both of these reactions occur through a series of decay processes and spontaneous fission (SF). Since so few atoms of unnilpentium (dubnium) are produced, and they have such a short half-life, its melting point, boiling point, and density cannot be determined. In addition, its valence and oxidation state are also unknown.

History

In 1967 the scientists of the Joint Institute of Nuclear Research in Dubna, Russia, bombarded americium-243 with neon-22 to produce two isotopes of unnilpentium, then known as hahnium. The reaction is as follows: $_{95}$Am-243 + $_{10}$Ne-22 → $_{105}$Hahnium-260 and 261 (Unp-260 and 261). In 1970 Albert Ghiorso and his team at Berkeley bombarded californium-249 with heavy nitrogen (N-15) in their Heavy Ion Linear Accelerator (HILAC). The reaction is as follows: $_{98}$Cf-249 + $_{7}$N-15 → $_{105}$Ha-260 (Unp-260).

After evidence of the formation of unnilpentium in 1970 by the Berkeley team, the American Chemical Society (ACS) changed the name "Unp" to "hahnium." To settle the confusing claims, the IUPAC recommended that scientists from the Russian and Berkeley groups get together to review and confirm conflicting data. This cooperative venture was never implemented. In 1997 the IUPAC decided to change the name "hahnium" ($_{105}$Ha) to "dubnium" ($_{105}$Db) after the location of the Russian nuclear research lab that first produced the new element. Dubnium is the name by which it is known by today.

Common Uses

There are no uses for unnilpentium (dubnium) except for research purposes.

Examples of Compounds

No compounds are published for unnilpentium (dubnium), but it could possibly form compounds similar to its homologue: element 71, tantalum, in group 5.

Hazards

The hazards of dubnium are similar to all radioactive heavy metals of the transactinide series, but there is no threat to the public given that such a small amount of the element is produced and exists.

SEABORGIUM (UNNILHEXIUM)

SYMBOL: Sg (Unh)   **PERIOD:** 7   **SERIES NAME:** Transactinide   **ATOMIC NO:** 106
ATOMIC MASS: ~263 amu   **VALENCE:** Unknown   **OXIDATION STATE:** Unknown
    NATURAL STATE: Solid (metal)
ORIGIN OF NAME: Named after and in honor of the nuclear chemist Glenn T. Seaborg.
ISOTOPES: There a total of 16 isotopes of unnilhexium (seaborgium) with half-lives rang-
    ing from 2.9 milliseconds to 22 seconds. All are artificially produced and radioactive, and
    they decay by spontaneous fission (SF) or alpha decay.

ELECTRON CONFIGURATION

Energy Levels/Shells/Electrons	Orbitals/Electrons
1-K = 2	s2
2-L = 8	s2, p6
3-M = 18	s2, p6, d10
4-N = 32	s2, p6, d10, f14
5-O = 32	s2, p6, d10, f14
6-P = 12	s2, p6, d4
7-Q = 2	s2

Properties and Characteristics

Seaborgium ($_{106}$Sg-266) has several chemical and physical properties similar (homologous) to molybdenum ($_{42}$Mo-96) and tungsten ($_{74}$W-184), the elements located just above seaborgium in group 6 (VIB). Scientists do not know why Sg has properties and characteristics similar to the elements above it in group 6. This factor seems odd considering that the two elements in the transactinide series preceding seaborgium ($_{104}$Rf-261 and $_{105}$Db-262) do not have homologues above them in groups 4 ad 5. Seaborgium-266 is the most stable isotope of seaborgium. It has a half-life of 21 seconds and decays into rutherfordium-262 through spontaneour fission (SF) and alpha emission.

The melting point, boiling point, and density of seaborgium are unknown. What is known is that its isotopes are radioactive metals with short half-lives and that these isotopes decay by fission and alpha emissions.

History

Similar to several other elements in both the actinide and the transactinide series, there was some question as to which laboratory first discovered unnilhexium (seaborgium). In June

1975 the nuclear research group in Dubna, Russia, claimed to have discovered the now-named seaborgium by bombarding isotopes of lead (lead-207 and lead-208) with high-energy ions of chromium-54 to produce seaborgium-259. Three months later, in September 1974, the Lawrence laboratory at Berkeley produced a few atoms of element 106 unnilhexium (element 106 seaborgium). The reaction is as follows: $_{98}Cf$-249 + $_8O$-18 → $_{106}Unh$-263 + 4 $_0n$-1 (4 free neutrons). The Berkeley nuclear reaction that produced Sg-106 was confirmed in 1993, whereas the Dubna reaction was not, resulting in the Berkeley group receiving credit for the discovery of element 106 unnilhexium, which was finally named "seaborgium" in March 1994 by the American Chemical Society, with the name accepted by the IUPAC. (Note: See the entry for element 104, unnilquadium, for more on the seaborgium controversy.)

Common Uses

None except for research purposes.

Examples of Compounds

None now known. However, because seaborgium is a metal, it might be possible for it to form compounds with some nonmetals such as the halogens.

Hazards

The hazards for seaborgium are the same as for any radioactive isotopes, but since only a few short-lived atoms are produced, there is no danger to the public.

BOHRIUM (UNNILSEPTIUM)

SYMBOL: Bh (Uns)   PERIOD: 7   SERIES NAME: Transactinide   ATOMIC NO: 107
ATOMIC MASS: ~272 amu   VALENCE: None   OXIDATION STATE: Unknown
NATURAL STATE: Solid
ORIGIN OF NAME: Named after the scientist Niels Bohr.
ISOTOPES: There are a total of 10 isotopes of unnilseptium (bohrium). Not all their half-lives are known. However, the ones that are known range from 8.0 milliseconds to 9.8 seconds for Bh-272, which is the most stable isotope of bohrium and which decays into dubnium-268 through alpha decay. Only one isotope, Uns-261, has a decay mode that involves both alpha decay and spontaneous fission. All the others decay by alpha emission.

ELECTRON CONFIGURATION

Energy Levels/Shells/Electrons	Orbitals/Electrons
1-K = 2	s2
2-L = 8	s2, p6
3-M = 18	s2, p6, d10
4-N = 32	s2, p6, d10, f14
5-O = 32	s2, p6, d10, f14
6-P = 13	s2, p6, d5
7-Q = 2	s2

Properties and Characteristics

Unnilseptium, or bohrium, is artificially produced one atom at a time in particle accelerators. In 1976 Russian scientists at the nuclear research laboratories at Dubna synthesized element 107, which was named unnilseptium by IUPAC. Only a few atoms of element 107 were produced by what is called the "cold fusion" process wherein atoms of one element are slammed into atoms of a different element and their masses combine to form atoms of a new heavier element. Researchers did this by bombarding bismuth-204 with heavy ions of chromium-54 in a cyclotron. The reaction follows: Bi-209 + Cr-54 + neutrons = (fuse to form) Uns-262 + an alpha decay chain.

The melting point, boiling point, and density of element 107, as well as some other properties, are not known because of the small number of atoms produced.

History

The Russian's claim of discovery of element 107 in 1976 was disputed by some chemists. In 1981 Peter Armbruster and Gottfried Munzenberg, from their nuclear laboratory in Darmstadt, Germany, claimed they had created element 107 (unnilseptium) in their Separator for Heavy Ion Products (SHIP) as well as some atoms of elements 108 and 109, thus confirming the Russian's claim for previously producing some atoms of element 107. The German scientists proposed the name nielsbohrium for element 107. In 1992 the IUPAC, along with the American Chemical Society (ACS), changed the name for element 107 to bohrium.

Common Uses

None, except for research purposes.

Examples of Compounds

None are known, but since unnilseptium is a metal, it may be possible for it to chemically combine with some nonmetals.

Hazards

Even though most isotopes of unnilseptium are alpha emitters, there is little radiation hazard because only a few atoms have been produced.

HASSIUM (UNNILOCTIUM)

SYMBOL: Hs (Uno) **PERIOD:** 7 **SERIES NAME:** Transactinide **ATOMIC NO:** 108
  ATOMIC MASS: ~277 amu **VALENCE:** Unknown **OXIDATION STATE:** Unknown
  NATURAL STATE: Solid
ORIGIN OF NAME: Named for the word *Hassias,* which is the Latin name for the German state of Hesse.
ISOTOPES: There are a total of eight isotopes for unniloctium (hassium) with atomic masses ranging from 263 to 277 and half-lives ranging from 0.8 milliseconds (Uno-264) to 12 minutes (Uno-277).

ELECTRON CONFIGURATION

Energy Levels/Shells/Electrons	Orbitals/Electrons
1-K = 2	s2
2-L = 8	s2, p6
3-M = 18	s2, p6, d10
4-N = 32	s2, p6, d10, f14
5-O = 32	s2, p6, d10, f14
6-P = 14	s2, p6, d6
7-Q = 2	s2

Properties and Characteristics

Most of the chemical and physical properties of unniloctium (hassium) are unknown. What is known is that its most stable isotope (hassium-108) has the atomic weight (mass) of about 277. Hs-277 has a half-life of about 12 minutes, after which it decays into the isotope seaborgium-273 through either alpha decay or spontaneous fission. Hassium is the last element located at the bottom of group 8, and like element 107, it is produced by a "cold fusion" process that in hassium's case is accomplished by "slamming" iron (Fe-58) into particles of the isotope of lead (Pb-209), along with several neutrons, as follows:

$$_{82}Pb\text{-}209 + {}_{26}Fe\text{-}58 + neutrons = {}_{108}Hs\text{-}256 + \alpha \text{ decay products.}$$

History

The physicists Peter Armbruster and Gottfried Munzenberg, using a linear accelerator in their laboratory in Darmstadt, Germany, produced a few atoms of the element 108 (unniloctium) in 1984. Their team also discovered elements 107, 109, 110, and 111. They proposed the name "hassium" for element 108 based on the word *hassia,* which is the Latin word for the province of Hesse in Germany. In 1992 the IUPAC and the ACS both accepted the name hassium as proposed by the German scientists to replace IUPAC's temporary name, "unniloctium," even though not much was known about this new synthetic element.

Common Uses

Considering that such a small amount of hassium is produced, it has no commercial uses beyond basic research in nuclear laboratories.

Examples of Compounds

No common compounds of hassium have been reported.

Hazards

Because only a few atom of hassium are produced at a time, radiation hazard is not a problem.

MEITNERIUM (UNNILENNIUM)

SYMBOL: Mt (Une) **PERIOD:** 7 **SERIES NAME:** Transactinide **ATOMIC NO:** 109
ATOMIC MASS: ¯276 amu **VALENCE:** Unknown **OXIDATION STATE:** Unknown
  NATURAL STATE: Solid
ORIGIN OF NAME: Named for the Austrian scientist Lise Meitner.
ISOTOPES: There are five isotopes of unnilennium (meitnerium), ranging from Une-266 to
  Une-276, and with half-lives ranging from 1.7 milliseconds to 0.72 seconds.

ELECTRON CONFIGURATION

Energy Levels/Shells/Electrons	Orbitals/Electrons
1-K = 2	s2
2-L = 8	s2, p6
3-M = 18	s2, p6, d10
4-N = 32	s2, p6, d10, f14
5-O = 32	s2, p6, d10, f14
6-P = 15	s2, p6, d7
7-Q = 2	s2

Properties and Characteristics

Not many chemical and physical properties of Une (or Mt) are known, but it is artificially produced by the basic process of combining the isotopes of two elements to produce a few atoms of a heavier isotope in linear accelerators. In this case, the creation of a few atoms of element 109 involves a similar nuclear process of fusion as was used for element 108. The reaction follows:

$$_{83}Bi\text{-}209 + {_{26}}Fe\text{-}58 \rightarrow {_{109}}Une\text{-}266 + {_{0}}n\text{-}1 \text{ (free neutron).}$$

The most stable isotope of unnilennium is meitnerium-276, which has a half-life of about 0.72 seconds. Une-276 decays into element 107 (Uns-272 or bohrium-272).

The melting point, boiling point, and density of unnilennium (meitneriumare unknown, as are many of its other chemical and physical properties.

History

In 1982 Peter Armbruster and Gottfried Munzenberg, who also discovered elements 107 and 108, discovered element 109 by using "cold fusion" that involved bombarding iron-58 into bismuth-209, producing just three atoms of unnilennium (meitnerium). These three atoms of Une-266 were produced in the scientists' linear accelerator (SHIP) located at the Heavy Ion Research Laboratory in Darmstadt, Germany. These atoms existed for just 3.4 thousandths of a second. The German scientists suggested the name "meitnerium" in honor of the Austrian-born physicist Lise Meitner, who was one of the founders of nuclear fission. In 1992 both the IUPAC and ACS accepted the name "meitnerium" for element 109, although the IUPAC's transition name (unnilennium) still applies.

The two scientists then traced the very short decay sequence of the three Une-266 atoms as they decayed into element 107 (unnilseptium or bohrium) and element number 105 (unnilpentium or dubnium). The decay sequence is as follows:

$$_{109}\text{Une-266} \rightarrow {}_{107}\text{Uns-262} + {}_0\text{n-1 (high-energy helium nucleus)}.$$

Unnilseptium-262 further decays as follows:

$$_{109}\text{Uns-262} \rightarrow {}_{105}\text{Unp-268} + {}_0\text{n-1}.$$

Unnilpentium-268 further decays as follows:

$$_{105}\text{Unp-268} + {}_{-1}\text{e-0 (high-energy electron)} \rightarrow {}_{104}\text{Unq-259 (rutherfordium)}.$$

Common Uses

None, except for research purposes.

Examples of Compounds

There are too few atoms of meitnerium produced at any one time to form compounds.

Hazards

None, except for radiation, which is not much of a risk given that only a few atoms exist in nuclear laboratories.

DARMSTADTIUM (UNUNNILIUM)

SYMBOL: Ds (Uun) PERIOD: 7 SERIES NAME: Transactinide ATOMIC NO: 110
ATOMIC MASS: ~281 amu VALENCE: Unknown OXIDATION STATE: Unknown NATU-
    RAL STATE: Solid (metal)
ORIGIN OF NAME: Named for the German city of Darmstadt.
ISOTOPES: There are a total of nine isotopes of Uun (Ds), ranging from Uun-267 to Uun-
    281, with half-lives ranging from 0.17 milliseconds to 1.1 minute.

ELECTRON CONFIGURATION

Energy Levels/Shells/Electrons	Orbitals/Electrons
1-K = 2	s2
2-L = 8	s2, p6
3-M = 18	s2, p6, d10
4-N = 32	s2, p6, d10, f14
5-O = 32	s2, p6, d10, f14
6- P = 16	s2, p6, d8
7-Q = 2	s2

Properties and Characteristics

The production and confirmation of elements 110 and higher required the development of new equipment. In 1994 Peter Armbruster's team used the Heavy Ion Research Laboratory's linear accelerator (SHIP) located in Darmstadt, Germany, to create a few atoms of element 110 that had an atomic mass of 267. The researchers bombarded a thin sheet of lead with high-energy ions of nickel. They "fired" over one billion (1×10^{18}) nickel ions at the lead target for seven days, resulting in the fusion of just one or two atoms of ununnilium 110. Uun's isotope, now known as darmstadtium-269, has a half-life of about 0.00017 seconds as it spontaneously decays into hassium-277 and four alpha particles in the process. Both the single atoms produced and their short half-lives have made it difficult to identify the element Uun and impossible to perform any chemistry or confirm the predicted chemical and physical properties and characteristics of element 110.

History

Germany, Russia, and the United States were all involved in the synthetic production of element 110 and those elements of higher atomic numbers. This search has been an ongoing international effort by Peter Armbruster's team that in 1994 claimed to have discovered element 110 in their laboratory. (See "Properties and Characteristics" section for more on this discovery.)

Several years before Armbruster's discovery, in 1991, Albert Ghiorso and others of the Berkeley group produced atoms of element 110. They did this by identifying the atoms of Uun by the products of the decay chain, using their new gas-filled Small Angle Separator System (SASSY-2). The reaction follows:

Bi-209 + Co-59 + neutron = $_{110}$Uun-267 + alpha particle.

In 1994 and 1995 Dr. Darleane Hoffman of LLNL in California and others from Germany used the Separator for Heavy Ion Reaction Products (SHIP) at the GSI laboratory in Darmstadt, Germany, to produce two new isotopes of element 110.

Pb-208 + Ni-62 + neutrons = $_{110}$Uun-269 + alpha particles, and
Pu-208 + Ni-64 = neutrons = $_{110}$Uun-271 + alpha particles.

Only a few atoms were produced by these processes.

A combined team of scientists from the Lawrence-Livermore National Laboratory (LLNL) in California, and from the laboratory in Dubna, Russia, reported the following "hot" fusion reaction:

Pu-244 + S-34 + neutrons = $_{110}$Uun-273 + alpha particles.

The Heavy Ion Reaction Separator (SHIP) located in the GSI laboratory in Germany was used to identify elements 107 (bohrium) through element 109 (meitnerium) during the years 1981 through 1984, and it was used again later, between 1994 and 1996, to verify elements 110 (Uun) through element 112 (Uub).

Common Uses

None are known except as an interest in nuclear laboratories.

Examples of Compounds

None have been reported.

Hazards

None, beyond minor risks from radiation because only a few atoms of darmstadtium are produced at a time, and all of its isotopes exist for only a small fraction of second.

RÖENTGENIUM (UNUNUNIUM)

SYMBOL: Rg and Uuu **PERIOD:** 7 **SERIES NAME:** Transactinide **ATOMIC NO:** 111
ATOMIC MASS: ~280 amu **VALENCE:** Unknown **OXIDATION STATE:** Unknown
ORIGIN OF NAME: Named in honor of the scientist Wilhelm Konrad Röentgen.
ISOTOPES: There are three isotopes of Uuu (Röentgenium). They are Rg-272 (half-life of 1.5 milliseconds), Rg-279 (half-life of 170 milliseconds), and Rg 280 (half-life of 3.2 seconds), the latter of which is röentgenium's most stable isotope. All of its isotopes are synthetic and unstable.

ELECTRON CONFIGURATION

Energy Levels/Shells/Electrons	Orbitals/Electrons
1-K = 2	s2
2-L = 8	s2, p6
3-M = 18	s2, p6, d10
4-N = 32	s2, p6, d10, f14
5-O = 32	s2, p6, d10, f14
6-P = 17	s2, p6, d9
7-Q = 2	s2

Properties and Characteristics

Not much is known about element 111, but it is assumed that it is a solid metal that has some of the properties of its homologues, gold and silver, located just above it in group 1 (IB) in the periodic table. The reaction that produces unununium is as follows:

Bi-209 + Ni-64 + neutrons = Uuu-272 + alpha decay particles.

Only a few atoms of Rg-272 have been produced. Its most stable isotope is Rg-280, with a half-life of 3.6 seconds.

History

Shortly after element 110 was discovered in 1994, the team led by Peter Armbruster and Sigurd Hofmann at the Institute for Heavy Ion Research at Darmstadt, Germany, discovered element 111 using their SHIP detection apparatus. One might speculate that the ancient alchemists would be impressed by the production of element 111 by the transmutation of two different elements. Very high-energy nickel ions were used to bombard bismuth atoms. This caused the nuclei of the two atoms to fuse together by the process off "cold fusion," producing unununium. Gold, located in period 6, group 11, just above unununium, is its homologue (element with some similar characteristics). The alchemist's ancient dream would be to produce gold by a similar process. Unfortunately, only three atoms of $_{111}$Rg were produced after an 18-day experiment, and their half-lives was extremely short.

Common Uses

None, except for research purposes.

Examples of Compounds

None are known at this time.

Hazards

Unununium (röentgenium) is not a public radiation hazard given that only a few atoms have been produced synthetically in nuclear laboratories and they do not exist over long periods of time.

UNUNBIUM ($_{112}$UUB-285)

SYMBOL: Uub   **PERIOD:** 7   **SERIES NAME:** Transactinide   **ATOMIC NO:** 112
ATOMIC MASS: ˜285 amu   **VALENCE:** Unknown   **OXIDATION STATE:** Unknown   **NATU-RAL STATE:** Assumed to be liquid metal
ORIGIN OF NAME: IUPAC transition name, "ununbium" (which literally means *112*), is used to name new elements until IUPAC's naming committee decides upon a permanent name.
ISOTOPES: There are four isotopes of ununbium ranging from Uub-277 to Uub-285. They have half-lives ranging from 0.24 milliseconds to 10 minutes for Uub-285. All are artificially produced, are radioactive, and are unstable.

ELECTRON CONFIGURATION

Energy Levels/Shells/Electrons	Orbitals/Electrons
1-K = 2	s2
2-L = 8	s2, p6
3-M = 18	s2, p6, d10
4-N = 32	s2, p6, d10, f14
5-O = 32	s2, p6, d10, f14
6-P = 18	s2, p6, d10
7-Q = 2	s2

Properties and Characteristics

Not much is known about the chemical and physical properties of ununbium ($_{112}$Uub-285) other than the properties that are assumed to be similar to its homologues. It is also known as "eka-mercury" since mercury is its major homologue located at the bottom of the column of group 12 (IIB). As a liquid metal, it is assumed to be more volatile than mercury. Uub is expected to also have some properties similar to zinc and cadmium, the other homologues above mercury in group 12. Ununbium's most stable isotope is Uub-285, which has a half-life of about 10 or 11 minutes, after which it decays into element 110, darmstadtium-281 (unnilpentium), through alpha decay.

History

The same international group of scientists that discovered elements 107, 108, 109, and 110—led by Peter Armbruster of the nuclear institute at Darmstadt, Germany—discovered element 112. They discovered ununbium on February 9, 1996, by accelerating to high-energy zinc atoms that fused with lead atoms. This was accomplished in their high-speed ion linear accelerator. The number of combined protons in the nuclei of the two atoms ($_{30}$Zn-70 and $_{82}$Pb-208) equals the 112 protons in the synthetic atoms of ununbium ($_{112}$Uub-277). This process produced atoms of the isotope $_{112}$ununbium-277, which in just 0.00024 of a second, decays into the element darmstadium-281 ($_{105}$Unp-281).

Common Uses

Because only a few atoms of ununbium have ever been synthesized, it currently has no uses besides basic research in nuclear laboratories.

Examples of Compounds

The formation of compounds involving ununbium has not been reported.

Hazards

None to the general public.

UNUNTRIUM ($_{113}$UUT-284)

SYMBOL: Uut   **PERIOD:** 7   **SERIES NAME:** Transactinide   **ATOMIC NO:** 113
ATOMIC MASS: ~284 amu   VALELNCE: Unknown   **OXIDATION STATE:** Unknown   **NATU-RAL STATE:** ~Solid
ORIGIN OF NAME: The IUPAC assigned the temporary name "ununtrium," which literally means *113*.
ISOTOPES: There are five isotopes of ununtrium, ranging from Uut-283 to Uut-287. Its most stable isotope is Uut-284, with a half-life of 0.48 seconds, which is used as its proposed standard mass number.

ELECTRON CONFIGURATION

Energy Levels/Shells/Electrons	Orbitals/Electrons
1-K = 2	s2
2-L = 8	s2, p6
3-M = 18	s2, p6, d10
4-N = 32	s2, p6, d10, f14
5-O = 32	s2, p6, d10, f14
6-P = 18	s2, p6, d10
7-Q = 3	s2, p1

Properties and Characteristics

Ununtrium is located on the periodic chart in group 13 (IIIA) just below thallium and indium. It is expected to have chemical and physical properties similar to these two homologues. Since only one or two unstable atoms of the isotopes of ununtrium have been synthesized, its melting point, boiling point, and density are not known.

History

In 2003 the Nuclear Research Laboratory in Dubna, Russia, and the Lawrence Livermore National Laboratory in California, collaborated in conducting a 27-day experiment that led to the discovery of ununtrium. They bombarded atoms of americium-243 with ions of calcium-48. This produced, among other particles, four atoms of ununpentium (element 115), which in less than 1/10 of a second decayed by alpha emission into atoms of ununtrium (element 113). Since no formal name has yet been proposed for element 113, IUPAC's temporary naming system was used to name element 113 "ununtrium" (*113*).

Following is one example of the nuclear reaction wherein an isotope of element 115 (ununpentium) decays within a few milliseconds into an isotope of element 113 (ununtrium-283), which in turn decays into element 111 (unununium):

$$_{115}\text{Uup-283} \rightarrow {}_{113}\text{Uut-283} + {}_{2}\text{He-4} \rightarrow {}_{111}\text{Uuu-279} + {}_{2}\text{He-4}.$$

(Note: This first reaction occurs in just 46.6 milliseconds, and the second reaction occurs in 147 milliseconds. Similar nuclear decay reactions of element 115 result in several other isotopes of $_{111}$Uut-284 with various fission decay rates into element 111.)

Common Uses

Since only a few atoms of ununtrium have been produced, there are currently no uses of element 113 outside of basic nuclear research.

Examples of Compounds

None have been reported.

Hazards

No hazards exist for the general public because only a few radioactive unstable atoms of ununtrium have been produced in nuclear laboratories.

The Superactinides (Super Heavy Elements) and Possible Future Elements

Introduction

There is some question regarding where the transactinide series ends and where the superactinide series and super heavy elements (also known as SHE) begin. Some references start the superactinides and SHEs at Z-114, Z-123, or Z-126, and others start at Z-141 and continue to Z-153, or even higher to Z-202. These series of elements are characterized by single electrons successively added to inner shells and orbital of the atom until they are filled. This is somewhat similar to the way electrons are added to the atoms of the lanthanide and actinide series.

In this book the superactinide elements begin at Z-114 because this is the first element that was recognized in what is known as the "**island of stability,**" also referred to as the "Island of Nuclear Stability." The stability of Z-114 is related to its exceptional long half-life of 30 seconds, which provides adequate time for detection and research on it. It also appears that the heavier the element, the shorter its half-life.

Scientists discovered some time ago that certain configurations of the nuclei of different atoms are more stable than the nuclei of other elements. Dr. Glenn T. Seaborg was the first to propose the concept of the "island of stability" wherein the nuclei of atoms are constructed as "shells," somewhat similar to the electron shells (energy levels) that surround the nuclei of atoms. The difference is that the shells in nuclei are very close together whereas electron shells are relatively distant from the nucleus. The energy level between shells of the nucleus maintains the shells' separation, just as the energy levels do that keep the electron shells separated, but the shells of the nuclei are much closer together. The importance of this hypothesis is that when a particular nucleus's shells (energy levels) are completely filled with protons and neutrons (nucleons), its shell configuration will have a longer half-life than will other configurations that leave shells of the nucleus incomplete. Thus, elements with stable nuclei seem to follow a pattern sometimes referred to as a "magic number." These magic numbers relate to the total of both protons and neutrons that, when combined in the nuclei, are referred to as nucleons. The elements that have these magic numbers include some of the superactinides and SHE elements, such as 114, 120, and 126, or even higher. Thus, element 114 was considered one of these in the "island of stability" for the superactinides, and the super heavy elements usually include elements Z-122 to Z-153. There is some speculation that some undiscovered isotopes of elements in the island of stability have half-lives of a billion or more years and that small traces of them might still be found in the crust and ancient rocks of the Earth since the Earth is only about 4.6 billion years old. There is also speculation that stable SHE elements of this island could be used to produce materials with exotic properties that have practical technological and industrial uses.

UNUNQUADIUM ($_{114}$UUQ)

SYMBOL: Uuq   **PERIOD:** 7   **SERIES NAME:** Superactinide   **ATOMIC NO:** 114

ATOMIC MASS: ~289 amu   **VALENCE:** Unknown   **OXIDATION STATE:** Unknown   **NATU-RAL STATE:** Assumed to be solid

ORIGIN OF NAME: IUPAC assigned the temporary name "ununquadium," literally meaning *114.*

ISOTOPES: There are five currently known isotopes of ununquadium, ranging from Uuq-285 to Uuq-289, with half-lives ranging from 0.85 milliseconds to 30 seconds. The most stable isotope is Uuq-289, which has a 30-second half-life.

ELECTRON CONFIGURATION

Energy Levels/Shells/Electrons	Orbitals/Electrons
1-K = 2	s2
2-L = 8	s2, p6
3-M = 18	s2, p6, d10
4-N = 32	s2, p6, d10, f14
5-O = 32	s2, p6, d10, f14
6-P = 18	s2, p6, d10
7-Q = 4	s2, p2

Properties and Characteristics

There is some question as to whether isotope 289 of $_{114}$ununquadium has been synthetically produced or identified, so it is sometimes called eka-lead because it is located at the bottom of the column of elements in group 14 (IVA). Lead and tin are located just above it in this column and are therefore Uuq's homologues; thus, Uuq has some properties similar to Pb.

Element 114 is the first element in the group referred to as "the island of stability," which consists of element 114 to the optimum element 184. The elements in this so-called island of stability all have a rather stable arrangement of neutrons and protons in their nuclei and a similar configuration of the electrons in their orbits (shells) that result in longer half-lives than some of the SHE elements not located in the "island."

As with Uuq's valence and oxidation state, its melting point, boiling point, and density are not known

History

In December 1998 the Lawrence Berkeley Laboratory sent a supply of plutonium-244 and calcium-48 to Russian scientists at the Joint Institute of Nuclear Research in Dubna, Russia. The Russians bombarded the plutonium with ions of calcium, which, after some time, produced a single atom of ununquadium-289. Uuq has a half-life of just 30 seconds, after which it decays successively into element 112 (ununbium), element 110, (darmstadtium), and element 108 (hassium). The Russian nuclear laboratory later synthesized several atoms of other isotopes of Ununquadium.

Common Uses

Not enough atoms of ununquadium have been produced to find any practical uses except in nuclear research laboratories.

Examples of Compounds

None are known.

Hazards

Atoms of ununquadium are produced in such small numbers that they are not a hazard to the general public.

UNUNPENTIUM ($_{115}$UUP)

SYMBOL: Uup **PERIOD:** 7 **SERIES NAME:** Superactinide **ATOMIC NO:** 115
ATOMIC MASS: ~288 amu **VALENCE:** Unknown **OXIDATION STATE:** Unknown **NATURAL STATE:** Assumed to be a solid metal
ORIGIN OF NAME: Ununpentium follows IUPAC's temporary naming system for element 115.
ISOTOPES: There are 5 known isotopes of the element ununpentium, ranging from Uup-287 to Uup-291. The first two isotopes synthetically produced and confirmed were Uup-287 and Uup-288. Ununpentium's most stable isotope is Uup-288, which has a half-life of 87 milliseconds. It decays by alpha emission into ununtrium (element 113).

ELECTRON CONFIGURATION

Energy Levels/Shells/Electrons	Orbitals/Electrons
1-K = 2	s2
2-L = 8	s2, p6
3-M = 18	s2, p6, d10
4-N = 32	s2, p6, d10, f14
5-O = 32	s2, p6, d10, f14
6-P = 18	s2, p6, d10
7-Q = 5	s2, p3

Properties and Characteristics

Ununpentium is also known as eka-bismuth because it is homologous to the element bismuth located at the bottom of Group 15 (VA). Its melting point, boiling point, and density are unknown as are many of its other properties. Several isotopes of element 115 were produced by the nuclear reaction that bombarded calcium into a target americium, resulting in the fusion of the calcium nuclei with the americium nuclei to form isotopes of element 115 (ununpentium).

Because only a few atoms of ununpentium have been produced and they exist for only a fraction of second, this metallic element's melting point, boiling point, density, and several other properties have not been determined.

History

A joint Russian and American team of physicists created two new super heavy elements—115 and 113—that provide more support for the "island of stability" concept. The experiments were conducted between July 14 and August 10 in 2003, but the results of the experiments were not published until February 2004. The experiments were conducted in the Joint Institute of Nuclear Research (JINR) in Dubna, Russia, and also involved the scientists of the Lawrence Berkely National Laboratory located at Berkeley, California. Only four atoms of two isotopes of ununpentium (element 115) were produced, and the results were published in a peer-reviewed scientific journal. Following is the nuclear reaction that produced these four atoms:

$$_{20}Ca\text{-}48 + \ _{95}Am\text{-}243 \rightarrow \ _{115}Uup\text{-}287 + 4 \text{ neutrons, and}$$
$$_{20}Ca\text{-}48 + \ _{95}Am\text{-}243 \rightarrow \ _{115}Uup\text{-}288 + 3 \text{ neutrons.}$$

It should be noted that these two isotopes of ununpentium (element 115), very soon after creation, decayed into isotopes of ununtrium (element 113), unununium (element 111), and so on as low as unnilpentium (element 105). All of these reactions were recorded in just fractions of a second.

It seems that element 115 (ununpentium) caught the interest of the UFO conspiracy-theory culture of pseudoscience. One advocate stated that the "sports model" of a flying disk used ununpentium-115 for fuel that "stepped up" to ununhexium. Further, Uuh-116's decay products included large quantities of "antimatter" that was milled to form discs of antimatter that could then be used to produce antigravity energy as a force to propel UFOs. One pseudoscientist suggested that the U.S. government had 500 pounds of element 115, even though only a few atoms have ever been produced.

Common Uses

None.

Examples of Compounds

None.

Hazards

Although the isotopes of ununpentium (element 115) are radioactive, there is not much risk of radiation poisoning since there are so few atoms produced.

UNUNHEXIUM ($_{116}$UUH)

SYMBOL: Uuh **PERIOD:** 7 **SERIES NAME:** Superactinide **ATOMIC NO:** 116 ATOMIC
MASS: ~292 amu **VALENCE:** Unknown **OXIDATION STATE:** Unknown **NATURAL
STATE:** Presumed to be a colorless gas
ORIGIN OF NAME: Its name, "ununhexium" (*116*), follows IUPAC's protocol.
ISOTOPES: The number of isotopes for Uuh is unknown, but the most stable one that
is known is ununhexium-292, with a half-life of about 0.6 milliseconds. It decays into
ununquadium-288 by alpha decay.

ELECTRON CONFIGURATION

Energy Levels/Shells/Electrons	Orbitals/Electrons
1-K = 2	s2
2-L = 8	s2, p6
3-M = 18	s2, p6, d10
4-N = 32	s2, p6, d10, f14
5-O = 32	s2, p6, d10, f14
6-P = 18	s2, p6, d10
7-Q = 6	s2, p4

Properties and Characteristics

Element 116 is the result of the decay process of element 118. (See entry for element 118 for details.)

Element 116 was also directly produced by bombarding atoms of curium-248 with ions of high-energy calcium-48 ions. At the bottom of group 6 (VIA) on the periodic table, Uuh is presumed to have some of the properties and characteristics of its homologues polonium and tellurium, located just above it in this group.

Because of the few atoms of element 116 produced and their extremely short half-lives, the melting point, boiling point, specific gravity, and most of their other properties are not known.

History

The nuclear reaction involving the bombardment of curium with calcium that directly produced element 116 occurred on December 6, 2000, at the Joint Institute for Nuclear Research in Dubna, Russia, in cooperation with personnel of the Lawrence-Livermore Berkeley Group. This nuclear reaction resulted in the production of a few atoms of the isotope ununhexium-292, which has a half-life of 0.6 milliseconds and emits four neutrons. Uuh-292 is also the most stable isotope of element 116 as it continues to decay into elements with Z numbers of 114, 112, 110, 108, and 106, plus emitting four alpha particles for each transmutation. (Z numbers are the number of protons in the nuclei of atoms.)

After the announcement and publication of the discovery of element 116 (and 118), controversy arose as to the actual production of these elements as reported. On July 27, 2001, not long after the announcement of their discoveries, the publication *Physical Review Letters* received correspondence that announced the retraction of the discovery of these elements. This retraction was the result of the Berkeley team's inability to reproduce the results. Since element 116 is part of the decay chain of element 118, its existence is also problematic. See the entry for element 118 for more detail on this issue.

Common Uses

Since isotopes of element 116 may not have been produced, there are no uses for it.

Examples of Compounds

None.

Hazards

There is no hazard to the general public because of the scarcity or nonexistence of ununhexium's isotopes, which are used only in nuclear laboratories.

UNUNSEPTIUM ($_{117}$UUS)

SYMBOL: Uus   **PERIOD:** 7   **SERIES NAME:** Superactinide   **ATOMIC NO:** 117
ATOMIC MASS: Unknown   **VALENCE:** Unknown   **OXIDATION STATE:** Unknown
    NATURAL STATE: Expected to be a solid silvery metal
ORIGIN OF NAME: Under the IUPAC's temporary system, it is assigned the name "ununseptium" (117).
ISOTOPES: Not yet produced.

ELECTRON CONFIGURATION

Energy Levels/Shells/Electrons	Orbitals/Electrons
1-K = 2	s2
2-L = 8	s2, p6
3-M = 18	s2, p6, d10
4-N = 32	s2, p6, d10, f14
5-O = 32	s2, p6, d10, f14
6-P = 18	s2, p6, d10
7-Q = 7	s2, p5

Properties and Characteristics

Because element 117 (ununseptium) has not yet been produced, its properties are not known. This does not hinder speculation as to what some of the properties and characteristics of $_{117}$Uus will be when it is discovered and where it will fit into the scheme of what is known of elements 116 and 118—if they are confirmed. One thing that can be assumed with a high degree of probability is that Uus does not have a "magic number" of protons and neutrons, and thus is not included as one of the elements in the "island of stability."

History

There is little history to report, except so far, its artificial production has not been reported. It is expected to be an extension of the halogens and will be located at the bottom of group 17(VIIA).

Common Uses

None, except as a "place filler" between Z-116 and Z-118 in the periodic table.

Examples of Compounds

None.

Hazards

None, because it does not seem to exist.

UNUNOCTIUM ($_{118}$UUO)

SYMBOL: Uuo   PERIOD: 7   SERIES NAME: Superactinide   ATOMIC NO: 118
ATOMIC MASS: Unknown   VALENCE: Unknown   OXIDATION STATE: Unknown   NATU-
   RAL STATE: Expected to be a gas
ORIGIN OF NAME: Ununoctium (118) follows the temporary naming system of IUPAC for
   elements that have not yet received proper names.
ISOTOPES: No isotopes of ununoctium have yet been produced.

ELECTRON CONFIGURATION

Energy Levels/Shells/Electrons	Orbitals/Electrons
1-K = 2	s2
2-L = 8	s2, p6
3-M = 18	s2, p6, d10
4-N = 32	s2, p6, d10, f14
5-O = 32	s2, p6, d10, f14
6-P = 18	s2, p6, d10
7-Q = 8	s2, p6

Properties and Characteristics

Not only is ununoctium expected to be a gas, but it should also be a nonmetal when discovered. It is located at the bottom of group 18 (VIIIA) in the periodic table and could be expected to have some of the characteristics of it neighbors above it in this group. When first and erroneously reported as being discovered, it was said to have 118 protons and 175 neutrons in its nucleus for an atomic mass number (amu) of ˜293, which would make it the heaviest of the yet-to-be discovered elements.

History

Element 118 has an interesting and fascinating history. In June 1999 the Lawrence Berkeley National Laboratory (LBNL) announced that its team working with the 88-inch cyclotron and related detection equipment had produced a few atoms of the heavy element Z-118. The team claimed it contained 118 protons and 175 neutrons in its nucleus for a total of 293 atomic mass units (amu) plus one neutron. One of its several decay products was element Z-116. Later LBNL's director announced that an internal investigation had resulted in the discipline for scientific misconduct of a member of the team working on element 118. Both internal and external investigating committees' examinations found that Victor Ninov, a member of the experimental team and major author of the first publication related to the discovery of element 118, "cooked" the data of the experiment. Nuclear research labs in Japan, German, and France could not replicate the experiment in order to confirm the artificial production of element 118. Even other scientists at Berkeley could not verify the results of Ninov's data. Ninov had developed his own computer program to track three decay sequences of the few atoms of element 118 through element 116 to the end of the chains as element 106. Other Berkeley scientists and the investigating committees could not locate his original data that claimed several decay chains for element 118. One visiting scientist from Poland suggested that element 118 could be fabricated by using a beam of $_{36}$Krypton-86 ions to bombard $_{82}$Lead-208. Using the Berkeley 88-inch cyclotron, a beam of ions of krypton-86 was aimed at thin sheets of lead-208. The proposed reaction follows:

$$_{36}Kr\text{-}86 > bombarded > \,_{82}Pb\text{-}208 \rightarrow \,_{118}Uuo\text{-}293 + one\ neutron.$$

The LBNL's sensitive gas-filled separator and silicon detector plate would detect any new particles produced by this bombardment, including element 118. These instruments would identify any alpha particles in a decay chain. Ninov's team reported that it had found three atoms of element 118 over a ten-day period of running the experiment. This resulted in three different alpha-decay sequences as reported as evidence of the discovery by Ninov. Also, the sequence of decay events for elements 118 and 116, if they really occurred, would be consistent with the theory for the "island of stability." The saga continues; several of Ninov's colleagues used their own computers and programs but could not verify the original data claimed, but never produced, by Ninov. The conclusion is that the original data was fabricated. Ninov is no longer employed at the LBNL, and the published papers of the discovery of 118 were retracted.

One conclusion that can be drawn from this is that even the best of scientific labs and investigators can be seduced into scientific misconduct. Fortunately, such misconduct is rare in the scientific community. Another conclusion is that because the decay chain of element 118 produced the following new elements—116, as it was the result of the decay process of the nonexistent element 118, 114, 112, 110, 108, and 106—it throws suspicion on the existence of element 116. Still, many of the elements identified by the suspect decay chain have also been independently produced synthetically.

Common Uses

None, except in theoretical research in nuclear physics.

Examples of Compounds

None.

Hazards

None, because it does not yet exist.

Possible Future Elements

For all practical purposes, the periodic table will end at or before element 120 (unbini-lium). In theory, however, elements with proton numbers up to and beyond element 126 may become possible according to the "island of stability" as proposed by Dr. Glenn T. Seaborg. Over the past 50 years, and particularly in the past 20 years, the elegance and sophistication of research equipment designed to synthetically produce and identify super heavy elements has been greatly improved. But as good as the equipment is and as elegant as the detection proce-dures are, they are still not adequate to produce some of the heavier elements. Improvements in instruments and procedures will need to be developed in order to continue researching the unknown elements. One large project intended to improve the situation was called the "Superconducting Super Collider" (SSC). It was a short-lived attempt to develop an advanced facility and equipment that would assist in the exploration of particle acceleration and discov-ery of new elements. In the early 1990s, a very large underground circular tunnel about 85 feet underground was being dug on a huge track of land south of Dallas, Texas, to house equip-ment for a very high-energy circular particle accelerator. When 14 miles of the tunnel was completed and some of the equipment was being installed and two billion dollars expended, the U.S. Senate and U.S. House of Representatives voted overwhelming in 1993 to cancel the funds for the project. This was a disaster for the scientific community and, in particular, particle physicists and chemists. The cancellation of SSC provided the opportunity for foreign countries to take the leadership in the field of high-energy particle physics and research related to the exploration of the nature of matter, including possible new heavy elements.

Some expanded periodic tables of chemical elements have been developed that include future elements up to the proton number 152 or 202. Someday the existence of new heavy elements may be realized if the United States once again takes the lead in particle physics and related research.

Glossary of Technical Terms

abrasive. A finely divided, hard, refractory material used to reduce, smooth, clean, or polish the surfaces of other, less hard substances.

absolute zero. In 1848 Lord Kelvin developed the Kelvin scale for measuring low temperatures starting at 0 degrees Kelvin, which eliminated the need for negative numbers, and thus is the absolute coldest it can become. It is the temperature at which all matter, such as a gas, possesses no thermal energy and at which all molecular motion ceases. It has never been reached. At this temperature, "entropy," or the state of disorder, no longer exists. This is an important concept for the third law of thermodynamics and quantum physics. Each Kelvin is equal to one degree, and this scale starts at 0 degrees Kelvin, which is equivalent to $-273.13°$ Celsius or $-459.4°$ Fahrenheit, which is the coldest temperature possible, and thus, the boiling point of water becomes 373 Kelvin.

absorption. In chemical terminology, the penetration of one substance into the inner structure of another. Also, the ability of the three forms of matter (solids, liquids, and gases) to "mix" and intersperse their atoms and molecules within the atoms or molecules of other or similar forms of matter.

acid. A substance that releases hydrogen ions when added to water. Strong acids, such as HCl, are proton donors and tend to dissociate in water and become ionized in solution—that is, they yield H^+ ions. They are sour to taste, turn litmus paper red, and react with some metals to release hydrogen gas—for example, $2HCl + 2Zn = 2ZnCl + H_2$.

actinides. The elements from actinium (atomic number 89) to lawrencium (atomic number 103). They are radioactive, and those beyond uranium (atomic number 92) are produced artificially, although neptunium and plutonium have recently been found in minute quantities in nature. They are all similar to $_{89}Ac$.

adsorption. Adherence or collection of atoms, ions, or molecules of a gas or liquid to the surface of another substance, which is called the adsorbent; for example, hydrogen gas collects (adsorbs) to the surface of several other elements, particularly metals. It is an important process for dyeing fabrics and is not to be confused with absorption, which is internal mixing or dispersion of one substance within another.

aliphatic. A major group of organic compounds in which the carbon atoms are connected in branched or straight chains rather than ring formation. There are three subgroups of aliphatic hydrocarbons: (1) alkanes (paraffins), (2) alkenes or alkadienes (olefins), and (3) alkynes (acetylenes).

alkali earth metals. Those metals found in group 2 (IIA) of the periodic table. They are beryllium ($_4$Be), magnesium ($_{12}$Mg), calcium ($_{20}$Ca), strontium ($_{38}$Sr), barium ($_{56}$Ba), and radium ($_{88}$Ra).

alkali metals. Those metals found in group 1 (IA) of the periodic table. They are hydrogen ($_1$H), lithium ($_3$Li), sodium ($_{11}$Na), potassium ($_{19}$K), rubidium ($_{37}$Rb), cesium ($_{55}$Cs), and francium ($_{87}$Fr).

allotropes. Elements that exist in two or more different forms in the same physical state. The chemical properties of each allotrope are the same, but the physical properties are different; for example, carbon has four allotropes: diamond, graphite, amorphous carbon, and fullerene. Although the chemical properties of allotropes of an element are the same, some physical properties, such as density, hardness, electrical conductivity, color, and even molecular structure, may differ. The concept of allotropes in chemistry is somewhat similar to the concept of different species of the same plant or animal in biology.

alloy. A mixture of two or more metals, or a mixture of metals and nonmetals. An alloy is a substance made by joining (fusing) a metallic substance with two or more other metals or nonmetals that results in a mixture with properties different from each of the former substances. Steel is an alloy of iron and carbon. If a small amount of nickel is added, it becomes stainless steel. Brass is an alloy of copper and zinc, and bronze an alloy of copper and tin. Sometimes other metals are also added.

alpha decay. The transformation of radioactive elements characterized by the emission of an alpha particle by a nuclide (atomic nucleus).

alpha particle. A nucleus of a helium atom (H^{++}), that is, two positive protons and two neutrons, without any electrons. Such positive particles are produced in nuclear reactions and result from the decay of radioactive elements. They are one of the three basic forms of radiation (alpha, beta, and gamma). Alpha particles, the heaviest of the three, have a range of penetration in air of five centimeters.

alpha (α) radiation. One of the three basic forms of radiation, the others being beta (β) and gamma (γ). Alpha (α) radiation consists of helium ($_2$He) nuclei.

amalgam. An alloy that contains mercury. It is a "solution" of various metals that can be combined without melting them together; for example, gold, silver, platinum, uranium, copper, lead, potassium, and sodium will form amalgams with mercury.

amorphous. Noncrystalline, having no molecular lattice structure that is characteristic of the solid state; for example, all liquids and some powders are amorphous. It is a state of matter having no orderly arrangements of its particles. Glass is an example of a solid amorphous solution.

analogous (analogue). In chemistry, refers to (1) an element that has similar properties and characteristics as another element on the periodic table or (2) a chemical compound similar to another compound but whose composition differs by one element.

anesthetic. A chemical compound that induces loss of sensation in a specific part or all of the body.

anhydrous. A substance that contains no water.

anion. An ion having a negative charge that is attracted to the anode (positive pole) in an electrolyte during the process known as electrolysis.

annealing. Treating a metal, an alloy, or a glass with heat and then allowing it to cool for the purposes of removing internal stress, thus making the material less brittle. Also referred to as *tempering*.

anode. The positively charged electrode in an electrolytic cell, electron tube, or storage battery. The anode is where oxidation takes place when electrons are lost in the process. Also, the collector of electrons.

antiknock agent. A substance added to engine fuel to prevent knocking—that is, a shuddering sound in engines caused by the uneven and premature burning of fuel.

antimatter. *See* antiparticle.

antioxidant. An organic compound added to rubber, fats, oils, food products, and lubricants to retard oxidation that causes deterioration and rancidity. This also refers to some foods and diet supplements that neutralize free radicals, such as the hydroxyl free radical ($-OH$) in human cells.

antiparticle. A subatomic particle, namely a positron, antiproton, or antineutron, with the identical mass of the ordinary particle to which it corresponds but opposite in electrical charge or in magnetic moment. Antiparticles make up antimatter, the mirror image of the particles of matter that make up ordinary matter as we know it on Earth. This is a theoretical concept devised to relate relativistic mechanics to the quantum theory.

antiproton. The antiparticle of the proton. It has the identical mass as the proton but is opposite in electrical charge; that is, it is a proton with a negative charge.

antiseptic. In chemistry, it refers to any substance that retards, stops, or destroys the growth of infectious microorganisms, such as alcohol, boric acid, hydrogen peroxide, and hexachlorophene, among many others.

aqua regia. A very corrosive, vaporizing yellow liquid that is a mixture of one part nitric acid and three or four parts hydrochloric acid. The ancient alchemists named it "aqua regia" (meaning "royal water") because it is the only acid that will dissolve platinum and gold.

aromatic. In chemistry, it refers to a group of unsaturated cyclic hydrocarbons characterized by the six–carbon ring arrangement. The most common is benzene. Substances composed of these particular hydrocarbons typically have a strong but not noxious odor, hence the term "aromatic."

asphyxiant. A gas that has little or no positive effect but can cause unconsciousness and death by replacing air, thus depriving an organism of oxygen (for example, carbon dioxide).

atom. The smallest part of an element that can exist and that is recognizably part of a particular element. A stray electron could belong to any element, but an atom, with its unique configuration of neutrons (neutral charge), protons (positive charge), and electrons (negative charge), can always be identified as a specific element.

atomic mass (atomic weight). The average (mean) weight of all the isotopes of each specific chemical element found in nature. Atomic *weight* is expressed in atomic mass units (amu), which is defined as exactly 1/12 the *mass* of the carbon-12 atom. It is the total weight of protons plus neutrons found in the nucleus of an atom.

atomic mass number. The total number of protons and neutrons in the nucleus of an atom. Also known as the nucleon number.

atomic number (proton number). The number of positively charged protons found in the nucleus of an atom, upon which its structure and properties depend. This number determines the location of an element in the periodic table. For a neutral atom, the number of electrons equals the number of protons. The proton number = the Z-number

atomic radius. One half the distance between two contiguous atoms in the crystals of elements.

atomic weight. The total number of protons plus neutrons in an atom. (*See also* atomic mass.)

Babbit metal. A soft alloy of metals such as tin, silver, arsenic, and cadmium combined with a lead base. It can be cast or used as a coating on steel bearings to form an oil-like coating that reduces friction. Used to make oil-less bearings.

base. An alkali substance that reacts with (neutralizes) an acid to form a salt—for example, $4HCl + 2Na_2O \rightarrow 4NaCl + 2H_2O$ [hydrochloric acid + sodium hydroxide yields sodium chloride (table sale) + water]. A base in water solution tastes bitter, feels slippery or soapy, turns litmus paper blue, and registers above 7 on the *p*H scale. A base can accept two electrons. In water, a base dissociates to yield hydroxyl ions (-OH). (*See also* *p*H.)

beta particle. In essence, a beta particle is a high-speed electron and is considered one of the three basic types of radiation—alpha, beta, and gamma. It is a negative particle expelled from the nucleus during the decay of a radioactive atom. If it has a negative charge, it is identical to an electron. If it has a positive charge, it is called a positron (*see* antiparticle). It can travel at up to 99% of the speed of light and can be stopped by several layers of paper.

beta radiation (β). One of the three basic types of radiation, the others being alpha (α) and gamma (γ). Beta (β) radiation can be harmful and can cause severe burns on the skin.

binary compound. A compound formed by two elements, such as HCl (hydrogen chloride), NaCl (sodium chloride), CO (carbon monoxide), and so forth.

boiling point. The temperature at which a liquid boils (vaporizes)—that is, the transition of a liquid to a gas under standard atmospheric pressure.

bolides. Two or more parts of a large meteor formed when the meteor, usually called a *fireball*, produces bright streaks of light and splits. Called bolides after the Greek word *bolis*, which means "missile." A loud hissing noise can sometimes be heard as one of these large meteors passes through the atmosphere.

brazing. A welding method in which a nonferrous filler alloy is inserted between the ends or edges of the metals to be joined.

carbon black. A high-volume commercial chemical product that is produced through the incomplete combustion (partial oxidation) of petroleum oil and used in the manufacture of rubber (tires), pigments, and printer's ink.

carbon dating. A method of dating very old organic matter by using the rate of decay of carbon-14, a naturally occurring radioactive form of carbon with a half-life of 5,580 years. It can be used to accurately calculate and confirm the date when an organic substance was living by comparing the amount of carbon-14 with the carbon-12 remaining in the substance.

carcinogen (carcinogenic). A substance that causes cancer.

catalyst. Any substance that affects the rate of a chemical reaction without itself being consumed or undergoing a chemical change. There are three characteristics of catalysts: (1) not much is needed to affect the chemical reaction; (2) although involved in the reaction, a catalyst is not changed or used up; and (3) each catalyst is unique for each specific reaction. Most catalysts accelerate change (positive catalysts), but a few can retard change (negative catalysts). Both types change the amount of energy or time required for the chemical reaction to take place. An example is platinum/palladium pellets in automobile catalytic converters that change toxic, harmful gases to less toxic gases and thus help control pollution. An enzyme is a biological catalyst that affects chemical reactions in living organisms.

catalytic converter. A device used in gasoline engines (internal combustion engines) whereby harmful gases, such as carbon monoxide and hydrocarbon pollutants, are converted to less harmful gases (e.g., carbon dioxide) as they pass through platinum converters. The CO_2 produced is not changed. The platinum/palladium pellets in the converters will last as long as the car because they are not consumed in the chemical reaction. (*See also* catalyst.)

cathode. A negatively charged electrode or plate, as in an electrolytic cell, storage battery, or electron tube similar to a TV. Also, the primary source of electrons in a cathode ray tube, such as the Crookes tube. Within the electrolyte, the positively charged ions, called cations, flow to the cathode during electrolysis.

cations. Positively charged ions that migrate to the cathode or negatively charged electrode in an electrolytic system. During electrolysis, cations are attracted to the cathode. Cations are usually positive ions of metals.

cell. *In chemistry:* A cell consists of two electrodes or plates in a liquid electrolyte that provides the pathway for electrons to migrate from one electrode to the other. *In biology:* The smallest unit of a biological structure or an organism capable of independent function. It is composed of cytoplasm, one or more nuclei, a number of organelles, and inanimate matter, all of which is surrounded by a semipermeable plasma membrane.

chain reaction (nuclear). Atomic nuclei are bombarded with neutrons and split to release more neutrons, a phenomenon known as fission. These neutrons then split further nuclei, with the release of more neutrons. The process keeps repeating itself exponentially, with the generation of enormous amounts of energy. Controlled nuclear chain reactions are the source of energy produced in nuclear power stations, and the nuclear (atomic) bomb must obtain "critical mass" to create an explosion. It takes 33 pounds of uranium-235, or just 10 pounds of plutonium-239, to attain a critical mass that can produce an uncontrolled fission chain reaction resulting in a nuclear explosion.

chemical. The definitive molecular structure of all matter, as well as of its properties, characteristics, and functions. A substance produced by or used in a chemical process.

chemical bond. Electrostatic force that holds together the elements that form molecules of compounds. This attractive force between atoms is strong enough to hold the compound together until a chemical reaction causes the substance to either form new bonds or break the bonds that form the molecule.

chemical reaction. A chemical change involving either endothermic (heat-absorbing) or exothermic (heat-releasing) reactions. Some common examples are oxidation, reduction, ionization, combustion, and hydrolysis.

chemical reduction. The process in which electrons are gained. The opposite of chemical oxidation.

chemistry. The science concerned with the physical and chemical properties, composition, structures, and interactions of the elements of the Earth. There are many branches and subbranches of chemistry, including (1) *analytical chemistry,* the study of "what it is" and "how much there is of it"; (2) *biochemistry,* the study of chemical reactions in living things; (3) *electrochemistry,* the science of the relationships of electricity and chemical changes; (4) *geochemistry,* the study of the composition of the Earth, water, atmosphere, and our natural resources; (5) *industrial chemistry,* or chemical engineering, concerned with the production and use of our resources; (6) *inorganic chemistry,* the study of all atoms and molecules except carbon; (7) *organic chemistry,* the study of carbon compounds and the chemistry of all living things; (8) *physical chemistry,* concerned with the physical properties of matter; (9) *polymer chemistry,* the study of very large molecules such as rubber and plastics; and (10) *nuclear chemistry,* the study of the atomic nucleus, including fission and fusion reactions and their results.

chlorofluorocarbons (CFCs). Compounds containing chlorine, fluorine, and carbon. They are used as refrigerants and as propellants in aerosols. In the upper atmosphere, they decompose to give chlorine atoms, which react with ozone molecules to form chlorine monoxide. Chlorine from the oceans also affects the ozone layer. Such a phenomenon may cause thinning of the ozone layer. They are also suspected, along with CO_2, of increasing the so-called greenhouse effect. Efforts are underway to reduce the use of CFCs.

chlorohydrocarbons (CHCs). Similar to CFCs, except they contain hydrocarbon molecules instead of just carbon. (*See also* **chlorofluorocarbons.**)

chromatography. Any of a group of techniques used to separate complex mixtures, that is, vapors, liquids, or solutions, by a process of selective adsorption (not to be confused with absorption), the result being that the distinct layers of the mixture can be identified. The most popular techniques are liquid, gas, column, and paper chromatography.

cladding. The process by which two different metals are rolled together under heat and pressure to form a bond between them. The interface between the two metals becomes an alloy, but the "outerface" of each maintains its original properties. An example would be when low-grade steel is "clad" with high-grade stainless steel or some other noncorrosive metal.

combustion. A chemical reaction in which a substance combines with a gas to give off heat and light (fire). It commonly involves an oxidation reaction that is rapid enough to produce heat and light.

compound. A substance in which two or more elements are joined by a chemical bond to form a substance different from the combining elements. The combining atoms do not vary their ratio in the new compound and can only be separated by a chemical reaction, not a physical force.

conductors. Substances that allow heat or electricity to flow through them.

control rods. Rods made of metals that have a high capacity to absorb neutrons in a nuclear reactor. Their purpose is to control the rate of fission in nuclear reactors; they absorb the neutrons that sustain chain reactions.

convulsion. A violent spasm.

corrosion. The electrochemical degradation of metals or alloys as a result of reaction with their environment, which is accelerated by the presence of acids or bases. Also, the destruction of body tissues by strong acids and bases. The rusting of iron is a form of corrosion called oxidation.

covalent bond (homopolar). Sharing of electrons by two or more atoms. A single covalent bond involves the sharing of two electrons with two different atoms. In a double covalent bond, four electrons are shared, and in a triple bond, six electrons are shared. In a nonpolar covalent bond, the electrons are not shared evenly, but in a polar bond, they are definitely shared evenly. This type of bond always produces a molecule. Also known as "electron pair bond."

cracking. The process used in the refining of petroleum, by which hydrocarbon molecules are broken down into smaller molecules, which are then used in the production of chemicals and fuels. Heat was originally used to crack petroleum, but today most hydrocarbon fuels are formed by catalytic cracking. Platinum pellets or silica alumina are usually used as catalysts.

critical mass. The minimum mass of fissionable material (^{235}U or ^{239}Pu) that will initiate an uncontrolled fission chain reaction, as in a nuclear (atomic) bomb. The critical mass of uranium-235 is about 33 pounds, and of plutonium-239, about 10 pounds.

crucible. A cone-shaped container having a curved base and made of a refractory (heat-resistant) material, used in laboratories to heat and burn substances. Crucibles may or may not have covers. A crucible is also a special type of furnace used in the steel industry that has a cavity for collecting the molten metal.

crust. The outermost layer of crystalline rocks and minerals that make up what lies directly beneath the Earth's continental soils and the ocean floors. The continental crust is that portion of the Earth where humans and other air-breathing life forms exist. It is distinct from the oceanic crust. Most of the Earth's elements are found in the crust (except those that are either artificially or synthetically produced).

cryogenic. Study of behavior of matter at very low temperatures, below –200°C. The use of the liquefied gases, oxygen, nitrogen, and hydrogen, at approximately –260°C is standard industrial practice. (*See also* superconductor.)

crystalline. A substance that has its atoms, molecules, or ions arranged in a regular three-dimensional structure (e.g., igneous or metamorphic rock and diamonds).

cupric. Form of the word "copper" used in naming copper compounds in which the copper has a valence of +2 (e.g., cupric oxide [CuO] and cupric chloride [$CuCl_2$]).

cuprous. Form of the word "copper" used in naming copper compounds in which the copper has a valence of +1 (e.g., cuprous oxide [Cu_2O] and cuprous chloride [$CuCl$]).

cyclotron. A particle accelerator made up of two hollow cylinders (similar to two opposing D structures) that are connected to a high-frequency alternating voltage source in a constant magnetic field. The charge particles, which are injected near the midpoint of the gap between these two hollow cylinders, are then accelerated in a spiral path of increasing expanse so that the path traveled by these accelerated particles increases with their speed, and a deflecting magnetic field deflects them to a target. (*See also* particle accelerator.)

decahydrate. A chemical compound that contains 10 molecules of water.

decomposition. The process whereby compounds break up into simpler substances. Decomposition usually involves the release of energy as heat.

dehydration. The removal of water, usually 95% or more, from a substance or a compound by exposure to a high temperature. Also, the excessive loss of water from the human body.

dehydrogenation. A chemical process that removes hydrogen from a compound. It is considered a form of oxidation.

deliquescence. The absorption of water from the atmosphere by a substance to such an extent that a solution is formed; for example, water-soluble chemical salts (powders) that dissolve in the water absorbed from the air.

density. The relative measure of how much material is in a substance as compared to a standard. It may be calculated by dividing weight (mass) by volume. Density of solids and liquids is compared to the density of 1 milliliter (ml or cc) of water at 4°C. The weight of 1 cc of water is exactly 1 gram, and thus the density of water = 1. Therefore, water is used as a relative standard for comparison of densities of solids and liquids (weight/volume or g/cc). For gases, the density is expressed as grams per liter (g/L); for air, it is 1.293. Thus, the specific gravity of air = 1.0 (1.293/1.293), and 1.0 is used as the standard for comparing the specific gravity (density) of gases. Specific gravity is the same as density for liquids and solids, but it is not exactly the same as density for gases. (*See also* specific gravity.)

depilatory. A substance (usually a sulfide) used to remove hair from skin of humans or the hides of cattle, pigs, and horses.

depression. A general feeling of uneasiness or distress about past or present circumstances or future uncertainties. It can be as mild as just a day or two of "feeling low" or can be clinical depression, which is a serious condition needing medical and/or psychological treatment.

desiccant. A hygroscopic substance (e.g., silica gel) that can adsorb water vapor from the air. Used to maintain a dry atmosphere in food containers, chemical reagents, and so forth.

deuteron. A nuclear particle, with a mass of 2 and a positive charge of 1. The deuterium atom is called heavy hydrogen (^2D) [hydrogen with an atomic weight of 2].

diamagnetic. Refers to a substance wherein an activated magnetic field is positioned at right angles to the magnetic line. In other words, it is repelled by the magnet.

diatomic molecule. A molecule that contains two atoms of the same element (e.g., H_2, O_2, and N_2).

die. A device with a cone-shaped hole through which soft metal is forced to form wire, pipe, or rods. Also, a form for cutting out or forming materials. (*See also* ductile; extrusion.)

dielectric. Refers to any material that either (1) is a nonconductor of electricity (insulator), or (2) has very low electrical conductivity (i.e., less than a millionth [10^{-6}] of a mho/cm, or siemens).

diode. An electronic device that restricts the flow of electrical current to one direction, such as a vacuum tube made up of two electrodes, one cathode and one anode, or a semiconductor used as a rectifier to change AC current to DC current.

distillation. A process used to purify and separate liquids. A liquid is said to have been distilled when it has been heated to a vapor, and then the vapor is condensed back into a liquid.

divalent. Having a valence of 2. Similar to bivalent.

ductile. Easily shaped. A ductile substance can be extruded through a die and easily drawn into wires. (*See also* die.)

eka. Eka means "first" in Sanskrit. It was used by Mendeleev as a prefix for yet-to-be-discovered elements that he predicted would "fit" into specific blank spaces in the next lowest position of the same groups in his periodic table. For example, he knew that the unknown element for the blank space following calcium would be closely related to boron, so he gave it the name of "eka-boron" until that unknown element, scandium, was definitely identified and named. Scandium's properties are almost exactly as Mendeleev predicted they would be. Several other eka elements were eka-aluminum (for gallium), eka-silicon (for germanium), and eka-manganese (technetium), although Mendeleev missed this latter one.

electrochemistry. Use of electricity as the energy source to break up the oxides of elements. Used to extract metals from ores.

electrode. A conductor that allows an electric current to pass through an electrolyte (solution). The current enters and leaves the electrolyte solution through anode and cathode electrodes. Electrodes are made of metals, such as copper, platinum, silver, zinc, lead, and sometimes carbon.

electroluminescence. Luminescence generated in crystals by electric fields or currents in the absence of bombardment or other means of excitation. It is observed in many crystalline substances, such as silicon carbide, germanium, and diamond.

electrolysis. A process in which an electric current is passed through a liquid, known as an electrolyte, producing chemical changes at each electrode. The electrolyte decomposes, thus enabling elements to be extracted from their compounds. Two examples are the production of chlorine gas by the electrolysis of sodium chloride (NaCl, or table salt), and the electrolysis of water (H_2O) to produce oxygen (O) and hydrogen (H).

electrolyte. A compound that, when molten or in solution, will conduct an electric current. The electric current decomposes the electrolyte. Some salts and acids make excellent electrolytic solutions.

electromagnetic spectrum. Electromagnetic waves, that is, the total range of the wavelengths of radiation, from the very short high-frequency waves, including cosmic rays, gamma rays, X-rays, and ultraviolet radiation, to wavelengths of visible light (violet to reds), and then the long infrared heat radiation, microwaves, and finally the longest electromagnetic radiation of the spectrum, such as radio waves and electric currents.

electron. An extremely small, negatively charged particle that moves around the nucleus of an atom. An electron's mass is only 1/1837 of the mass of a proton. The number of electrons in a neutral atom is always equal to the number of positively charged protons in the nucleus. Ions are similar to atoms with an electrical charge. For ions, the number of electrons does not equal the number of protons. The electrons surrounding the nuclei of atoms are partly responsible for the chemical properties of elements. The interaction of the electrons of atoms is the chemistry of our Earth's elements.

electron affinity. The energy required to remove an electron from a negative ion, the result of which is the restoration of the atom's or molecule's neutrality.

electron capture. Either of two processes: (1) an atom or ion passing through a filtering substance either gains or loses one or more of its orbital electrons; or (2) the radioactive transformation of the atomic nucleus (nuclide) in which the electron attaches to its nucleus, also called electron attachment.

electronegative. The positive nucleus can exert attraction for additional electrons, even though, as a whole, the atom is electrically neutral. The nuclei of metals have a very weak attraction for additional electrons. Thus, they exhibit weak electronegativity (by giving up electrons to become positive ions), whereas the nuclei of nonmetals have a tendency to attract electrons and, thus, are highly electronegative (gain electrons to become negative ions). When atoms "share" electrons, the atom that "keeps" the electrons for most of the time is the electronegative partner. The most highly electronegative elements are fluorine (4.0), oxygen (3.5), chlorine (3.0), nitrogen (3.0), and bromine (2.8).

electroplating. Sending an electrical charge from one electrode through an electrolyte to deposit a thin coating of a metal on an object at the opposite electrode.

electropositive. The opposite of electronegative—for example, metals.

element. A pure substance composed of atoms. Each specific element's atoms have the same number of protons in their nuclei. Elements cannot be broken down into simpler substances by normal chemical means. Each element has a unique proton number and contains an equal number of electrons and protons. Thus, the atoms of elements have no electrical charge. All substances and materials are composed of about one hundred elements. Many of the heavier elements are man-made, unstable, and radioactive and decay by fission into other elements. (*See also* atom.)

emetic. A chemical agent that induces vomiting.

energy. The capacity for doing work. There are many forms of energy. Two representative classifications are potential energy and kinetic energy. The word "potential" is derived from the Latin word *potens,* meaning "having power." The word "kinetic" is derived from the Greek word *kinetos,* which means "moving." Several forms of energy include the following: (1) mechanical energy (e.g., using a sledgehammer); (2) chemical energy (e.g., burning fuel, rusting or oxidation, and photosynthesis); (3) electrical energy (e.g., a generator sending electrons through wires, and lightning); (4) radiant energy (e.g., sunlight and X-rays); and (5) nuclear energy (e.g., the fission or fusion of atomic nuclei and $E = mc^2$). There are many sources of energy. Most of them, except nuclear energy, involve the electrons of atoms, namely chemistry.

enzyme. Any of a number of proteins or conjugated proteins produced by living organisms that act as biochemical catalysts in those organisms.

eutectic system. The composition of specified mixtures that are solidified at the lowest possible temperature (e.g., alloys).

evaporation. The process whereby a liquid changes to a gas. The "interface" between the liquid state of a substance and the gaseous state of that substance. (*See also* liquid; gas.)

extrude. The process whereby a material is forced through a cone-shaped hole in a metal die, followed by cooling and hardening.

femtosecond. One-quadrillionth (10^{-15}) of a second. Laser pulses can be measured in femtoseconds.

fermentation. A chemical change induced by bacteria in yeast, molds, or fungi that involves the decomposition of sugars and starches to ethanol and carbon dioxide. Used in the preparation of breads and the manufacture of beer, wine, and other alcoholic beverages.

ferromagnetic. Characteristic of iron and other metals and alloys of iron, nickel, cobalt, and various materials, which possess a high magnetic permeability. Also, the ability to acquire high magnetization in a weak magnetic field.

filament. A continuous, finely extruded wire, usually tungsten, gold, or a metal carbide, or synthetic material that is heated electrically to incandescence in electric lamps. Also used to form cathodes in electron tubes.

fission (nuclear). The splitting of an atom's nucleus with the resultant release of enormous amounts of energy, additional neutrons, and the production of smaller atoms of different elements. Fission occurs spontaneously in the nuclei of unstable radioactive elements, such as uranium-235 or plutonium-239. This process is used in the generation of nuclear power, as well as in nuclear bombs.

flammable. Easily bursting into flame. Combustible.

fluorescent. Consisting of a gas-filled tube with an electrode at each end. Passing an electric current through the gas produces ultraviolet radiation, which is converted into visible light by a phosphor coating on the inside of the tube. This emission of light by the phosphor coating is called fluorescence. (*See also* phosphor.)

flux. A chemical that assists in the fusing of metals while preventing oxidation of the metals, such as in welding and soldering. Also, lime added to molten metal in the furnace to absorb impurities and form a slag that can be removed.

forge. A furnace or hearth where metals are heated and softened (wrought)—for example, wrought iron.

fossil fuel. Fuel resulting from an organic source in past ages that is buried within the earth (e.g., petroleum, natural gas, and coal) and that is considered nonrenewable.

fractional crystallization. The separation of ores by using their unique rates of crystallization. Different minerals crystallize at different temperatures and rates. Thus, they can be separated if these factors are controlled.

fractional distillation. A process used to obtain a substance in a form as pure as possible. The evaporated vapors (fractions) are "boiled" off and collected at different temperatures, which separates the fractions by either their different boiling points or their specific condensation temperatures.

fractionation. The process of separating and isolating a mixture of either gases or liquids. The separation may be accomplished by distillation, crystallization, electrophoresis, particle filtration, centrifugation, or gel filtration or chromatography. This procedure is based on the fact that the different components of the mixture have different physical and chemical properties.

free electron. An electron not attached to any one atom and not restricted from flowing.

free element. An element found in nature, uncombined with other elements.

free radical. A molecular fragment having one or more unpaired electrons and charges, usually short-lived and highly reactive—for example, the hydroxyl radical, -OH, which is thought to contribute to the aging process of our bodies.

fuel cell. A cell in which the chemical energy from oxidation of a gas (fuel) is directly converted into electricity. An electrochemical device for continuously converting chemicals (a fuel, such as H_2, and an oxidant, such as O_2) into direct-current electricity and water. Once the device is operating, a "redox" or reduction-plus-oxidation reaction takes place, releasing electrical energy. Also, an aircraft fuel tank made of or lined with an oil-resistant synthetic rubber.

fungicide. A chemical compound or solution that will destroy or inhibit the growth of fungi.

fusion (thermonuclear reaction). An endothermic nuclear reaction yielding large amounts of energy in which the nuclei of light atoms (e.g., forms of heavy hydrogen, such as deuterium and tritium) join or fuse to form helium (e.g., energy of the sun or the hydrogen bomb). Might be thought of as the opposite of fission nuclear reactions.

galvanization. The process of coating iron or steel with a layer of zinc, making it more resistant to corrosion than the underlying metal.

gamma (γ) radiation. The most intense of the three forms of radiation, the others being alpha (α) and beta (β). Somewhat similar to X-rays, but more penetrating.

gamma rays. A type of radiation emitted by radioactive elements. Gamma rays are capable of traveling through 15 centimeters of lead. Because of their ability to kill body cells, they are used in the treatment of cancer. Gamma rays have the shortest wavelength and most energy of the three major types of radiation (alpha, beta, and gamma).

gas. A state of matter in which atoms and molecules are linked by very few bonds and are, therefore, given great freedom of movement. Gases have a fixed mass (weight), but no fixed shape or volume.

gastrointestinal. Relating to the stomach and intestines.

Geiger counter. Also known as a scintillation counter. A device used to detect, measure, and record radiation. The instrument gets its name from one of its parts, the Geiger tube, which is a gas-filled tube containing coaxial cylindrical electrodes.

getter. Also called a scavenger. Any substance added to a mixture of substances used to reduce or inactivate any traces of impurities. A metal added to alloys to reduce impurities in the alloys. Also, small amounts of metals inserted into vacuum tubes to absorb any remaining trace gases from the vacuum. (*See also* misch metal.)

graphite. The crystalline allotrope of carbon, which is steel gray to black. Used in lead pencils, lubricants, and coatings. Also fabricated forms, such as molds, bricks, crucibles, and electrodes. Used to moderate radiation in nuclear reactors.

group. Refers to families of elements that exhibit similar characteristics and that are "grouped" in vertical columns on the periodic table. Elements in the same group, in general, exhibit similar combining powers (valence), but do not exhibit the same degree of reactivity to other elements.

half-life. The time required for one-half of the atoms of heavy radioactive elements to decay or disintegrate by fission into lighter elements. There is no method of determining when or if a particular atom in a radioactive substance will decay given that the act is random, but an average count of atoms decaying into atoms of lighter elements can be measured as the half-life of all atoms of the radioactive substance.

halides. Binary compounds of the halogens (e.g., fluorides, chlorides, iodides) and bromides (e.g., LiF, NaCl, KI).

halogens. Electronegative monovalent nonmetallic elements of group 17 (VIIA) of the periodic table (fluorine, chlorine, iodine, bromine, astatine). In pure form, they exist as diatomic molecules (e.g., Cl_2). Fluorine is the least dense and most active, whereas iodine is the most dense and least active (excluding radioactive astatine).

heat. The energy produced by the kinetic motion of molecules. Heat may be thought of as the energy that is transferred from one body to another by one of the following three forms of heat transfer: radiation, conduction, or convection. Because all molecular motion ceases at absolute zero on the Kelvin scale, no heat will be evident at the equivalent temperatures of –273.13°C or –459.4°F. Temperature may be thought of as an average of the total molecular motion (heat) one body has that may be transferred to other bodies. Temperature is measured by thermometers and thermocouples that react to heat.

herbicide. A substance, either organic or inorganic, used to destroy unwanted vegetation—such as, weeds, grasses, and woody plants.

homologue. Homogenous (Latin, "the same kind"), meaning the elements in a group have very similar characteristics and properties. Thus, they react in much the same way.

hydration. Combining with water to form a hydrate, that is, a compound containing water combined in a definite ratio, the water being retained or thought of as being retained in its molecular state.

hydride. An inorganic compound (either binary or complex) of hydrogen with another element. A hydride may be either ionic or covalent. Most common are hydrides of sodium, lithium, aluminum, and potassium, such as LiH, KH, NaH, and AlCHO (a complex hydride compound of aluminum + the carbonal –CHO group).

hydrocarbon. An organic compound whose molecules consist of *only* carbon and hydrogen. Hydrocarbons can form chains of molecules known as "aromatic compounds" or form stable rings known as "aliphatic compounds." All petroleum products and their derivatives are hydrocarbons.

hydrocracking. The cracking of petroleum and/or petroleum products using hydrogen. (*See also* cracking and hydrocarbon.)

hydrogasification. Production of gaseous or liquid fuels by direction addition of hydrogen to coal.

hydrogenation. Also known as hardening. It involves the conversion of an unsaturated compound into a saturated compound by the addition of hydrogen—for example, converting liquid oils to solids by adding hydrogen. An example is the addition of hydrogen to corn oil to form margarine.

hydrous. Describes any substance that contains an indeterminate amount of water.

hydroxide. A chemical compound containing the hydroxyl group of –OH, such as bases, certain acids, phenols, and alcohols, which form –OH ions in solution. Most metal hydroxides are bases—for example, the alkali metals.

hygroscopic. Descriptive of a substance that can absorb water from the atmosphere, but not enough to form a solution.

iatrochemistry. The treatment of specific diseases and illness with specific chemicals, first practiced by Paracelsus in the sixteenth century. *Iatros* is the Greek word for physician.

igneous rock. Rock formed by the cooling of magma (molten rock).

incandescent. Giving off a white light after being heated to a particularly high temperature. For example, household electric light bulbs produce incandescent light due to electricity heating the bulb's filaments in a vacuum or inert atmosphere.

inert. Unreactive and does not readily form compounds.

inert gases (noble gases). Extremely unreactive gases found in group 18 (VIIIA) of the periodic table.

infrared light. Radiation of electromagnetic wavelengths that are longer than the visible light wavelengths and are invisible to the naked eye. Similar to "heat" rays.

ingested. Eaten and swallowed into the gastrointestinal system.

ingot. A mass of metal that is shaped into a bar or block. Also, a casting mold for metal. Ingot iron is a highly refined steel with a high degree of ductility and resistance to corrosion.

inhaled. Breathed into the lungs.

inorganic. Involving neither organic life nor the products of organic life or compounds of carbon (with a few exceptions).

inorganic chemistry. The study of all elements and compounds other than organic carbon compounds.

insecticide. A chemical used to kill insects.

insoluble. Cannot be dissolved in a liquid or made into a solution.

insulator. Any substance or mixture that has an extremely low level of thermal or electrical conductivity, such as, glass, wood, polyethylene, and polystyrene.

intoxication. A state that occurs from inhaling industrial fumes—for example, zinc smelting or ingesting too much of a substance such as zinc, alcohol, or drugs—the results of which are lethargy, depression, possible addiction, and other health problems.

ion. An atom or group of atoms that have gained or lost electron(s) and, thus, have acquired an electrical charge. The loss of electrons gives positively charged ions. The gain of electrons gives negatively charged ions. If the ion has a net positive charge in a solution, it is a "cation." If it has a net negative charge in solution, it is an "anion." An ion often has different chemical properties from the atoms from which it originated.

ion-exchange process. A reversible chemical reaction in which the mobile hydrated ions of a solid (the ion exchanger) are exchanged equivalently for the ions of a fluid (usually a water solution). This process is used in water softening, electroplating, chromatography, recovery of metals, and separation of radioactive isotopes from atomic fission, to name a few.

ionic bond (electrovalent bond). The bond formed by transfer of one or more electrons from one atom to another during a chemical reaction. It is similar to a chemical bond where there is an attraction between atoms that is strong enough to form a new chemical unit (compound) that differs in its properties from the bonding atoms.

ionization energy. The chemical process for producing ions in which a neutral atom or molecule either gains or loses electrons, thus attaining a net charge and becoming an ion.

island of stability. On the periodic table, it is part of the superactinide series of super heavy elements (SHE) starting with Z-114 ($_{114}$Uuq) to Z-126 ($_{126}$Ubh) and beyond. Refer to text for more detailed explanation. Z = number of protons in the nucleis.

isomer. In chemistry, chemical compounds with the same molecular composition but different chemical structures. For example, butane has two isomers, C_4H_{10} and $C_2H_4(CH_3)_2$. In nuclear physics, isomers refer to the existence of atomic nuclei with the same atomic number and the same mass number but different energy states.

isotopes. Atoms of the same element with different numbers of neutrons. All atoms of an element always contain the same number of protons. Thus, their proton (atomic) number remains the same. However, an atom's nucleon number, which denotes the total number of protons and neutrons, can be different. These atoms of the same element with different atomic weights (mass) are called isotopes. Isotopes of a given element all have the same chemical characteristics (electrons and protons), but they may have slightly different physical properties.

lanthanide elements. Also known as rare-earth elements. Elements in period 6, starting at group 3, of the periodic table. The lanthanide elements start with proton number 57 (lanthanum, $_{57}$La) and continue through number 71 (lutetium, $_{71}$Lu).

laser. A device that produces a continuous single-colored (frequency or wavelength) beam of light. Laser light is very intense, can travel long distances, moves in one direction, and does not spread. Lasers are used in surgery, in supermarket bar code scanners, and in the production of holograms. The word "laser" is derived from the phrase Light Amplification by the Stimulated Emission of Radiation.

lathe. A powerful machine that shapes a piece of metal, wood, or other material, by means of a gripping apparatus that allows the material being "worked" to be rotated against a cutting device, by which process it is formed into a specified implement, such as, metal screws, wooden bowls.

leach. To remove or dissolve soluble material from a composite using an insoluble solid—for example, a liquid solvent. In industrial applications, the solvent is percolated (dripped, filtered, or penetrated) through a sieve that holds the leached material. Salt and sugar can be extracted using this process. Minerals and metals are often leached out of various ores via the percolation of water, acids, or chemical compounds.

legume. A plant that takes N_2 from the atmosphere and incorporates it into nodules on its roots, such as bean and pea plants.

lepton. In particle physics, any light particle. Leptons have a mass smaller than the proton mass and do not experience the strong nuclear force. They interact with electromagnetic and gravitational fields and essentially interact only through weak interactions.

liquid. A state of matter in which the bonds between particles are weaker than those found in solids, but stronger than those found in gases. Liquids always take the shape of the vessels in which they are contained. They have fixed masses and volumes but no fixed shape.

lubricant. An oily substance, in either a liquid or solid form, that reduces friction, heat, and wear when applied as a surface coating to moving parts or between solid surfaces.

luminescence. Emitting light without giving off heat. Also known as "cold light." (*See also* phosphor.)

machinable. The ability to be rolled, pounded, and cut on a lathe.

magnet. A body or an object that has the ability to attract certain substances (e.g., iron). This is due to a force field caused by the movement of electrons and the alignment of the magnet's atoms.

magnetic field. The space around a magnet or an electrical current where the existence of magnetic force is detectable at every point.

malleable. The ability to be beaten or hammered into different shapes. Malleable materials are usually ductile and machinable as well.

mass. The quantity (amount) of matter contained in a substance. Mass is constant regardless of its location in the universe. Mass should not be confused with weight. The weight of a body is dependent on the size (mass) of the planet plus the square of the distance between the object and the planet. The weight of an object on Earth can be thought of as the gravitational attraction between the object and the Earth. The greater the mass of an object, the more it will weigh on the Earth, but its mass will be the same on the moon even though the moon's gravity is $1/6^t$ that of Earth's.

matter. Everything that has mass and takes up space, regardless of where it is located in space. Modern physics has shown that matter can be transformed into energy, and vice versa; $E = mc^2$. (*See also* mass.)

melting point. Temperature at which a solid changes to a liquid.

metabolism. A chemical transformation that occurs in organisms when nutrients are ingested, utilized, and finally eliminated (e.g., digestion and absorption), followed by a complicated series of degradations, syntheses, hydrolyses, and oxidations that utilize enzymes, bile acids, and hydrochloric acid. Energy is an important by-product of the metabolizing of food.

metal. An element that is shiny, ductile, malleable, a good conductor of heat and electricity, has high melting and boiling points, and tends to form positive ions during chemical reactions. Metals exhibit very low electronegativity and become cations that are attracted to the negative electrode or cathode in electrolytic solutions. Different metals possess these properties in varying degrees. Most of the approximately 118 elements so far discovered are classified as metals. Most metals, other than those in groups 1 and 2, are called transition metals or super heavy metals. Metals exhibit oxidation states of +1 (the alkali metals) to +7 (represented by some of the transition metals).

metallics. Refers to those elements with an iridescent luster characteristic of metals.

metalloid. An element that is neither a metal nor a nonmetal. It has properties that are characteristic of both groups (e.g., silicon, selenium, and arsenic). The term "metalloid," although descriptive, is an old word used for these types of elements, which are now referred to as semiconductors.

metallurgy. The process of extracting and purifying metals from their ores, resulting in the creation of usable items and products.

mineral. An inorganic substance that occurs in nature (e.g., limestone, sand, clay, coal, and many ores).

misch metal. A commercial mixture of rare-earth metals that is used as an alloy material for aluminum and other metals. The rare-earth mixtures may be compacted into various forms for a variety of uses. Misch metals create sparks with only slight friction and are used for flints in cigarette lighters and other "spark" igniters. They are also used as "getters" to absorb gases in vacuum tubes and to remove impurities from metal alloys.

moderator. A substance of low atomic weight, such as, beryllium, carbon (graphite), deuterium, or ordinary water, which is used in nuclear reactors to control the neutrons that sustain chain reactions.

molecule. The smallest particle of an element containing more than one atom (e.g., O_2) or a compound that can exist independently. It is usually made up of a group of atoms of different elements joined by ionic or covalent bonds.

monatomic molecule. A molecule containing only one atom. Only the noble gases have monatomic molecules.

monomer. A simple molecule or chemical compound that is capable of combining with other like or unlike molecules to form a polymer. (*See also* polymer.)

monovalent (univalent). An atom having a valence of 1.

mordant. A fixative used in the textile industry. A substance that binds or fixes a dye to a fiber to prevent streaking or "running" when moisture is applied.

muon. The semistable second-generation lepton, with a mass 207 times that of an electron. It has a spin of one-half and a mass of approximately 105 MeV, and a mean lifetime of approximately 2.2×10^{-6} second. Also known as a *mu-meson*.

mutation. The process of being changed or altered, that is, the alteration of inherited genes or chromosomes of an organism.

natural gas. A mixture of hydrocarbon and combustible gases produced within the Earth and extracted through oil and gas wells. Although its composition varies, depending on where it was formed in the Earth, natural gas is composed of about 90% methane (CH_4). It is moved from place to place through extensive pipelines in the United States and can be liquefied for easy storage and shipping.

neutrino. An electrically neutral, stable fundamental particle in the lepton family of sub-atomic particles. It has a spin of one-half and a small or possibly a zero at-rest mass, with a weak interaction with matter. Neutrinos are believed to account for the continuous energy distribution of beta particles and are believed to protect the angular momentum of the beta decay process.

neutron. A fundamental particle of matter with a mass of 1.009 (of a proton) and having no electrical charge. It is a part of the nucleus of all elements except hydrogen.

neutron absorber. A material in which a large number of neutrons that pass through bond with nuclei and are not redischarged.

noble elements (gases). Also called inert gases, found in group 18 (VIIIA) of the periodic table. They are extremely stable and unreactive.

noble elements (metals). Unreactive metals (e.g., platinum and gold), not easily dissolved by acids and not oxidized by heating in air. Not a precise descriptive chemical term.

noncombustible. A material, either a solid, liquid, or gas, that will not ignite or burn despite extreme high temperatures (e.g., water, carbon dioxide).

nonmetal. A classification of 25 elements. One group has moderate electrical and thermal conductivity (e.g., semiconductors). They are also called metalloids. The other group has very low thermal and electrical conductivity but high electronegativity—for example, insulators and any material not classed as a metal (e.g., plastics, oil, paper). Nonmetals form ionic and covalent compounds by either gaining or sharing electrons with metals, and thus becoming negative ions. In electrolytic solutions, nonmetals form anions, which are attracted to the positive electrode (anode).

nuclear reaction. A change in the atomic nucleus of an element involving the processes known as fission, fusion, neutron capture, or radioactive decay, as opposed to a chemical reaction that involves the transformation of the electron structure of the element.

nuclear reactor. An apparatus in which controlled nuclear fission chain reactions take place. The heat produced by such reactions is used to generate electricity. Fissionable elements used in nuclear reactors can be used to make nuclear weapons. Nuclear reactors can also be used to produce numerous radioisotopes.

nucleon. A general term for either the neutron or the proton, in particular as a constituent of the nucleus.

nucleus. The core of an atom, which provides almost all of the atom's mass. It contains protons, neutrons, and quarks held together by gluons (except hydrogen's nucleus, which is a single proton) and has a positive charge equal to the number of protons. This charge is balanced by the negative charges of the orbital electrons.

orbital. The shell (space) around an atom or molecule where the likelihood of locating electrons is greatest.

orbital electrons. Electrons found in shells outside the nucleus of an atom.

orbital theory. A quantum theory of matter dealing with the nature and behavior of electrons in either a single atom or combined atoms. The theory describes the part of the atom where electrons are most likely to be found as energy levels.

ore. A metal compound found in nature. Also, a mineral from which it is possible to extract the metal.

organic chemistry. The study of carbon compounds (e.g., alcohols, ethers, foods, animals). It does not include the carbonates, hydrogen carbonates, or oxides of carbon. Organic chemistry includes the chemistry of living things and their products as well as carbon chemistry.

oxidation. A reaction in which oxygen combines chemically with another substance and undergoes one of three processes: the gaining of oxygen, the loss of hydrogen, or the loss of electrons. In oxidation reactions the element being oxidized loses electrons to form positive ions, whereas in complementary chemical reactions called reduction, electrons are gained to form negative ions.

oxide. A compound formed when oxygen combines with one other element, either a metal or nonmetal (e.g., magnesium oxide, copper oxide, and carbon dioxide).

oxidizer. A substance that supports combustion in a flammable material, such as fuel or propellant. The oxidizer need not contain oxygen in order to effect the oxidation process.

oxidizing agent. An agent that induces oxidation in substances. The agent itself is reduced or gains electrons (e.g., peroxides, chlorates, nitrates, perchlorates, and permanganates).

ozonosphere (ozone layer). The layer found in the upper atmosphere, between 10 and 30 miles in altitude. This thin layer contains a high concentration of ozone gas (O_3), which partially absorbs solar ultraviolet (UV) radiation and prevents it from reaching the Earth. It is mostly formed over the equator and drifts toward the North and South Poles. It seems to have a cyclic nature.

paramagnetic. Characteristic of an element in which an induced magnetic field is in the same direction and greater in strength than the actual magnetizing field, but much weaker than in ferromagnetic materials.

particle. A very small piece of a substance that maintains the characteristics of that substance. Also known as fundamental particles found in the nuclei of atoms and referred to as subatomic particles.

particle accelerator. A machine designed to speed up the movement of atomic particles and electrically charged subatomic particles that are directed at a target. These subatomic particles, also called "elementary particles," cannot be further divided. They are used in high-energy physics to study the basic nature of matter as well as the ultimate origin of life, nature, and the universe. Particle accelerators are also used to synthesize elements by "smashing" subatomic particles into nuclei to create new, heavy, unstable elements, such as the superactinides. (*See also* cyclotron.)

period. Refers to the tabular arrangement of elements with consecutive atomic numbers arranged in rows running from left to right in the periodic table of the elements. Elements in a given period have similar chemical and physical properties. Periods in the table follow the octet rule.

petrochemicals. Organic chemical compounds containing carbon and hydrogen that are derived from petroleum or natural gas.

pH. A scale of numbers (1 to 14) used to measure the acidity or alkalinity of a substance or solution. A substance with a pH of 8 to 14 is alkaline or basic. Technically it is defined as the logarithm of the reciprocal of the hydrogen +ion as concentrated in a solution.

pharmaceutical. Relating to the manufacture and sale of drugs as well as medicinal, dietary, and personal hygiene products.

phlogiston. The hypothetical substance believed at one time to be the volatile component of combustible material. It was used to explain the principle of fire before oxidation and reduction were known and prior to the discovery of the principle of combustion by Lavoisier.

phosphor. A substance, either inorganic or organic, or in liquid or crystal form, that absorbs energy and emits light (luminescence). If the substance glows for only a very short time after being excited, it is called fluorescent (e.g., a fluorescent light tube). If it continues to glow for a long time, it is called phosphorescent (e.g., the afterglow of a television screen when it is turned off and glow-in-the-dark clock faces. (*See also* luminescence; phosphorescence.)

phosphorescence. The emission of light after a substance has ceased to absorb radiation—that is, it glows in the dark after exposure to light. Electrons of the atoms that absorb energy are moved to a higher energy level (shell). After the source of energy is removed, the energized electrons jump back to their original level, or lower shell. In doing so, the excited electrons continue to emit light energy many hours or days later. (*See also* phosphor.)

photoelectric cells. There are two types of these cells: photoconductive and photovoltaic. Electrical conductivity of photoconductive cells increases as light grows brighter. Used in photographic light meters to control light conditions and correct exposures. Also used in burglar alarms. Photovoltaic cells convert light into electricity.

photon. The quantum unit of electromagnetic radiation or light that can be thought of as a particle. Photons are emitted when electrons are excited in an atom and move from one energy level (orbit) to another. According to Albert Einstein's theory, light can be thought of as either waves or particles or as both. This concept is difficult to understand because it is described only in mathematical terms.

photosynthesis. Process by which chlorophyll-containing cells in plants and bacteria convert carbon dioxide and water into carbohydrates, resulting in the simultaneous release of energy and oxygen. The process of photosynthesis is responsible for all food on Earth.

photovoltaic. Describes a substance or apparatus that is capable of generating a voltage when it is exposed to electromagnetic radiant energy (e.g., light).

piezoelectricity. The generation of electric energy by the application of mechanical pressure, usually on a crystalline substance such as quartz. The term is named after the Greek word *piezo,* meaning "to squeeze."

pigment. A substance, usually a dry powder, used to color another substance or mixture.

pitchblende. A brownish-black mineral of uraninite and uranium oxide with small amounts of water and uranium decay products. The essential ore of uranium.

plasma. An electrically neutral, highly ionized gas, which is made up of ions, electrons, and neutral particles. A plasma is produced by heating a gas to a high temperature, which causes all the electrons to separate from their nuclei, thus forming ions. These ion nuclei can be forced together under great pressure to form larger nuclei and produce tremendous amounts of energy (e.g., nuclear fusion, the sun, the stars, and the hydrogen bomb).

platinum group. Elements that tend to be found together in nature (e.g., ruthenium, osmium, rhodium, iridium, palladium, and platinum).

polarize. To effect the complete or partial polar separation of positive and negative electric charges in a nuclear, atomic, molecular, or chemical system; or to reduce the glare of sunlight by changing the orientation of the light waves passing through the glass (used in windows and eyeglasses).

polyatomic molecule. A chemical molecule containing three or more atoms.

polymer. A natural or synthetic compound of high molecular weight. Examples of natural polymers: polysaccharide, cellulose, and vegetable gums. Examples of synthetic polymers: polyvinyl chloride, polystyrene, and polyesters.

polyunsaturated. Relating to long-chain carbon compounds, those having two or more double bonds (unsaturated) per molecule (e.g., linoleic and linolenic acids, commonly known as corn oil and safflower oil, respectively).

positron. The positively charged antiparticle of an electron; e+ or p+. (*See also* antiparticle.)

prism. A homogeneous, transparent solid, usually with a triangular base and rectangular sides, used to produce or analyze a continuous spectrum of light.

proton. A positively charged particle found in the nucleus of an atom. For neutral atoms, the number of protons is always equal to the number of negatively charged electrons, and this number is the same as the proton number in neutral atoms. Z = proton number.

proton number. Also known as the atomic number. The number of protons found in the nucleus of an atom. Every element has a unique proton number, and every atom of that same element has the same proton number. Elements are arranged in order of their proton number in the periodic table. The atomic number of an element may be represented by the symbol Z. For example, Z-12 represents the element carbon twelve. Used for heavy elements.

p-type. An abbreviation for positive-type. It describes the electrical conductivity of certain semiconductors, crystal rectifiers, and silicon.

pyrophoric. Any material that can ignite spontaneously in air at just about room temperature (e.g., phosphorus, lithium hydride, and titanium dichloride); or a metal or alloy that will spark when slight friction is applied to it (e.g., barium or misch metal that is used as flint in cigarette lighters).

pyrotechnics. The manufacture and/or release of fireworks, flares, and warning equipment. Fireworks are composed of oxidizers (e.g., potassium nitrate and ammonium perchlorate) and fuels (e.g., aluminum, dextrin, sulfur, magnesium). Colors produced by some of the compounds of a few elements that are added to fireworks are as follows: strontium = red, barium = green, copper = blue, sodium = yellow, and magnesium and aluminum = bright white.

quantum. The basic unit of electromagnetic energy that is not continuous but occurs in discrete bundles called "quanta." For example, the photon is a small packet (quantum) of light with both particle and wave-like characteristics. A quantum unit for radiation is the frequency v to the product hv, where h is Planck's constant. The quantum number is the basic unit used to measure electromagnetic energy. To simplify, a very small bit of something. The quantum theory of energy ushered in the era known as "modern physics" and quantum mechanics. (It is also one of the most misused words in the lexicon of popular culture, with the word commonly being used to connote its *opposite* meaning—"very large.") (*See also* photon.)

quantum mechanics. The modern theory of physics that relates electromagnetic radiation with matter and describes the interaction between radiation and matter, as follows: (1) the structure and behavior of matter (atoms and molecules) on the subatomic level, (2) the theory of electromagnetic radiation, and (3) the interaction between matter and radiation.

quark. A hypothetical subnuclear particle having an electric charge one-third to two-thirds that of the electron. Also known as the "fundamental subatomic particle," which is one of the smallest units of matter. (*See also* subatomic particle.)

radiation. The emission of energy in the form of electromagnetic waves (e.g., light [photons], radio waves, X-rays). Also used to denote the energy itself.

radical. Also known as "free radical." A group of atoms having one unpaired electron.

radioactive. Describes the uncontrolled discharge of radiation from the decay of unstable atomic nuclei or from a nuclear reaction, involving alpha particles, nucleons, electrons, and gamma rays.

radioactive elements. Those elements subject to spontaneous disintegration (fission) of their atomic nuclei, with resultant emission of energy in the form of radiation. Radiation takes the form of alpha or beta particles or gamma rays or subatomic particles. (*See also* radiation.)

radiography. The process whereby an image is produced on a radiosensitive surface, such as, photographic film, using radiation. An example is an X-ray film.

radioisotope. The isotopic form of a natural of synthetic element that exhibits radioactivity. The same as a radioactive isotope of an element.

rare-earth elements. Elements found in the lanthanide series in period 6, starting in group 3 of the periodic table, beginning with atomic number 57 and ending with 71. They are usually found as oxides in minerals and ores. Some are produced as by-products from the radioactive decay of uranium and plutonium. They are not really "rare" or scarce but were just difficult to separate and identify. The rare-earth classification is sometimes extended to the actinide series, which includes elements from Z-89 to Z-103.

reagent. A substance used in chemical reactions to detect, measure, examine, or analyze other substances.

rectifier. A device (diode) that converts alternating current (ac) to direct current (dc).

reducing agent. A substance that induces reduction, such as acceptance of one or more electrons by an atom or ion, the removal of oxygen from a compound, or the addition of hydrogen to a compound. It is the opposite of oxidation, although the agent itself is oxidized (loses electrons).

refractory material. A material that is resistant to extremely high temperatures (e.g., carbon, chrome-ore magnesite, fire clay [aluminum silicates]). Primarily used for lining steel furnaces and coke ovens.

resonance structure. In chemistry, it refers to the mathematic concept dealing with the wave functions of electrons that asserts that a compound is capable of two or more possible structures. In other words, their geometry may be identical, but the arrangement of their paired electrons is different. This concept is applied to aromatic compounds, such as benzene.

salt. A compound formed when some or all of the hydrogen in an acid is replaced by a metal; for example, the reaction of zinc with hydrochloric acid results in hydrogen gas and the salt zinc chloride ($2HCl + Zn \rightarrow ZnCl_2 + H_2$).

saturated compounds. Organic compounds containing only single bonds (e.g., straight-chain paraffins). Because it does not contain double or triple bonds, it cannot add elements or compounds.

saturated hydrocarbons. A compound of hydrogen and carbon in which all valence bonds are filled; that is, there are no double or triple bonds.

scintillation counter. *See* Geiger counter.

semiconducting elements. *See* semiconductor.

semiconductor. Usually a "metalloid" (e.g., silicon) or a compound (e.g., gallium arsenide), which has conductive properties greater than those of an insulator but less than those of a conductor (metal). It is possible to adjust their level of conductivity by changing the temperature or adding impurities. Semiconductors are important in the manufacture of computers and many other electronic devices.

shells. Theoretical spherical spaces surrounding the nucleus of an atom. They contain electrons. Also referred to as orbits or energy levels. They might be thought of as somewhat similar to the layer around the core of an onion.

sintering. The process whereby powdered iron and other powdered metals or rare-earths are "pressed" together to fit a mold, using minimum heat below the melting point of the materials. Sintering is used to produce homogeneous (uniform throughout) metal parts. It increases the strength, conductivity, and density of the product.

slag. The vitreous (that is, glassy) mass or residue that is the leftover result of smelting of metallic ores.

smelting. Heating of an ore or mineral to separate the metallic component from the other substances in the ore by chemical reduction. Also known as "roasting."

solar evaporation. The evaporation of seawater by the sun in shallow ponds that results in the reclamation of sea salt and other compounds. The evaporated water, which is desalinated, can be condensed and collected as a source of drinking water.

soldering flux. A chemical flux that assists in the flow of solder during the welding process and removes and prevents the formation of oxides on the welded site.

solders. Alloys, usually containing lead and tin, which melt readily and are used to join metals.

solid. A state of matter in which the atoms or molecules of a substance are rigidly fixed in position. A solid has a fixed shape, a fixed mass, and a fixed volume.

solution. Usually considered a liquid mixture produced by dissolving a solid, liquid, or gas in a solvent. A solution is an even mixture at the molecular or ionic level of one or more substances. It can be a solution of solids as well as liquids or gases. Therefore, by definition, glass is a solid solution.

solvent. A substance that can dissolve another substance. Usually a liquid (the solvent) in which a solid, liquid, or gas (the solute) has dissolved to form a uniform mixture (the solution).

specific gravity. The density of a substance as related to the density of some standard (e.g., water = 1.0). The specific gravity of solids and liquids is equivalent to their densities (g/cc). The specific gravity of gases, and to some extent liquids, is dependent on the temperature and pressure of the gas, because these factors affect the gas's volume. Therefore, the number representing the specific gravity of a gas may or may not be the same as the number for its density (g/L). (*See also* density.)

specific heat. The ratio of the amount of heat required to raise the temperature of a substance 1°C to the amount of heat required to raise an equal mass of a substance (usually water) 1°C in temperature.

spectrograph. A spectroscope equipped to photograph spectra. (For more detail, *see* spectrum)

spectrometer. *See* spectroscope.

spectroscope. An instrument designed to analyze the wavelength of light emitted or absorbed by the excitation of elements. Each element, when heated, exhibits a unique line or color in the electromagnetic (visible light) spectrum.

spectroscopic analysis. *See* spectroscopy

spectroscopy. The range of all electromagnetic radiation, arranged progressively according to wavelength. The visible and near-visible spectrum of radiation ranges from the invisible rays called infrared, which we cannot see but can feel as heat, to the range of visual light from deep red to violet, then to ultraviolet, whose wavelengths are too short to see but are just on the edge of the visual spectrum. (*See also* electromagnetic spectrum.)

steel. An alloy whose main constituent is iron. By the adding of different elements to form alloys of the iron, different steels can be produced—for example, adding nickel to produce stainless steel.

subatomic particle. A component of an atom whose reactions are characteristic of the atom (e.g., electrons, protons, and neutrons), whereas subnuclear particles come from the nucleus during nuclear reactions and in cyclotrons (e.g., quarks, mesons).

sublimation. The direct passage of a substance from a solid to a vapor without going through the liquid state.

superactinide elements. A group of heavy, radioactive, synthetic, unstable elements, some of which have not yet been produced in the laboratory. They have atomic numbers ranging from Z-114 to as high as Z-168 or Z-186, with very short half-lives, and are produced in extremely small amounts by particle accelerators.

superconductor. A metal, alloy, or compound that at temperatures near absolute zero loses both electrical resistance and magnetic permeability (is strongly repelled by magnets), thereby having infinite electrical conductivity.

synthesize. To create a new and complex product by combining separate elements.

synthetic. Man-made; not of natural origin. Artificial.

tarnish. A type of corrosion caused by the reaction of a metal with hydrogen sulfide in the atmosphere (e.g., black silver sulfide forming on silver).

tau. The 19th letter of the Greek alphabet, τ. In particle physics, it is the name for the tau meson (formerly called the κ meson) and the tau particle. The negative tau ($-\tau$), along with its neutrino, is one of the three major leptons.

tensile. The ability to be stretched. Tensile strength is how much a material or object will resist being pulled apart.

tetravalent. Having a valance of 4.

thermochemical. A chemical reaction involving heat; for example, endothermic (absorbs heat) and exothermic (gives off heat).

thermocouple. An instrument made up of dissimilar metals or semiconducting materials used to measure extremely high temperatures that are beyond the range of liquid-in-glass thermometers. Also, a bimetal used in home thermostats to control heat and cooling in furnaces and air conditioners.

thermoelectrical. The characteristic of electricity generated by a temperature difference between two dissimilar materials. Usually semiconductors or metallic conductors. (*See also* thermocouple.)

tincture. A dilute solution of a chemical or pharmaceutical, such as a tincture of iodine, which is less volatile than the original substance.

toxic. Damaging or harmful to the body or its organs and cells (e.g., poison).

tracer. A radioactive entity (e.g., radioisotope) that is added to the reacting elements and/or compounds in a chemical process and that can be traced and detected through that process. Some examples of tracers are carbon-14 and, in medicine, radioactive forms of iodine and sodium.

tractable. Easily worked; malleable.

transactinide elements. A number of unstable radioactive elements that extend beyond the transuranic elements—that is, beyond lawrencium, with atomic numbers of 104 through 114. They do not exist in nature and have very short half-lives; some exist for only a fraction of a second and are synthesized in very small amounts.

transistor. A device that overcomes the resistance when a current of electricity passes through it. Transistors are important in the manufacture of many types of electronics.

transition elements. Elements located in periods 4, 5, and 6, ranging from group 3 of the periodic table through group 12. They are metals that represent a gradual shift from those having a very weak electronegativity (form positive ions) to the strongly electronegative nonmetallic elements that form negative ions. That is, they are elements in "transition" from metals to nonmetals. The lanthanide and actinide series of elements are also considered "transition" elements because they use electron shells inside the outer shell as their valence electron shells and are classed as metals.

transmutation. The transformation, either artificially or naturally, of the atoms of one element into the atoms of a different element as the result of nuclear reactions.

transuranic element. A radioactive element of the actinide series with a higher atomic number than uranium ($_{92}$U). Thus, they are a subseries of the actinides that are not only produced artificially by nuclear bombardment but are also radioactive.

triatomic molecule. A molecule that contains three atoms—for example, ozone (O_3) or magnesium chloride ($MgCl_2$).

trivalent. Having a valence of 3. For example, elements in group 13 of the periodic table.

ultraviolet (UV). The radiation wavelength in the electromagnetic spectrum from 100 to 3,900 angstroms, between the X-ray region and visible violet light.

unsaturated hydrocarbons. Carbon compounds containing atoms that share more than one valence bond and are capable of dissolving more of a solute at a given temperature. In other words, they have at least one double or triple carbon-to-carbon bond and are not in the aromatic group of elements (e.g., ethylene, acetylene).

vacuum. A space devoid of matter; that is, it contains no or very, very few atoms or molecules.

valence. The whole number that represents the combining power of one element with another element. Valence electrons are usually, but not always, the electrons in the outermost shell.

vasodilator. A chemical agent that causes the dilation of blood vessels.

viscosity. The internal resistance with which a liquid resists flowing when pressure is applied. The lower the viscosity, the easier the substance will flow.

vulcanization. A process in which sulfur is added to rubber, resulting in the increased hardness of the rubber.

water. A colorless liquid, which is a compound consisting of the atoms of hydrogen and oxygen. It is found in the atmosphere, in the Earth, and in all living creatures. It is essential to sustain life. Pure water has a pH of 7 (neutral).

X-ray diffraction. A method of spectroanalysis of substances using X-rays instead of light waves. It involves scattering of the X-radiation at angles that are unique to the substance being analyzed, thus yielding a spectrum of the atomic or molecular substance being analyzed.

Z-Number. The number of protons in a nucleus. Used mainly for heavier elements.

Discoverers of the Chemical Elements

Element	Symbol	Date/ Place	Name(s)/ Nationality	Reference Page #
Actinium	Ac	1899 (Paris, France)	Andre Louis Debierne (French)	307
Aluminum	Al	1825 (Copenhagen, Denmark)	Hans Christian Öersted (Danish)	178
Americium	Am	1944 (Chicago, Illinois)	Glenn Seaborg, Ralph James, Leon Morgan, Albert Ghiorso (all American)	321
Antimony	Sb	c. 2000 BCE	Unknown	217
Argon	Ar	1894 (London and Bristol, England)	John William Strutt,aka Lord Rayleigh (British) and Sir William Ramsay (Scottish)	267
Arsenic	As	c.1250 (Germany)	Albertus Magnus (German)	215
Astatine	At	1940 (Berkeley, California)	Dale Corson, Kenneth Mackenzie (both American) and Emilio Segrè (Italian-American)	257
Barium	Ba	1808 (London, England)	Sir Humphry Davy (British)	78
Berkelium	Bk	1949 (Berkeley, California)	Stanley Thompson, Glenn Seaborg, and Albert Ghiorso (all American)	324
Beryllium	Be	1798 (Paris, France)	Louis Vauquelin (French)	65
Bismuth	Bi	c.500 BCE	Unknown	220
Boron	B	1808 (London England and Paris, France)	Sir Humphry Davy (British) and Joseph Gay-Lussac, Louis Thenard (both French)	175

Element	Symbol	Date/ Place	Name(s)/ Nationality	Reference Page #
Bromine	Br	1826 (Montpellier, France)	Antoine Balard (French)	225
Cadmium	Cd	1817 (Gottingen, Germany)	Friedrich Strohmeyer (German)	143
Calcium	Ca	1808 (London, England)	Sir Humphry Davy (British)	173
Californium	Cf	1950 (Berkeley, California)	Stanley Thompson, Glenn Seaborg, Albert Ghiorso, Kenneth Street, Jr. (all American)	326
Carbon	C	pre-history	Unknown	189
Cerium	Ce	1803 (Västmanland, Sweden) and Berlin, German	Jons Jakob Berzelius, Wilhelm Hisinger (both Swedish) and Martin Klaproth (German)	279
Cesium	Cs	1860 (Heidelberg, German)	Robert Bunsen and Gustav Kirchhoff (both German)	59
Chlorine	Cl	1774 (Uppsala, Sweden) 1810 (London, England)	Carl Wilhelm Scheele (Swedish) Sir Humphry Davy (British)	248
Chromium	Cr	1797 (Paris, France)	Louis Vauquelin (French)	95
Cobalt	Co	1735(Stockholm Sweden)	Georg Brandt (Swedish)	105
Copper	Cu	c.5000 BC	unknown	111
Curium	Cm	1944 (Berkeley, California)	Glenn Seaborg, Ralph James, Albert Ghiorso (all American)	323
Dysprosium	Dy	1886 (Paris, France)	Paul de Boisbaudran (both French)	294
Einsteinium	Es	1952 (Berkeley, California)	Gregory Choppin, Stanley Thompson, Albert Ghiorso (all American) and Bernard Harvey (British-American)	328
Erbium	Er	1842-43 (Stockholm, Sweden)	Carl Mosander (Swedish)	297
Europium	Eu	1896, 1901 (Paris, France)	Eugène Demarçay (French)	289
Fermium	Fm	1952 (Berkeley, California)	Gregory Choppin, Stanley Thompson, Albert Ghiorso (all American) and Bernard Harvey (British-American)	330
Fluorine	F	1886 (Paris, France)	Henri Moissan (French)	245
Francium	Fr	1939 (Paris, France)	Marguerite Perey (French)	62
Gadolinium	Gd	1886 (Geneva, Switzerland) and (Paris, France)	Jean-Charles de Marignac (Swiss) and Paul de Boisbaudran (French)	290

Element	Symbol	Date/ Place	Name(s)/ Nationality	Reference Page #
Gallium	Ga	1875 (Paris, France)	Paul de Boisbaudran (French)	181
Germanium	Ge	1886 (Freiberg, Germany)	Clemens Winkler (German)	198
Gold	Au	c.4000 BC	Unknown	165
Hafnium	Hf	1923 (Copenhagen, Denmark)	Dirk Coster (Dutch) and Karl von Hevesy (Hungarian-Swedish)	147
Helium	He	1895 (London, England)	Sir William Ramsay (Scottish)	261
Holmium	Ho	1879 (Uppsala, Sweden)	Per Teodor Cleve (Swedish)	295
Hydrogen	H	1766 (London, England)	Henry Cavendish (British)	40
Indium	In	1863 (Freiberg, Germany)	Ferdinand Reich and Hieronymous Richter (both German)	159
Iodine	I	1811 (Paris, France)	Bernard Courtois (French)	254
Iridium	Ir	1803 (London, England)	Smithson Tennant (British)	159
Iron	Fe	c.3000 BCE	Unknown	100
Krypton	Kr	1898 (London, England)	Sir William Ramsay (Scottish) and Morris Travers (British)	268
Lanthanum	La	1839 (Stockholm, Sweden)	Carl Mosander (Swedish)	277
Lawrencium	Lr	1961 (Berkeley, California)	Almon E. Larsh, Torbjorn Sikkeland, Albert Ghiorso, and Robert Latimer (all American)	335
Lead	Pb	c.5000 BCE	Unknown	203
Lithium	Li	1817-18 (Stockholm, Sweden)	Johann Arfvedson (Swedish)	47
Lutetium	Lu	1907-08 (Paris, France) and (Vienna, Austria)	Georges Urbain (French) and Carl von Welsbach (Austrian)	302
Magnesium	Mg	1755 (Edinburgh, Scotland) 1808 (London, England)	Joseph Black (Scottish) Sir Humphry Davy (British)	69
Manganese	Mn	1774 (Stockholm, Sweden)	Johan Gahn (Swedish)	97
Mendelevium	Md	1955 (Berkeley, California)	Gregory Choppin, Stanley Thompson, Albert Ghiorso, Glenn Seaborg (all American) and Bernard Harvey (British)	332
Mercury	Hg	c.1500 BCE	Unknown	168
Molybdenum	Mo	1781 (Uppsala, Sweden)	Peter Hjelm (Swedish)	127
Neodymium	Nd	1885 (Vienna, Austria)	Carl von Welsbach (Austrian)	283

Element	Symbol	Date/ Place	Name(s)/ Nationality	Reference Page #
Neon	Ne	1898 (London, England)	Sir William Ramsay (Scottish) and Morris Travers (British)	265
Neptunium	Np	1940 (Berkeley, California)	Edwin McMillan and Philip Abelson (both American)	316
Nickel	Ni	1751 (Stockholm, Sweden)	Baron Axel Cronstedt (Swedish)	108
Niobium	Nb	1801 (London, England) 1844 (Berlin, Germany)	Charles Hatchett (British) Heinrich Rose (German)	124
Nitrogen	N	1772 (Edinburgh, Scotland)	Daniel Rutherford (Scottish)	207
Nobelium	No	1958 (Berkeley, California)	Torbjorn Sikkeland, John Walton, Albert Ghiorso, and Glenn Seaborg (all American)	333
Osmium	Os	1803 (London, England)	Smithson Tennant (British)	157
Oxygen	O	1772-74 (Uppsala, Sweden) and London, England)	Carl Scheele (Swedish) and Joseph Priestley (British)	223
Palladium	Pd	1803 (London, England)	William Wollaston (British)	137
Phosphorus	P	1669 (Hamburg, Germany)	Hennig Brand (German)	212
Platinum	Pt	c.1300 BCE 1735 (South America)	unknown Antonio de Ulloa (Spanish)	162
Plutonium	Pu	1941 (Berkeley, California)	Glenn Seaborg, Arthur Wahl, Joseph Kennedy (all American)	318
Polonium	Po	1898 (Paris, France)	Marie Curie (Polish)	214
Potassium	K	1807 (London, England	Sir Humphry Davy (British)	53
Praseodymium	Pr	1885 (Vienna, Austria)	Carl von Welsbach (Austrian)	281
Promethium	Pm	1946 (Oak Ridge, Tennessee)	Jacob Marinsky, Lawrence Glendenin, and Charles Coryell (all American)	285
Protactinium	Pa	1913 (Berlin, Germany) 1918 (Berlin, Germany)	Kasimir Fajans (Polish) and Otto Göhring (German) Otto Hahn (German), Lise Meitner (Austrian), and John Cranston and Frederick Soddy (both British)	311
Radium	Ra	1898 (Paris, France)	Marie Curie (Polish) and Pierre Curie (French)	81
Radon	Rn	1900 (Halle, Germany)	Friedrich Dorn (German)	272

Element	Symbol	Date/ Place	Name(s)/ Nationality	Reference Page #
Rhenium	Re	1925 (Berlin, Germany)	Walter Noddack, Ida Tacke Noddack and Otto Berg (all German)	155
Rhodium	Rh	1803 (London, England)	William Wollaston (British)	135
Rubidium	Rb	1861 (Heidelberg, German)	Robert Bunsen and Gustav Kirchhoff (both German)	57
Ruthenium	Ru	1827 or 28 (Russia) 1844 (Russia)	Gottfried Osann (German) Karl Klaus (Estonian)	133
Samarium	Sm	1879 (Paris, France)	Paul de Boisbaudran (French)	287
Scandium	Sc	1879 (Uppsala, Sweden)	Lars Nilson (Swedish)	87
Selenium	Se	1817 (Stockholm, Sweden)	Jöns Berzelius and Wilhelm von Hisinger (both Swedish)	237
Silicon	Si	1824 (Stockholm, Sweden)	Jöns Berzelius (Swedish)	194
Silver	Ag	c.5000 BCE	unknown	140
Sodium	Na	1807 (London, England)	Sir Humphry Davy (British)	50
Strontium	Sr	1790 (Edinburgh, Scotland) 1808 (London, England)	Adair Crawford (Irish) Sir Humphry Davy (British)	76
Sulfur	S	pre-history	unknown	234
Tantalum	Ta	1802 (Uppsala, Sweden)	Anders Ekeberg (Swedish)	150
Technetium	Tc	1925 (Berlin, Germany) 1937 (Berkeley, California)	Walter Noddack and Ida Tacke Noddack (both German) Emilio Segré (Italian-American) and Carlo Perrier (Italian)	130
Tellurium	Te	1782-83 (Sibiu, Romania)	Franz Muller (Austrian)	239
Terbium	Tb	1843 (Stockholm, Sweden)	Carl Mosander (Swedish)	292
Thallium	Tl	1861 (London, England)	Sir William Crookes (British)	186
Thorium	Th	1815, 1829 (Stockholm, Sweden)	Jöns Berzelius (Swedish)	309
Thulium	Tm	1879 (Uppsala, Sweden)	Per Cleve (Swedish)	299
Tin	Sn	c.3500 BCE	unknown	200
Titanium	Ti	1791 (Cornwall, England) 1793 (Berlin, Germany)	Rev. William Gregor (British) Martin Klaproth (German)	90

Element	Symbol	Date/ Place	Name(s)/ Nationality	Reference Page #
Tungsten	W	1783 (Vergara, Spain)	Don Juan José de Elhuyar and Don Fausto de Elhuyar (both Spanish)	153
Uranium	U	1789 (Berlin, German)	Martin Klaproth (German)	312
Vanadium	V	1801 (Mexico City, Mexico)	Andrés del Rio (Mexican)	93
		1830 (Uppsala, Sweden)	Nils Sefstrom (Swedish)	
Xenon	Xe	1898 (London, England)	Sir William Ramsay (Scottish) and Morris Travers (British)	270
Ytterbium	Yb	1878 (Geneva, Switzerland)	Jean Charles de Marignac (Swiss)	300
Yttrium	Y	1789 or 94 (Abo, Finland or Stockholm, Sweden)	Johann Gadolin (Finnish)	119
Zinc	Zn	c.1000 BCE	Unknown	114
		c.1526 (Basel, Germany)	Paracelsus (German)	
		1746 (Berlin, German)	Andreas Marggraf (German)	
Zirconium	Zr	1789 (Berlin, Germany)	Martin Klaproth (German)	122

Transactinides, Superactinides, Super Heavy Elements*

Element	Symbol	Date/ Place	Name(s)/ Nationality	Reference Page #
Rutherfordium (Unnilquadium)	Rf or Uuq	1964 (Dubna, Russia)	Georgy Flerov, Ivo Zvara (both Russian)	342
		Berkeley, California	Albert Ghiorso (American)	
Dubnium (Unnilpentium)	Db or Unp	1967 (Dubna, Russia) Berkeley, California	Georgy Flerov (Russian) Albert Ghiorso (American)	343
Seaborgium (Unnilhexium)	Sg or Unh	1974 (Berkeley, California)	Albert Ghiorso (American)	345
Bohrium (Unnilseptium)	Bh or Uns	1981 (Darmstadt, Germany)	Peter Armbruster and Gottfried Műnzenberg (both German)	346
Hassium (Unniloctium)	Uno or Hs	1984 (Darmstadt, Germany)	Peter Armbruster and Gottfried Műnzenberg (both German)	347
Meitnerium (Unnilennium)	Mt or Une	1982 (Darmstadt, Germany)	Peter Armbruster and Gottfried Műnzenberg (both German)	349
Darmstadtium (Ununnilium)	Ds or Uun	1994-95 (Berkeley, California)	Darleane Hoffman (American)	350
		(Darmstadt, Germany)	Peter Armbruster, Sigurd Hofmann (both German)	
		Dubna, Russia	Russian Team @ JINR	

Element	Symbol	Date/ Place	Name(s)/ Nationality	Reference Page #
Röentgenium (Unununium)	Rg or Uuu	1994-95 (Darmstadt, Germany)	Peter Armbruster and Sigurd Hofmann (both German)	352
Ununbium	Uub	1996 (Darmstadt, Germany)	Peter Armbruster and Sigurd Hofmann (both German)	353
Ununtrium	Uut	2003 (Dubna, Russia) and (Berkeley, California)		354
Ununquadium	Uuq	1998 (Berkeley, California and Dubna, Russia)		358
Ununpentium	Uup	2003 (Berkeley, California and Dubna, Russia)		359
Ununhexium	Uuh	2000 (Berkeley, California and Dubna, Russia) (in dispute)		361
Ununseptium	Uus	Undiscovered		362
Ununoctium	Uuo	In dispute		363

*Transactinide, Superactinide, and Super Heavy Elements beyond Lawrencium 103 are all made artificially and in minute amounts. All are unstable and radioactive with very short half-lives.

Chemical Elements in Order of Abundance in Earth's Crust

Rank	Element	Symbol & Atomic Number	Reference Page #
1	Oxygen	$_8O$	223
2	Silicon	$_{14}Si$	194
3	Aluminum	$_{13}Al$	178
4	Iron	$_{26}Fe$	100
5	Calcium	$_{20}Ca$	73
6	Sodium	$_{11}Na$	50
7	Magnesium	$_{12}Mg$	69
8	Potassium	$_{19}K$	53
9	Titanium	$_{22}Ti$	90
10	Hydrogen	$_1H$	40
11	Manganese	$_{25}Mn$	97
12	Phosphorus	$_{15}P$	212
13	Fluorine	$_9F$	245
14	Carbon	$_6C$	189
15	Sulfur	$_{16}S$	234
16	Strontium	$_{38}Sr$	76
17	Barium	$_{56}Ba$	78
18	Zirconium	$_{40}Zr$	122
19	Vanadium	$_{23}V$	93
20	Chlorine	$_{17}Cl$	248
21	Chromium	$_{24}Cr$	95

Rank	Element	Symbol & Atomic Number	Reference Page #
22	Rubidium	$_{37}$Rb	57
23	Nickel	$_{28}$Ni	108
24	Zinc	$_{30}$Zn	114
25	Cerium	$_{58}$Ce	279
26	Copper	$_{29}$Cu	111
27	Yttrium	$_{39}$Y	119
28	Neodymium	$_{60}$Nd	283
29	Lanthanum	$_{57}$La	277
30	Nitrogen	$_{7}$N	207
31	Lithium	$_{3}$Li	47
32	Cobalt	$_{27}$Co	105
33	Niobium	$_{41}$Nb	124
34	Gallium	$_{31}$Ga	181
35	Lead	$_{82}$Pb	203
36	Beryllium	$_{4}$Be	65
37	Thorium	$_{90}$Th	309
38	Boron	$_{5}$B	175
39	Samarium	$_{62}$Sm	287
40	Gadolinium	$_{64}$Gd	290
41	Praseodymium	$_{59}$Pr	281
42	Scandium	$_{21}$Sc	89
43	Dysprosium	$_{66}$Dy	294
44	Uranium	$_{92}$U	312
45	Ytterbium	$_{70}$Yb	300
46	Erbium	$_{68}$Er	297
47	Hafnium	$_{72}$Hf	147
48	Cesium	$_{55}$Cs	279
49	Tin	$_{50}$Sn	200
50	Holmium	$_{67}$Ho	295
51	Tantalum	$_{73}$Ta	150
52	Germanium	$_{32}$Ge	198
53	Arsenic	$_{33}$As	215
54	Molybdenum	$_{42}$Mo	127
55	Europium	$_{63}$Eu	289
56	Argon	$_{18}$Ar	269
57	Terbium	$_{65}$Tb	292

Rank	Element	Symbol & Atomic Number	Reference Page #
58	Tungsten	$_{74}$W	153
59	Thallium	$_{81}$Tl	186
60	Lutetium	$_{71}$Lu	302
61	Thulium	$_{69}$Tm	299
62	Bromine	$_{35}$Br	225
63	Antimony	$_{51}$Sb	217
64	Iodine	$_{53}$I	254
65	Cadmium	$_{48}$Cd	143
66	Silver	$_{47}$Ag	140
67	Selenium	$_{34}$Se	237
68	Mercury	$_{80}$Hg	168
69	Indium	$_{49}$In	157
70	Bismuth	$_{83}$Bi	220
71	Tellurium	$_{52}$Te	239
72	Gold	$_{79}$Au	165
73	Helium	$_{2}$He	216
74	Ruthenium	$_{44}$Ru	133
75	Platinum	$_{78}$Pt	162
76	Technetium	$_{43}$Tc	130
77	Palladium	$_{46}$Pd	137
78	Rhenium	$_{75}$Re	155
79	Rhodium	$_{45}$Rh	135
80	Osmium	$_{76}$Os	157
81	Krypton	$_{36}$Kr	268
82	Neon	$_{10}$Ne	265
83	Iridium	$_{77}$Ir	159
84	Xenon	$_{54}$Xe	270
85	Radium	$_{88}$Ra	81
86	Radon	$_{86}$Rn	272
87	Astatine	$_{85}$At	257
88	Francium	$_{87}$Fr	62
89	Promethium	$_{61}$Pm	285
90	Actinium	$_{89}$Ac	307
91	Protactinium	$_{91}$Pa	311
92	Polonium	$_{84}$Po	241
93	Plutonium	$_{94}$Pu	318

Rank	Element	Symbol & Atomic Number	Reference Page #
94	Neptunium	$_{93}$Np	316
95	Americium	$_{95}$Am	321
96	Curium	$_{96}$Cm	323
97	Berkelium	$_{97}$Bk	324
98	Californium	$_{98}$Cf	326
99	Einsteinium	$_{99}$Es	328
100	Fermium	$_{100}$Fm	330
101	Mendelevium	$_{101}$Md	332
102	Nobelium	$_{102}$No	333
103	Lawrencium	$_{103}$Lr	335
104	Unnilquadium (Rutherfordium)	$_{104}$Unq or $_{104}$Rf	342
105	Unnilpentium (Dubnium)	$_{105}$Unp or $_{105}$Db	343
106	Unnilhexium (Seaborgium)	$_{106}$Unh or $_{106}$Sg	345
107	Unnilseptium (Bohrium)	$_{107}$Uns or $_{107}$Bh	346
108	Unniloctium (Hassium)	$_{108}$Uno or $_{108}$Hs	347
109	Unnilennium (Meitnerium)	$_{109}$Une or $_{109}$Mt	349
110	Ununnilium (Darmstadtium)	$_{110}$Uun or $_{110}$Ds	350
111	Unununium (Röentgenium)	$_{111}$Uuu or $_{111}$Rg	352
112	Ununbium	$_{112}$Uub	353
113	Ununtrium	$_{113}$Uut	354
114	Ununquadium	$_{114}$Uuq	358
115	Ununpentium	$_{115}$Uup	359
116	Ununhexium (in dispute)	$_{116}$Uuh	361
117	Ununseptium	Uus (undiscovered)	362
118	Ununoctium (in dispute)	118Uuo	363

Nobel Laureates in Chemistry (1901–2005)

Nobel laureates are listed consecutively by year starting with 1901 and ending with 2005. The country indicates where the chemist did major work. If the chemist's country of birth is different than where major work is done, this is indicated after the "b." In many years, several chemists are recognized. When this occurs, it may be for work in similar areas of chemistry, for working collaboratively or independently, or for work in entirely different areas. When several laureates have been recognized for work in similar areas, the work is cited for the first laureate on the list.

Year	Recipient(s)	Country	Work
1901	Jacobus van't Hoff	Germany, b. Netherlands	laws of chemical dynamics and osmotic pressure in solutions
1902	Herman Emil Fischer	Germany	sugar and purine syntheses
1903	Svante August Arrhenius	Sweden	electrolytic dissociation theory
1904	William Ramsay	Great Britain	discovery of noble gases
1905	Adolf von Baeyer	Germany	organic dyes and hydroaromatic compounds
1906	Henri Moissan	France	isolation of fluorine and electric furnace
1907	Eduard Buchner	Germany	fermentation in absence of cells and biochemistry
1908	Ernest Rutherford	Great Britain, b. New Zealand	radioactive decay
1909	Wilhelm Ostwald	Germany, b. Russia	chemical equilibrium, kinetics, and catalysis
1910	Otto Wallach	Germany	pioneering work with alicyclic compounds
1911	Marie Curie	France, b. Poland	discovery of radium and polonium

Year	Recipient(s)	Country	Work
1912	Victor Grignard	France	discovery of Grignard's reagent
	Paul Sabatier	France	hydrogenation with metal catalysts
1913	Alfred Werner	Switzerland, b. Germany	bonding of inorganic compounds
1914	Theodore Richards	United States	determination of atomic weights
1915	Richard Willstätter	Germany	studies of plant pigments, especially chlorophyll
1918	Fritz Haber	Germany	synthesis of ammonia
1920	Walter H. Nernst	Germany	thermochemistry
1921	Frederick Soddy	Great Britain	radioactive substances and isotopes
1922	Francis W. Aston	Great Britain	mass spectrography and discovery of isotopes
1923	Fritz Perl	Austria	organic microanalysis
1925	Richard A. Zsigmondy	Germany, b. Austria	colloid chemistry
1926	Theodore Svedberg	Sweden	disperse systems
1927	Heinrich Otto Wieland	Germany	bile acids
1928	Adolf Windaus	Germany	sterols relationship with vitamins
1929	Arthur Harden	Great Britain	fermentation of sugar and sugar enzymes
	Hans von Euler-Chelpin	Sweden, b. Germany	
1930	Hans Fischer	Germany	synthesis of hemin
1931	Carl Bosch	Germany	high-pressure chemical processing
	Friedrich Bergins	Germany	
1932	Irving Langmuir	United States	surface chemistry
1934	Harold Urey	United States	discovery of heavy hydrogen
1935	Frédéric Joliot	France	synthesis of new radioactive elements
	Irène Joliot-Curie	France	
1936	Peter Debye	Germany, b. Netherlands	dipole moments and x-ray diffraction
1937	Walter Hayworth	Great Britain	carbohydrates and vitamin C
	Paul Karrer	Switzerland	vitamins A and B12
1938	Richard Kuhn	Germany, b. Austria	carotenoids and vitamins
1939	Adolf F.J. Butenandt	Germany	sex hormones
	Leopold Ruzicka	Switzerland, b. Hungary	polymethylenes and terpenes
1943	George DeHevesy	Sweden, b. Hungary	isotope tracers
1944	Otto Hahn	Germany	fission of heavy nuclei

Year	Recipient(s)	Country	Work
1945	Atturi I. Virtanen	Finland	agricultural and food chemistry and preservation of fodder
1946	James Sumner;	United States	crystallization of enzymes
	John H. Northrop	United States	preparation of proteins and enzymes in
	Wendell Stanley	United States	pure form
1947	Robert Robinson	Great Britain	alkaloids
1948	Arne W.K. Tiselius	Sweden	electrophoresis and serum proteins
1949	William F. Giauque	United States	low temperature thermodynamics
1950	Otto Diels	Germany	diene synthesis
	Kurt Alder	Germany	
1951	Edwin McMillan	United States	chemistry of transuranium elements
	Glenn Seaborg	United States	
1952	Archer J.P. Martin	Great Britain	invention of partition chromatography
	Richard L.M. Synge	Great Britain	
1953	Hermann Staudinger	Germany	macromolecular chemistry
1954	Linus Pauling	United States	chemical bonding and molecular structure of proteins
1955	Vincent Du Vigneaud	United States	sulfur compounds of biological importance, synthesis of polypeptide hormone
1956	Cyril Hinshelwood	Great Britain	mechanisms of chemical reaction
	Nikolay Semenov	USSR	
1957	Alexander Todd	Great Britain	nucleotides and their co-enzymes
1958	Frederick Sanger	Great Britain	protein structure, insulin
1959	Jaroslav Heyrovsky	Czechoslovakia	polarographic methods of analysis
1960	Willard Libby	United States	carbon-14 dating
1961	Melvin Calvin	United States	CO_2 assimilation in plants
1962	Max Preutz	Great Britain, b. Austria	structure of globular proteins
	John Kendrew	Great Britain	
1963	Giulio Natta	Italy	high polymers
	Karl Ziegler	Germany	
1964	Dorothy Crowfoot Hodgkin	Great Britain	x-ray techniques of the structures of biochemical substances
1965	Robert Woodward	United States	organic synthesis
1966	Robert Mulliken	United States	chemical bonds and electronic structure of molecules
1967	Manfred Eigen	Germany	study of very fast chemical reactions
	Ronald Norrish	Great Britain	
	George Porter	Great Britain	

Year	Recipient(s)	Country	Work
1968	Lars Onsager	United States, b. Norway	thermodynamic of irreversible processes
1969	Derek Barton	Great Britain	conformation
	Odd Hensel	Norway	
1970	Luis Leloir	Argentina	sugar nucleotides and carbohydrate bio-synthesis
1971	Gerhard Herzberg	Canada, b. Germany	structure and geometry of free radicals
1972	Christian Anfinsen	United States	ribonuclease, amino acid sequencing and biological activity;
	Stanford Moore	United States	chemical structure and catalytic activity of ribonuclease
	William Stein	United States	
1973	Ernst Fischer	Germany	organometallic sandwich compounds
	Geoffrey Wilkinson	Great Britain	
1974	Paul Flory	United States	macromolecules
1975	John Cornforth	Great Britain	stereochemistry of enzyme-catalyzed reactions
	Vladmir Prelog	Switzerland, b. Bosnia	stereochemistry of organic molecules
1976	William Lipscomb	United States	structure of borane and bonding
1977	Ilya Prigogine	Belgium, b. Russia	theory of dissipative structures
1978	Peter Mitchell	Great Britain	chemiosmotic theory
1979	Herbert Brown	United States, b. Great Britain	organic synthesis of boron and phospho-rus compound
	George Wittig	Germany	
1980	Paul Berg	United States	recombinant DNA
	Walter Gilbert	United States	nucleic acid base sequences
	Frederick Sanger	Great Britain	
1981	Kenichi Fukui	Japan	chemical reactions and orbital theory
	Roald Hoffman	United States, b. Poland	
1982	Aaron Klug	Great Britain	crystallographic electron microscopy applied to nucleic acids and proteins
1983	Henry Taube	United States, b. Canada	electron transfer in metal complexes
1984	Robert Merrifield	United States	chemical synthesis
1985	Herbert Hauptman	United States	crystal structures
	Jerome Karle	United States	
1986	Dudley Herschbach	United States	chemical elementary processes

Year	Recipient(s)	Country	Work
	Yuan Lee	United States, b. Taiwan	
	John Polanyi	Canada	
1987	Donald Cram	United States	development of molecules with highly
	Jean-Marie Lehn	France	selective structure specific interactions
	Charles Pedersen	United States, b. Korea (Norwegian)	
1988	Johann Deisenhofer	Germany and United States, b. Germany	photosynthesis
	Robert Huber	Germany	
	Michel Hartmut	Germany	
1989	Sidney Altman	United States, b. Canada	catalytic properties of RNA
1990	Elias J. Corey	United States	organic synthesis
1991	Richard Ernst	Switzerland	nuclear resonance spectroscopy
1992	Rudolph Marcus	United States-, b. Canada	electron transfer in chemical systems
1993	Kary Mullis	United States	invention of PCR method
	Michael Smith	Canada, b. Great Britain	mutagenesis and protein studies
1994	George Olah	United States, b. Hungary	carbocation chemistry
1995	Paul Crutzen	Germany, b. Netherlands	atmospheric chemistry
	Mario Molina	United States, b. Mexico	stratospheric ozone depletion
	Rowland F. Sherwood	United States	
1996	Robert Kurl	United States	discovery of fullerenes
	Harold Kroto	Great Britain	
	Richard Smalley	United States	
1997	Paul Boyer	United States	enzyme mechanism of ATP;
	John Walker	Great Britain	discovery of ion transport enzyme Na+, K+-ATPase
	Jens Skou	Denmark	
1998	Walter Kohn	United States	density function theory;
	John Pople	Great Britain	computational methods in quantum chemistry
1999	Ahmed Zewail	United States, b. Egypt	transition states using femto spectroscopy

Year	Recipient(s)	Country	Work
2000	Alan J. Heeger	United States	discovery of conductive polymers
	Alan MacDiarmid	United States	
	Hideki Shirakawa	Japan	
2001	William Knowles	United States	chirally catalyzed hydrogenation reactions
	Ryoji Noyori	Japan	
	K. Barry Sharpless	United States	
2002	John B. Fenn	United States	identification of biological macromolecules;
	Koichi Tanaka	Japan	
	Kurt Wüthrich	Switzerland	NMR analysis of biological macromolecules
2003	Peter Agre	United States	discovery of water channels;
	Roderick MacKinnon	United States	structural and mechanistic studies of ion channels
2004	Aaron Ciechanover	Israel	discovery of ubiquitin-mediated protein degradation
	Avram Hershko	Israel, b. Hungary	
	Irwin Rose	United States	
2005	Yves Chauvin	France	development of the metathesis method in organic synthesis
	Robert H. Grubbs	United States	
	Richard R. Schrock	United States	

Sources: Myers, Richard. *The Basics of Chemistry.* Westport, CT: Greenwood Publishing Group, 2003; Nobelprize.org.

Selected Bibliography

Print Media

Asimov, Isaac. *Asimov's Biographical Encyclopedia of Science and Technology.* New York: Doubleday, 1964.

———. *Asimov's Chronology of Science and Discovery.* New York: Harper & Row, 1989.

———. *Beginnings, The Story of Origins—of Mankind, Life, the Earth, the Universe.* New York: A Berkeley Book, 1989.

Barnes-Svarney, Patricia, ed. *The New York Public Library Science Desk Reference.* New York: Simon & Schuster, Macmillian, 1995.

Boorstin, Daniel J. *The Discoverers.* New York: Vintage Books, 1983.

Campbell, Norman. *What Is Science?* New York: Dover, 1953.

Cromer, Alan. *Uncommon Sense.* New York: Oxford University Press, 1993.

Emsley, John. *The Elements,* 3rd ed. New York: Oxford University Press,1998.

Ferris, Timothy. *Coming of Age in the Milky Way.* New York: Doubleday, 1988.

Gamow, George. *Mr. Tompkins in Paperback.* Cambridge, England: Cambridge University Press, 1994.

The Handy Science Answer Book. Carnegie Library of Pittsburgh. Detroit: Visible Ink Press, 1994.

Heiserman, David L. *Exploring Chemical Elements and Their Compounds.* New York: Tab Books, Division of McGraw-Hill,1992.

Holmyard, E. J. *Alchemy.* New York: Dover, 1990. (first published Middlesex, England: Penguin Books, 1957.)

Ihde, Aaron J. *The Development of Modern Chemistry.* New York: Dover, 1984.

Lewis, Richard J., Sr. *Hawley's Condensed Chemical Dictionary,* 12th ed. New York: Van Nostrand Reinhold, 1993.

McLellan, C. R., Marion C. Day, and Roy W. Clark. *Concepts of General Chemistry.* Philadelphia: F. A. Davis, 1966.

Newton, David E. *Recent Advances and Issues in Chemistry.* Phoenix, AZ: Oryx Press, 1999.

Parker, Sybil P., ed. *McGraw-Hill Dictionary of Scientific and Technical Terms,* 5th ed. New York: McGraw-Hill, 1994.

Pauling, Linus. *General Chemistry.* New York: Dover, 1970.

Read, John. *From Alchemy to Chemistry.* Toronto, Ontario: General Publishing, 1995.

Samuels, David. "Buried Suns: The past and possible future of America's nuclear-testing program." *Harper's Magazine,* June 2005, 66.

Schneider, Herman, and Leo Schneider. *The Harper Dictionary of Science in Everyday Language.* New York: Harper & Row, 1988.

Singh, Simon. *Big Bang, The Origin of the Universe.* New York: HarperCollins, 2004.

Stwertka, Albert. *A Guide to the Elements,* 2nd ed. New York: Oxford University Press, 2001.

Trefil, James. *1001 Things Everyone Should Know about Science.* New York: Doubleday 1992.

Upshall, Michael, ed. *Hutchinson Dictionary of Chemistry.* London: Brockhampton Press, 1997.

Digital Media

Atomic Structure. Chemistry Concepts On-Line Series for Grades 9–12, MS-DOS. Thornwood, NY: Edunetics, 1996.

Banks, Alton J., and Jon L. Holmes. *Periodic Table CD Software,* Journal of Chemical Education Software. Madison: University of Wisconsin–Madison, 1995.

The Carbon Cycle. DOS software for Windows. Redmond, WA: Software Labs, 1992.

Chemicals for Windows. DOS Software for Windows. Redmond, WA: Software Labs, 1992.

Chemistry. High School Level Interactive Multimedia CD Software for Windows. San Jose, CA: Sequoia Business, 1994.

Chemistry. Interactive CD for Windows. Saratoga, CA: Stanford Studyware, 1995.

Elements. Science and Nature Interactive CD-ROM Software Series. London: Mentorom Multimedia, YITM, 1995.

Name That Element. Computer Graphics Game I and II, DOS. Redmond, WA: Software Labs, 1992.

The Periodic Table and Chemical Formulas. Chemistry Concepts On-Line for Grades 9–12, DOS. Thornwood, NY: Edunetics, 1996.

Online Media

AAE Periodic Table.com. http://periodtable.com/page/AAE-Rejects.Html (accessed October 24, 2005).

Abundance of the Chemical Elements. Wikipedia. http://en.wikipedia.org/wiki/abundance_of_the-chemical_elements (accessed December 2, 2005).

Aliens on Earth. http://www.ufomind.com/misc/1999/jun/d10–001.shtml (accessed October 24, 2005).

Answers.com. http://www.answers.com/topic/creating-elements?hl = element&hl = 112 (accessed October 20, 2005).

Answers.com. http:///www.answers.com/topic/island-of-stability.html (accessed October 23, 2005).

Atomic Electron Configuration Table. http://www.en.wikipedia.org/wiki/Atomic_electron_configuration_table (accessed November 2, 2005).

Chemistry: Wed Elements Periodic Table. http://www.webelements.com/webelements/elements/text.html (accessed October 24, 2005).

Discovery and Isolation of Plutonium. http://science.kennesaw.edu/˜mhermes/nuclear/nc-04.htm (accessed November 2, 2005).

CMS—Discovery of Elements 113 and 115. http://www-cms.11n1.gov/e113_115/about.html (accessed October 23, 2005).

Atoms—The Inside Story. http://www.schoolscience.co.uk/content/5physics/particles/ partich2pg1.html (accessed October 5, 2005).

Element 113. http://www.radiochemistry.org/periodictable/elements/113.html (accessed October 12, 2005).

Element 115. http://www.gravitywarpdrive.com/Element-115.htm (accessed October 24, 2005).

Everything: Lawrencium. http://www.everything2.com/index.pl?node_id = 365883 (accessed November 2, 2005).

The Free Dictionary. http://www.encyclopedia.thefreedictionary.com (accessed October 20, 2005).

The Free Dictionary: Island of Stability. http://encyclopedia.thefreedictionary.com/ Island+of+stability (accessed October 20, 2005).

The Free Dictionary: Periodic Table. http://columbia.thefreedictionary.com/ Periodic+table+of+the+chemical+elements (accessed October 13, 2005).

The Free Dictionary: Roentgenium. http://columbia.thefreedictionary.com/element+111 (accessed October 20, 2005).

The Free Dictionary: Unbihexium. http://www.encyclopedia.thefreedictionary.com/unbihexium (accessed October 20, 2005).

The Free Dictionary: Untriseptium. http://encyclopedia.thefreedictionary.com/untriseptium (accessed October 20, 2005).

The Free Dictionary: Ununhexium. http://www.encyclopedia.thefreedictionary.com/ununhexium (accessed October 20, 2005).

The Free Dictionary: Ununoctium. http://www.encyclopedia.thefreedictionary.com/ununoctium (accessed October 20, 2005).

The Free Dictionary: Ununseptium. http://www.encyclodedia.thefreedictionary.com/ununseptium (accessed October 20, 2005).

Infoplease: Actinide Series. http://www.infoplease.com/ce6/sci/A0802378.html (accessed September 6, 2005).

Infoplease: Lanthanide Series. http://www.infoplease.com/ce6/sciA0828840.html (accessed September 6, 2005).

Infoplease: Synthetic Elements. http://www.infoplease.com/ce6/sci/A0847507.html (accessed September 6, 2005).

Infoplease: Transactinide Series. http://www.infoplease.com/ce6/sci/A0849243.html (accessed September 6, 2005).

Infoplease: Transuranium Elements. http://www.infoplease.com/ce6/sci/A0849293.html (accessed September 6, 2005).

International Union of Pure and Applied Chemistry (IUPAC). htttp://www.iupac.org/reports/ periodic_table/ (accessed October 9, 2005).

Island of Stability. http://www.rzuser.uni-heidelbert.de/˜q61/el118.html (accessed October 23, 2005).

The Island of Stability. http://absoluteastronomy.com/encyclopedia/i/is/island_of_stability.html (accessed October 23, 2005).

History of Chemistry. http://www.en.wikipedia.org/wiki/History_of_chemistry.html (accessed October 9, 2005).

It's Elemental. Jefferson Lab. http://education.jbab.org/itselemental.html (accessed October 9, 2005).

Jefferson Lab: Questions and Answers: How do I read an electron configuration table? http://education.jlab.org/qa/electron_config.html (accessed November 2, 2005).

List of Elements of the Periodic Table—Sorted by Abundance in Earth's crust. http://www .science.co .ilPTelements.asp?s=Earth (accessed December 2, 2005).

Kennesaw State University, ChemCase.com: Nuclear Chemistry, 4. The Discovery and Isolation of Plutonium. http://science.kennesaw.edu/˜mhermes/nuclear/nc-04.htm (accessed November 2, 2005).

Names of Scientists Associated with Discoveries of Elements of Periodic Table, by Dr. John Andraos. http://www.careerchem.com/NAMED/Elements-Discoverers.pdf >

New York Times Full Document. http://www.sanacacio.net/118-saga/story.html (accessed November 2, 2005).

Nit-Picking—I say pnicogen, you say pnictogen. http://www.chm.bris.ac.uk/motm/ pinctogen/pnictogenh.htm (accessed June 8, 2005).

Nobel Prize in Chemistry. http://en.wikipedia.org/wiki/Noble_Prize_in_chemistry (accessed October 9, 2005).

Nuclear Power. http://en.wikipedia.org/wiki/Nuclear_power (accessed October 4, 2005).

Organic Chemistry—Revision Notes. http://www.revision-notes.co.uk/revision/130.html (accessed October 5, 2005).

Periodic Table: Actinium. http://www.chemicalelements.com/elements/ac.html (accessed September 6, 2005).

Periodic Table: Americium. http://www.chemicalelements.com/elements/am.html (accessed September 6, 2005).

Periodic Table: Plutonium. http://www.chemicalelements.com/elements/pu.html (accessed September 6, 2005).

Periodic Table: Protactinium. http://www.chemicalelements.com/elements/pa.html (accessed September 6, 2005).

Periodic Table: Thorium <http://www.chemicalelements.com/elements/th.html> (accessed September 6, 2005).*Periodic Table: Uranium* <http://www.chemicalelements.com/elements/u.html> (accessed September 6, 2005).*Periodic Table: Electron Configuration.* http://www.chemicalelements.com/ show/electronconfig.html (accessed November 12, 2005).

Periodic Table (extended). http//en.wikipedia.org/w/index.php?title=Periodic_table_%28extended%29&printable=yes (accessed November 2, 2005).

Periodic Table of The Elements. http://www.radiochemistry.org/periodictable/elements.html (accessed October 25, 2005).

Physical Sciences Information Gateway. http://www.psigate.acuk/roads/cgi-bin/serach_webcatalogue2.pl?limit=2006&term1=ele… (accessed October 12, 2005).

Physics Today. http://www.aip.org/pt/vol-55/iss-9/p15.html (accessed October 12, 2005).

Physics Web. http://www.physicsweb.org/articles/news/8/2/1 (accessed October 24, 2005).

Priestley Metal Address. http://www.pubs.acs.org/hotartcl/cenear/032700/print7813address.html (accessed October 13, 2005).

Radiochemistry Society: Periodic Table of the Elements. http://www.radiochemistry.org/ periodictable/element/115.html (accessed October 12, 2005).

Rare Earths. http://www.du.edu/˜jcalvert/phys/rare.html (accessed August 2, 2005).

Research News. http://www.enews.lbl.gov/Science-Articles/Archive/elements-116–118.html (accessed October 24, 2005).

Science Help Online Chemistry, Lesson 3–6. Electron Configuration. http://www.fordhamprep .org/gcurran/sho/lessons/lesson36.htm (accessed November 2, 2005).

Spontaneous Fission. http://enwikipedia.org/wiki/Spontaneours_fission (accessed October 4, 2005).

Superactinides—Wikipedia, the Free Encyclopedia. http://en.wikipedia.org/wiki/Superactinides (accessed October 9, 2005).

The Superconducting Super Collider Project. http://www.hep.net/scc/ (accessed November 1, 2005).

Superheavy Elements 114, 115, and 118 Discovered http://www.wealth4freedom.com/ truth12/115.htm> (accessed October 12, 2005).

Superheavy Elements "Island of Stability." LBNL Image Library. http://www.imglib.lbl.gov .ImgLib/COLLECTIONS/BERKELEY-LAB/SEABORG-ARCHIVE (accessed October 23, 2005).

Transactinides. http://www.kernchemie.uni-mainz.dejvkratz/TRANSaeng.HTML (accessed October 9, 2005).

Uranium. http://www.ist-socrates.berkeley.edu/~eps2wisc/u.html (accessed September 6, 2005).

Victory News Magazine. http://www.victorynewsmagazine.com/ScienceAndAfterLife.htm

Western Oregon University: A Brief History of the Periodic Table. http://www.wou.edu/las/ physics/ch412/prehist.htm (accessed October 9, 2005).

Electronic Resources

Chemical Heritage Foundation
315 Chestnut Street
Philadelphia, PA 19106
Phone: 215-925-2222. Fax: 215-925-1954
www.chemheritage.org

Chemical Heritage Foundation (CHF) serves the community of the chemical and molecular sciences. CHF maintains a collection of materials that document the history and heritage of the chemical and molecular sciences, technologies, and industries.

About the Author

ROBERT E. KREBS has written seven books for Greenwood Press. He has taught chemistry, biology, and other sciences at several schools and universities. Dr. Krebs has served as a science specialist in the federal government and a research administrator for four universities. He retired as Associate Dean for Research in the Graduate College at the Medical Center of the University of Illinois at Chicago.

PERIODIC TABLE OF

TRANSITION ELEMENTS

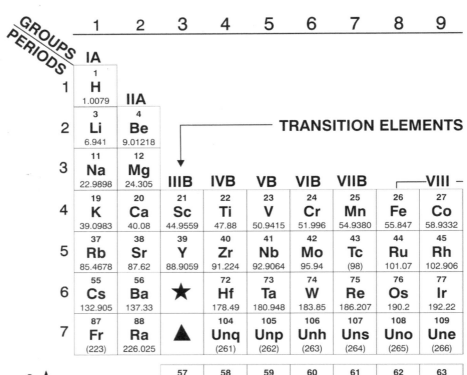

GROUPS / PERIODS	1	2	3	4	5	6	7	8	9
	IA	IIA	IIIB	IVB	VB	VIB	VIIB	VIII	VIII
1	1 **H** 1.0079								
2	3 **Li** 6.941	4 **Be** 9.01218							
3	11 **Na** 22.9898	12 **Mg** 24.305							
4	19 **K** 39.0983	20 **Ca** 40.08	21 **Sc** 44.9559	22 **Ti** 47.88	23 **V** 50.9415	24 **Cr** 51.996	25 **Mn** 54.9380	26 **Fe** 55.847	27 **Co** 58.9332
5	37 **Rb** 85.4678	38 **Sr** 87.62	39 **Y** 88.9059	40 **Zr** 91.224	41 **Nb** 92.9064	42 **Mo** 95.94	43 **Tc** (98)	44 **Ru** 101.07	45 **Rh** 102.906
6	55 **Cs** 132.905	56 **Ba** 137.33	★	72 **Hf** 178.49	73 **Ta** 180.948	74 **W** 183.85	75 **Re** 186.207	76 **Os** 190.2	77 **Ir** 192.22
7	87 **Fr** (223)	88 **Ra** 226.025	▲	104 **Unq** (261)	105 **Unp** (262)	106 **Unh** (263)	107 **Uns** (264)	108 **Uno** (265)	109 **Une** (266)

6 ★ Lanthanide Series (RARE EARTH)

57 **La** 138.906	58 **Ce** 140.12	59 **Pr** 140.908	60 **Nd** 144.24	61 **Pm** (145)	62 **Sm** 150.36	63 **Eu** 151.96

7 ▲ Actinide Series (RARE EARTH)

89 **Ac** 227.028	90 **Th** 232.038	91 **Pa** 231.036	92 **U** 238.029	93 **Np** 237.048	94 **Pu** (244)	95 **Am** (243)